Statistics and Probability for Engineering Applications
With Microsoft® Excel

Statistics and Probability for Engineering Applications

With Microsoft® Excel

by
W.J. DeCoursey
*College of Engineering,
University of Saskatchewan
Saskatoon*

Amsterdam   Boston   London   New York   Oxford   Paris
San Diego   San Francisco   Singapore   Sydney   Tokyo

Newnes is an imprint of Elsevier Science.

Copyright © 2003, Elsevier Science (USA). All rights reserved.

No part of this publication may be reproduced, stored in a retrieval system, or transmitted in any form or by any means, electronic, mechanical, photocopying, recording, or otherwise, without the prior written permission of the publisher.

 Recognizing the importance of preserving what has been written, Elsevier Science prints its books on acid-free paper whenever possible.

Library of Congress Cataloging-in-Publication Data

ISBN: 0-7506-7618-3

British Library Cataloguing-in-Publication Data
A catalogue record for this book is available from the British Library.

The publisher offers special discounts on bulk orders of this book.
For information, please contact:

Manager of Special Sales
Elsevier Science
200 Wheeler Road
Burlington, MA 01803
Tel: 781-313-4700
Fax: 781-313-4880

For information on all Newnes publications available, contact our World Wide Web home page at: http://www.newnespress.com

10 9 8 7 6 5 4 3 2 1

Printed in the United States of America

Contents

Preface xi

What's on the CD-ROM? xiii

List of Symbols xv

1. Introduction: Probability and Statistics 1
    1.1   Some Important Terms 1
    1.2   What does this book contain? 2

2. Basic Probability 6
    2.1   Fundamental Concepts 6
    2.2   Basic Rules of Combining Probabilities 11
        2.2.1   Addition Rule 11
        2.2.2   Multiplication Rule 16
    2.3   Permutations and Combinations 29
    2.4   More Complex Problems: Bayes' Rule 34

3. Descriptive Statistics: Summary Numbers 41
    3.1   Central Location 41
    3.2   Variability or Spread of the Data 44
    3.3   Quartiles, Deciles, Percentiles, and Quantiles 51
    3.4   Using a Computer to Calculate Summary Numbers 55

4. Grouped Frequencies and Graphical Descriptions 63
    4.1   Stem-and-Leaf Displays 63
    4.2   Box Plots 65
    4.3   Frequency Graphs of Discrete Data 66
    4.4   Continuous Data: Grouped Frequency 66
    4.5   Use of Computers 75

5. Probability Distributions of Discrete Variables 84

- 5.1 Probability Functions and Distribution Functions 85
  - (a) Probability Functions 85
  - (b) Cumulative Distribution Functions 86
- 5.2 Expectation and Variance 88
  - (a) Expectation of a Random Variable 88
  - (b) Variance of a Discrete Random Variable 89
  - (c) More Complex Problems 94
- 5.3 Binomial Distribution 101
  - (a) Illustration of the Binomial Distribution 101
  - (b) Generalization of Results 102
  - (c) Application of the Binomial Distribution 102
  - (d) Shape of the Binomial Distribution 104
  - (e) Expected Mean and Standard Deviation 105
  - (f) Use of Computers 107
  - (g) Relation of Proportion to the Binomial Distribution 108
  - (h) Nested Binomial Distributions 110
  - (i) Extension: Multinomial Distributions 111
- 5.4 Poisson Distribution 117
  - (a) Calculation of Poisson Probabilities 118
  - (b) Mean and Variance for the Poisson Distribution 123
  - (c) Approximation to the Binomial Distribution 123
  - (d) Use of Computers 125
- 5.5 Extension: Other Discrete Distributions 131
- 5.6 Relation Between Probability Distributions and Frequency Distributions 133
  - (a) Comparisons of a Probability Distribution with Corresponding Simulated Frequency Distributions 133
  - (b) Fitting a Binomial Distribution 135
  - (c) Fitting a Poisson Distribution 136

6. Probability Distributions of Continuous Variables 141

- 6.1 Probability from the Probability Density Function 141
- 6.2 Expected Value and Variance 149
- 6.3 Extension: Useful Continuous Distributions 155
- 6.4 Extension: Reliability 156

7. The Normal Distribution 157

- 7.1 Characteristics 157
- 7.2 Probability from the Probability Density Function 158
- 7.3 Using Tables for the Normal Distribution 161
- 7.4 Using the Computer 173
- 7.5 Fitting the Normal Distribution to Frequency Data 175
- 7.6 Normal Approximation to a Binomial Distribution 178
- 7.7 Fitting the Normal Distribution to Cumulative Frequency Data 184
- 7.8 Transformation of Variables to Give a Normal Distribution 190

8. Sampling and Combination of Variables 197

- 8.1 Sampling 197
- 8.2 Linear Combination of Independent Variables 198
- 8.3 Variance of Sample Means 199
- 8.4 Shape of Distribution of Sample Means: Central Limit Theorem 205

9. Statistical Inferences for the Mean 212

- 9.1 Inferences for the Mean when Variance Is Known 213
  - 9.1.1 Test of Hypothesis 213
  - 9.1.2 Confidence Interval 221
- 9.2 Inferences for the Mean when Variance Is Estimated from a Sample 228
  - 9.2.1 Confidence Interval Using the t-distribution 232
  - 9.2.2 Test of Significance: Comparing a Sample Mean to a Population Mean 233
  - 9.2.3 Comparison of Sample Means Using Unpaired Samples .. 234
  - 9.2.4 Comparison of Paired Samples 238

10. Statistical Inferences for Variance and Proportion 248

- 10.1 Inferences for Variance 248
  - 10.1.1 Comparing a Sample Variance with a Population Variance 248
  - 10.1.2 Comparing Two Sample Variances 252
- 10.2 Inferences for Proportion 261
  - 10.2.1 Proportion and the Binomial Distribution 261

    10.2.2 Test of Hypothesis for Proportion 261
    10.2.3 Confidence Interval for Proportion 266
    10.2.4 Extension 269

11. Introduction to Design of Experiments 272
    11.1 Experimentation vs. Use of Routine Operating Data 273
    11.2 Scale of Experimentation 273
    11.3 One-factor-at-a-time vs. Factorial Design 274
    11.4 Replication 279
    11.5 Bias Due to Interfering Factors 279
        (a) Some Examples of Interfering Factors 279
        (b) Preventing Bias by Randomization 280
        (c) Obtaining Random Numbers Using Excel 284
        (d) Preventing Bias by Blocking 285
    11.6 Fractional Factorial Designs 288

12. Introduction to Analysis of Variance 294
    12.1 One-way Analysis of Variance 295
    12.2 Two-way Analysis of Variance 304
    12.3 Analysis of Randomized Block Design 316
    12.4 Concluding Remarks 320

13. Chi-squared Test for Frequency Distributions 324
    13.1 Calculation of the Chi-squared Function 324
    13.2 Case of Equal Probabilities 326
    13.3 Goodness of Fit 327
    13.4 Contingency Tables 331

14. Regression and Correlation 341
    14.1 Simple Linear Regression 342
    14.2 Assumptions and Graphical Checks 348
    14.3 Statistical Inferences 352
    14.4 Other Forms with Single Input or Regressor 361
    14.5 Correlation 364
    14.6 Extension: Introduction to Multiple Linear Regression 367

15. Sources of Further Information **373**
    15.1 Useful Reference Books 373
    15.2 List of Selected References 374

Appendices **375**
    Appendix A: Tables 376
    Appendix B: Some Properties of Excel Useful
        During the Learning Process 382
    Appendix C: Functions Useful Once the
        Fundamentals Are Understood 386
    Appendix D: Answers to Some of the Problems 387

Engineering Problem-Solver Index **391**

Index ***393***

Preface

This book has been written to meet the needs of two different groups of readers. On one hand, it is suitable for practicing engineers in industry who need a better understanding or a practical review of probability and statistics. On the other hand, this book is eminently suitable as a textbook on statistics and probability for engineering students.

Areas of practical knowledge based on the fundamentals of probability and statistics are developed using a logical and understandable approach which appeals to the reader's experience and previous knowledge rather than to rigorous mathematical development. The only prerequisites for this book are a good knowledge of algebra and a first course in calculus. The book includes many solved problems showing applications in all branches of engineering, and the reader should pay close attention to them in each section. The book can be used profitably either for private study or in a class.

Some material in earlier chapters is needed when the reader comes to some of the later sections of this book. Chapter 1 is a brief introduction to probability and statistics and their treatment in this work. Sections 2.1 and 2.2 of Chapter 2 on Basic Probability present topics that provide a foundation for later development, and so do sections 3.1 and 3.2 of Chapter 3 on Descriptive Statistics. Section 4.4, which discusses representing data for a continuous variable in the form of grouped frequency tables and their graphical equivalents, is used frequently in later chapters. Mathematical expectation and the variance of a random variable are introduced in section 5.2. The normal distribution is discussed in Chapter 7 and used extensively in later discussions. The standard error of the mean and the Central Limit Theorem of Chapter 8 are important topics for later chapters. Chapter 9 develops the very useful ideas of statistical inference, and these are applied further in the rest of the book. A short statement of prerequisites is given at the beginning of each chapter, and the reader is advised to make sure that he or she is familiar with the prerequisite material.

This book contains more than enough material for a one-semester or one-quarter course for engineering students, so an instructor can choose which topics to include. Sections on use of the computer can be left for later individual study or class study if so desired, but readers will find these sections using Excel very useful. In my opinion a course on probability and statistics for undergraduate engineering students should include at least the following topics: introduction (Chapter 1), basic probability (sections 2.1 and 2.2), descriptive statistics (sections 3.1 and 3.2), grouped frequency (section 4.4), basics of random variables (sections 5.1 and 5.2), the binomial distribution (section 5.3) (not absolutely essential), the normal distribution (sections 7.1, 7.2, 7.3), variance of sample means and the Central Limit Theorem (from Chapter 8), statistical inferences for the mean (Chapter 9), and regression and correlation (from Chapter 14). A number of other topics are very desirable, but the instructor or reader can choose among them.

It is a pleasure to thank a number of people who have made contributions to this book in one way or another. The book grew out of teaching a section of a general engineering course at the University of Saskatchewan in Saskatoon, and my approach was affected by discussions with the other instructors. Many of the examples and the problems for readers to solve were first suggested by colleagues, including Roy Billinton, Bill Stolte, Richard Burton, Don Norum, Ernie Barber, Madan Gupta, George Sofko, Dennis O'Shaughnessy, Mo Sachdev, Joe Mathews, Victor Pollak, A.B. Bhattacharya, and D.R. Budney. Discussions with Dennis O'Shaughnessy have been helpful in clarifying my ideas concerning the paired t-test and blocking. Example 7.11 is based on measurements done by Richard Evitts. Colleagues were very generous in reading and commenting on drafts of various chapters of the book; these include Bill Stolte, Don Norum, Shehab Sokhansanj, and particularly Richard Burton. Bill Stolte has provided useful comments after using preliminary versions of the book in class. Karen Burlock typed the first version of Chapter 7. I thank all of these for their contributions. Whatever errors remain in the book are, of course, my own responsibility.

I am grateful to my editor, Carol S. Lewis, for all her contributions in preparing this book for publication. Thank you, Carol!

W.J. DeCoursey
Department of Chemical Engineering
College of Engineering
University of Saskatchewan
Saskatoon, SK, Canada
S7N 5A9

What's on the CD-ROM?

Included on the accompanying CD-ROM:
- a fully searchable eBook version of the text in Adobe pdf form
- data sets to accompany the examples in the text
- in the "Extras" folder, useful statistical software tools developed by the Statistical Engineering Division, National Institute of Science and Technology (NIST). Once again, you are cautioned not to apply any technique blindly without first understanding its assumptions, limitations, and area of application.

Refer to the Read-Me file on the CD-ROM for more detailed information on these files and applications.

List of Symbols

$\bar{A}$ or A'	complement of A
$A \cap B$	intersection of A and B
$A \cup B$	union of A and B
$B \mid A$	conditional probability
$E(X)$	expectation of random variable X
$f(x)$	probability density function
f_i	frequency of result x_i
i	order number
n	number of trials
$_nC_r$	number of combinations of n items taken r at a time
$_nP_r$	number of permutations of n items taken r at a time
p	probability of "success" in a single trial
$\hat{p}$	estimated proportion
$p(x_i)$	probability of result x_i
Pr [...]	probability of stated outcome or event
q	probability of "no success" in a single trial
$Q(f)$	quantile larger than a fraction f of a distribution
s	estimate of standard deviation from a sample
s^2	estimate of variance from a sample
s_c^2	combined or pooled estimate of variance
$s_{y\mid x}^2$	estimated variance around a regression line
t	interval of time or space. Also the independent variable of the t-distribution.
X (capital letter)	a random variable
x (lower case)	a particular value of a random variable
$\bar{x}$	arithmetic mean or mean of a sample
z	ratio between $(x - \mu)$ and σ for the normal distribution
α	regression coefficient
β	regression coefficient
λ	mean rate of occurrence per unit time or space
μ	mean of a population
σ	standard deviation of population
$\sigma_{\bar{x}}$	standard error of the mean
σ^2	variance of population

CHAPTER 1

Introduction: Probability and Statistics

Probability and statistics are concerned with events which occur *by chance*. Examples include occurrence of accidents, errors of measurements, production of defective and nondefective items from a production line, and various games of chance, such as drawing a card from a well-mixed deck, flipping a coin, or throwing a symmetrical six-sided die. In each case we may have some knowledge of the likelihood of various possible results, but we cannot predict with any certainty the outcome of any particular trial. Probability and statistics are used throughout engineering. In electrical engineering, signals and noise are analyzed by means of probability theory. Civil, mechanical, and industrial engineers use statistics and probability to test and account for variations in materials and goods. Chemical engineers use probability and statistics to assess experimental data and control and improve chemical processes. It is essential for today's engineer to master these tools.

1.1 Some Important Terms

(a) *Probability* is an area of study which involves predicting the relative likelihood of various outcomes. It is a mathematical area which has developed over the past three or four centuries. One of the early uses was to calculate the odds of various gambling games. Its usefulness for describing errors of scientific and engineering measurements was soon realized. Engineers study probability for its many practical uses, ranging from quality control and quality assurance to communication theory in electrical engineering. Engineering measurements are often analyzed using statistics, as we shall see later in this book, and a good knowledge of probability is needed in order to understand statistics.

(b) *Statistics* is a word with a variety of meanings. To the man in the street it most often means simply a collection of numbers, such as the number of people living in a country or city, a stock exchange index, or the rate of inflation. These all come under the heading of *descriptive statistics*, in which items are counted or measured and the results are combined in various ways to give useful results. That type of statistics certainly has its uses in engineering, and we will deal with it later, but another type of statistics will engage our attention in this book to a much greater extent. That is *inferential statistics* or statistical inference. For example, it is often not practical to measure all the items produced by a process. Instead, we very frequently take a sample and measure the relevant quantity on each member of the sample. We infer something about all the items of interest from our knowledge of the sample. A particular characteristic of all the items we are interested in constitutes a *population*. Measurements of the diameter of all possible bolts as they come off a production process would make up a particular population. A *sample* is a *chosen part* of the population in question, say the measured diameters of twelve bolts chosen to be representative of all the bolts made under certain conditions. We need to know how reliable is the information inferred about the population on the basis of our measurements of the sample. Perhaps we can say that "nineteen times out of twenty" the error will be less than a certain stated limit.

(c) *Chance* is a necessary part of any process to be described by probability or statistics. Sometimes that element of chance is due partly or even perhaps entirely to our lack of knowledge of the details of the process. For example, if we had complete knowledge of the composition of every part of the raw materials used to make bolts, and of the physical processes and conditions in their manufacture, in principle we could predict the diameter of each bolt. But in practice we generally lack that complete knowledge, so the diameter of the next bolt to be produced is an unknown quantity described by a random variation. Under these conditions the distribution of diameters can be described by probability and statistics. If we want to improve the quality of those bolts and to make them more uniform, we will have to look into the causes of the variation and make changes in the raw materials or the production process. But even after that, there will very likely be a random variation in diameter that can be described statistically.

Relations which involve chance are called *probabilistic* or *stochastic* relations. These are contrasted with deterministic relations, in which there is no element of chance. For example, Ohm's Law and Newton's Second Law involve no element of chance, so they are deterministic. However, measurements based on either of these laws do involve elements of chance, so relations between the measured quantities are probabilistic.

(d) Another term which requires some discussion is randomness. A *random* action cannot be predicted and so is due to chance. A *random sample* is one in which every member of the population has an equal likelihood of appearing. Just which items appear in the sample is determined completely by chance. If some items are more likely to appear in the sample than others, then the sample is not random.

1.2 What does this book contain?

We will start with the basics of probability and then cover descriptive statistics. Then various probability distributions will be investigated. The second half of the book will be concerned mostly with statistical inference, including relations between two or more variables, and there will be introductory chapters on design and analysis of experiments. Solved problem examples and problems for the reader to solve will be important throughout the book. The great majority of the problems are directly applied to engineering, involving many different branches of engineering. They show how statistics and probability can be applied by professional engineers.

Some books on probability and statistics use rigorous definitions and many derivations. Experience of teaching probability and statistics to engineering students has led the writer of this book to the opinion that a rigorous approach is not the best plan. Therefore, this book approaches probability and statistics without great mathematical rigor. Each new concept is described clearly but briefly in an introductory section. In a number of cases a new concept can be made more understandable by relating it to previous topics. Then the focus shifts to examples. The reader is presented with carefully chosen examples to deepen his or her understanding, both of the basic ideas and of how they are used. In a few cases mathematical derivations are presented. This is done where, in the opinion of the author, the derivations help the reader to understand the concepts or their limits of usefulness. In some other cases relationships are verified by numerical examples. In still others there are no derivations or verifications, but the reader's confidence is built by comparisons with other relationships or with everyday experience. The aim of this book is to help develop in the reader's mind a clear *understanding* of the ideas of probability and statistics and of the ways in which they are used in practice. The reader must keep the assumptions of each calculation clearly in mind as he or she works through the problems. As in many other areas of engineering, it is *essential* for the reader to do many problems and to understand them thoroughly.

This book includes a number of computer examples and computer exercises which can be done using Microsoft Excel®. Computer exercises are included because statistical calculations from experimental data usually require many repetitive calculations. The digital computer is well suited to this situation. Therefore a book on probability and statistics would be incomplete nowadays if it did not include exercises to be done using a computer. The use of computers for statistical calculations is introduced in sections 3.4 and 4.5.

There is a danger, however, that the reader may obtain only an incomplete understanding of probability and statistics if the fundamentals are neglected in favor of extensive computer exercises. The reader should certainly perform several of the more basic problems in each section before doing the ones which are marked as computer problems. Of course, even the more basic problems can be performed using a spreadsheet rather than a pocket calculator, and that is often desirable. Even if a spreadsheet is used, some of the simpler problems which do not require repetitive

Chapter 1

calculations should be done first. The computer problems are intended to help the reader apply the fundamental ideas in conjunction with the computer: they are not "black-box" problems for which the computer (really that means the original programmer) does the thinking. The strong advice of many generations of engineering instructors applies here: *always* show your work!

Microsoft Excel has been chosen as the software to be used with this book for two reasons. First, Excel is used as a general spreadsheet by many engineers and engineering students. Thus, many readers of this book will already be familiar with Excel, so very little further time will be required for them to learn to apply Excel to probability and statistics. On the other hand, the reader who is not already familiar with Excel will find that the modest investment of time required to become reasonably adept at Excel will pay dividends in other areas of engineering. Excel is a very useful tool.

The second reason for choosing to use Excel in this book is that current versions of Excel include a good number of special functions for probability and statistics. Version 4.0 and later versions give at least fifty functions in the Statistical category, and we will find many of them useful in connection with this book. Some of these functions give probabilities for various situations, while others help to summarize masses of data, and still others take the place of statistical tables. The reader is warned, however, that some of these special functions fall in the category of "black-box" solutions and so are not useful until the reader understands the fundamentals thoroughly.

Although the various versions of Excel all contain tools for performing calculations for probability and statistics, some of the detailed procedures have been modified from one version to the next. The detailed procedures in this book are generally compatible with Excel 2000. Thus, if a reader is using a different version, some modifications will likely be needed. However, those modifications will not usually be very difficult.

Some sections of the book have been labelled as Extensions. These are very brief sections which introduce related topics not covered in detail in the present volume. For example, the binomial distribution of section 5.3 is covered in detail, but subsection 5.3(i) is a brief extension to the multinomial distribution.

The book includes a large number of engineering applications among the solved problems and problems for the reader to solve. Thus, Chapter 5 contains applications of the binomial distribution to some sampling schemes for quality control, and Chapters 7 and 9 contain applications of the normal distribution to such continuous variables as burning time for electric lamps before failure, strength of steel bars, and pH of solutions in chemical processes. Chapter 14 includes examples touching on the relationship between the shear resistance of soils and normal stress.

Introduction: Probability and Statistics

The general plan of the book is as follows. We will start with the basics of probability and then descriptive statistics. Then various probability distributions will be investigated. The second half of the book will be concerned mostly with statistical inference, including relations between two or more variables, and there will be introductory chapters on design and analysis of experiments. Solved problem examples and problems for the reader to solve will be important throughout the book.

A preliminary version of this book appeared in 1997 and has been used in second- and third-year courses for students in several branches of engineering at the University of Saskatchewan for five years. Some revisions and corrections were made each year in the light of comments from instructors and the results of a questionnaire for students. More complete revisions of the text, including upgrading the references for Excel to Excel 2000, were performed in 2000-2001 and 2002.

CHAPTER 2

Basic Probability

Prerequisite: A good knowledge of algebra.

In this chapter we examine the basic ideas and approaches to probability and its calculation. We look at calculating the probabilities of combined events. Under some circumstances probabilities can be found by using counting theory involving permutations and combinations. The same ideas can be applied to somewhat more complex situations, some of which will be examined in this chapter.

2.1 Fundamental Concepts

(a) *Probability* as a specific term is a measure of the likelihood that a particular event will occur. Just how likely is it that the outcome of a trial will meet a particular requirement? If we are certain that an event will occur, its probability is 1 or 100%. If it certainly will not occur, its probability is zero. The first situation corresponds to an event which occurs in every trial, whereas the second corresponds to an event which never occurs. At this point we might be tempted to say that probability is given by relative frequency, the fraction of all the trials in a particular experiment that give an outcome meeting the stated requirements. But in general that would not be right. Why? Because the outcome of each trial is determined by chance. Say we toss a fair coin, one which is just as likely to give heads as tails. It is entirely possible that six tosses of the coin would give six heads or six tails, or anything in between, so the relative frequency of heads would vary from zero to one. If it is just as likely that an event will occur as that it will not occur, its true probability is 0.5 or 50%. But the experiment might well result in relative frequencies all the way from zero to one. Then the relative frequency from a small number of trials gives a very unreliable indication of probability. In section 5.3 we will see how to make more quantitative calculations concerning the probabilities of various outcomes when coins are tossed randomly or similar trials are made. If we were able to make an *infinite* number of trials, then probability would indeed be given by the relative frequency of the event.

Basic Probability

As an illustration, suppose the weather man on TV says that for a particular region the probability of precipitation tomorrow is 40%. Let us consider 100 days which have the same set of relevant conditions as prevailed at the time of the forecast. According to the prediction, precipitation the next day would occur at any point in the region in about 40 of the 100 trials. (This is what the weather man predicts, but we all know that the weather man is not always right!)

(b) Although we cannot make an infinite number of trials, in practice we can make a moderate number of trials, and that will give some useful information. The *relative frequency* of a particular event, or the proportion of trials giving outcomes which meet certain requirements, will give an *estimate* of the probability of that event. The larger the number of trials, the more reliable that estimate will be. This is the *empirical* or *frequency approach* to probability. (Remember that "empirical" means based on observation or experience.)

Example 2.1

260 bolts are examined as they are produced. Five of them are found to be defective. On the basis of this information, estimate the probability that a bolt will be defective.

Answer: The probability of a defective bolt is approximately equal to the relative frequency, which is $5 / 260 = 0.019$.

(c) Another type of probability is the *subjective estimate*, based on a person's experience. To illustrate this, say a geological engineer examines extensive geological information on a particular property. He chooses the best site to drill an oil well, and he states that on the basis of his previous experience he estimates that the probability the well will be successful is 30%. (Another experienced geological engineer using the same information might well come to a different estimate.) This, then, is a subjective estimate of probability. The executives of the company can use this estimate to decide whether to drill the well.

(d) A third approach is possible in certain cases. This includes various gambling games, such as tossing an unbiased coin; drawing a colored ball from a number of balls, identical except for color, which are put into a bag and thoroughly mixed; throwing an unbiased die; or drawing a card from a well-shuffled deck of cards. In each of these cases we can say before the trial that a number of possible results are *equally likely*. This is the *classical* or "a priori" approach. The phrase "a priori" comes from Latin words meaning coming from what was known before. This approach is often simple to visualize, so giving a better understanding of probability. In some cases it can be applied directly in engineering.

Chapter 2

Example 2.2

Three nuts with metric threads have been accidentally mixed with twelve nuts with U.S. threads. To a person taking nuts from a bucket, all fifteen nuts seem to be the same. One nut is chosen randomly. What is the probability that it will be metric?

Answer: There are fifteen ways of choosing one nut, and they are equally likely. Three of these equally likely outcomes give a metric nut. Then the probability of choosing a metric nut must be 3 / 15, or 20%.

Example 2.3

Two fair coins are tossed. What is the probability of getting one heads and one tails?

Answer: For a fair or unbiased coin, for each toss of each coin

$$\Pr[\text{heads}] = \Pr[\text{tails}] = \frac{1}{2}$$

This assumes that all other possibilities are excluded: if a coin is lost that toss will be eliminated. The possibility that a coin will stand on edge after tossing can be neglected.

There are two possible results of tossing the first coin. These are heads (H) and tails (T), and they are equally likely. Whether the result of tossing the first coin is heads or tails, there are two possible results of tossing the second coin. Again, these are heads (H) and tails (T), and they are equally likely. The possible outcomes of tossing the two coins are HH, HT, TH, and TT. Since the results H and T for the first coin are equally likely, and the results H and T for the second coin are equally likely, the four outcomes of tossing the two coins must be equally likely. These relationships are conveniently summarized in the following tree diagram, Figure 2.1, in which each branch point (or node) represents a point of decision where two or more results are possible.

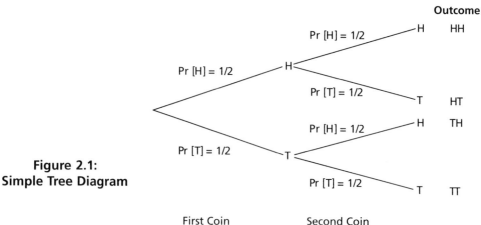

Figure 2.1: Simple Tree Diagram

Basic Probability

Since there are four equally likely outcomes, the probability of each is $\frac{1}{4}$. Both HT and TH correspond to getting one heads and one tails, so two of the four equally likely outcomes give this result. Then the probability of getting one heads and one tails must be $\frac{2}{4} = \frac{1}{2}$ or 0.5.

In the study of probability an event is a set of possible outcomes which meets stated requirements. If a six-sided cube (called a die) is tossed, we define the outcome as the number of dots on the face which is upward when the die comes to rest. The possible outcomes are 1,2,3,4,5, and 6. We might call each of these outcomes a separate event—for example, the number of dots on the upturned face is 5. On the other hand, we might choose an event as those outcomes which are even, or those evenly divisible by three. In Example 2.3 the event of interest is getting one heads and one tails from the toss of two fair coins.

(e) Remember that the probability of an event which is certain is 1, and the probability of an impossible event is 0. Then no probability can be more than 1 or less than 0. If we calculate a probability and obtain a result less than 0 or greater than 1, we know we must have made a mistake. If we can write down probabilities for all possible results, the sum of all these probabilities must be 1, and this should be used as a check whenever possible.

Sometimes some basic requirements for probability are called the axioms of probability. These are that a probability must be between 0 and 1, and the simple addition rule which we will see in part (a) of section 2.2.1. These axioms are then used to derive theoretical relations for probability.

(f) An alternative quantity, which gives the same information as the probability, is called the fair odds. This originated in betting on gambling games. If the game is to be fair (in the sense that no player has any advantage in the long run), each player should expect that he or she will neither win nor lose any money if the game continues for a very large number of trials. Then if the probabilities of various outcomes are not equal, the amounts bet on them should compensate. The fair odds in favor of a result represent the ratio of the amount which should be bet against that particular result to the amount which should be bet for that result, in order to give fairness as described above. Say the probability of success in a particular situation is 3/5, so the probability of failure is 1 − 3/5 = 2/5. Then to make the game fair, for every two dollars bet on success, three dollars should be bet against it. Then we say that the odds in favor of success are 3 to 2, and the odds against success are 2 to 3. To reason in the other direction, take another example in which the fair odds in favor of success are 4 to 3, so the fair odds against success are 3 to 4. Then

$$\Pr[\text{success}] = \frac{4}{4+3} = \frac{4}{7} = 0.571.$$

Chapter 2

In general, if Pr [success] = p, Pr [failure] = $1 - p$, then the fair odds *in favor* of success are $\frac{p}{1-p}$ to 1, and the fair odds *against* success are $\frac{1-p}{p}$ to 1. These are the relations which we use to relate probabilities to the fair odds.

> **Note for Calculation: How many figures?**
>
> How many figures should be quoted in the answer to a problem? That depends on how precise the initial data were and how precise the method of calculation is, as well as how the results will be used subsequently. It is important to quote enough figures so that no useful information is lost. On the other hand, quoting too many figures will give a false impression of the precision, and there is no point in quoting digits which do not provide useful information.
>
> Calculations involving probability usually are not very precise: there are often approximations. In this book probabilities as answers should be given to not more than three significant figures—i.e., three figures other than a zero that indicates or emphasizes the location of a decimal point. Thus, "0.019" contains two significant figures, while "0.571" contains three significant figures. In some cases, as in Example 2.1, fewer figures should be quoted because of imprecise initial data or approximations inherent in the calculation.
>
> It is important *not to round off figures before the final calculation*. That would introduce extra error unnecessarily. Carry more figures in intermediate calculations, and then *at the end* reduce the number of figures in the answer to a reasonable number.

Problems

1. A bag contains 6 red balls, 5 yellow balls and 3 green balls. A ball is drawn at random. What is the probability that the ball is: (a) green, (b) not yellow, (c) red or yellow?

2. A pilot plant has produced metallurgical batches which are summarized as follows:

	Low strength	**High strength**
Low in impurities	2	27
High in impurities	12	4

   If these results are representative of full-scale production, find estimated probabilities that a production batch will be:
   i) low in impurities
   ii) high strength
   iii) both high in impurities and high strength
   iv) both high in impurities and low strength

Basic Probability

3. If the numbers of dots on the upward faces of two standard six-sided dice give the score for that throw, what is the probability of making a score of 7 in one throw of a pair of fair dice?

4. In each of the following cases determine a decimal value for the probability of the event:
   a) the fair odds against a successful oil well are 10-to-1.
   b) the fair odds that a bid will succeed are 1-to-6.

5. Two nuts having U.S. coarse threads and three nuts having U.S. fine threads are mixed accidentally with four nuts having metric threads. The nuts are otherwise identical. A nut is chosen at random.
   a) What is the probability it has U.S. coarse threads?
   b) What is the probability that its threads are not metric?
   c) If the first nut has U.S. coarse threads, what is the probability that a second nut chosen at random has metric threads?
   d) If you are repairing a car engine and accidentally replace one type of nut with another when you put the engine back together, very briefly, what may be the consequences?

6. (a) How many different positive three-digit whole numbers can be formed from the four digits 2, 6, 7, and 9 if any digit can be repeated?
   (b) How many different positive whole numbers less than 1000 can be formed from 2, 6, 7, 9 if any digit can be repeated?
   (c) How many numbers in part (b) are less than 680 (i.e. up to 679)?
   (d) What is the probability that a positive whole number less than 1000, chosen at random from 2, 6, 7, 9 and allowing any digit to be repeated, will be less than 680?

7. Answer question 7 again for the case where the digits 2, 6, 7, 9 can not be repeated.

8. For each of the following, determine (i) the probability of each event, (ii) the fair odds against each event, and (iii) the fair odds in favour of each event:
   (a) a five appears in the toss of a fair six-sided die.
   (b) a red jack appears in draw of a single card from a well-shuffled 52-card bridge deck.

2.2 Basic Rules of Combining Probabilities

The basic rules or laws of combining probabilities must be consistent with the fundamental concepts.

2.2.1 Addition Rule

This can be divided into two parts, depending upon whether there is overlap between the events being combined.

(a) If the events are *mutually exclusive*, there is no overlap: if one event occurs, other events can not occur. In that case the probability of occurrence of one

Chapter 2

or another of more than one event is the *sum* of the probabilities of the separate events. For example, if I throw a fair six-sided die the probability of any one face coming up is the same as the probability of any other face, or one-sixth. There is no overlap among these six possibilities. Then Pr [6] = 1/6, Pr [4] = 1/6, so Pr [6 or 4] is $\frac{1}{6} + \frac{1}{6} = \frac{1}{3}$. This, then, is the probability of obtaining a six or a four on throwing one die. Notice that it is consistent with the classical approach to probability: of six equally likely results, two give the result which was specified. The Addition Rule corresponds to a logical **or** and gives a **sum** of separate probabilities.

Often we can divide all possible outcomes into two groups without overlap. If one group of outcomes is event A, the other group is called the *complement* of A and is written $\overline{A}$ or A'. Since A and $\overline{A}$ together include all possible results, the sum of Pr [A] and Pr [$\overline{A}$] must be 1. If Pr [$\overline{A}$] is more easily calculated than Pr [A], the best approach to calculating Pr [A] may be by first calculating Pr [$\overline{A}$].

Example 2.4

A sample of four electronic components is taken from the output of a production line. The probabilities of the various outcomes are calculated to be: Pr [0 defectives] = 0.6561, Pr [1 defective] = 0.2916, Pr [2 defectives] = 0.0486, Pr [3 defectives] = 0.0036, Pr [4 defectives] = 0.0001. What is the probability of at least one defective?

Answer: It would be perfectly correct to calculate as follows:

Pr [at least one defective] = Pr [1 defective] + Pr [2 defectives] +
    Pr [3 defectives] + Pr [4 defectives]
        = 0.2916 + 0.0486 + 0.0036 + 0.0001 = 0.3439.
but it is easier to calculate instead:
Pr [at least one defective] = 1 − Pr [0 defectives]
        = 1 − 0.6561
        = 0.3439 or 0.344.

(b) If the events are *not mutually exclusive*, there can be *overlap* between them. This can be visualized using a *Venn diagram*. The probability of overlap must be subtracted from the sum of probabilities of the separate events (i.e., we must not count the same area on the Venn Diagram twice).

The circle marked A represents the probability (or frequency) of event A, the circle marked B represents the probability (or frequency) of event B, and the whole rectangle represents all possibilities, so a probability of one or the total frequency. The set consisting of all possible outcomes of a particular experiment is called the *sample space* of that experiment. Thus, the rectangle on the Venn diagram

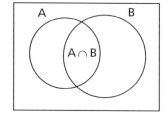

Figure 2.2: Venn Diagram

Basic Probability

corresponds to the sample space. An event, such as A or B, is any subset of a sample space. In solving a problem we must be very clear just what total group of events we are concerned with—that is, just what is the relevant sample space.

Set notation is useful:

Pr [A ∪ B) = Pr [occurrence of A **or** B **or** both], the *union* of the two events A and B.

Pr [A ∩ B) = Pr [occurrence of both A **and** B], the *intersection* of events A and B.

Then in Figure 2.2, the intersection A ∩ B represents the overlap between events A and B.

Figure 2.3 shows Venn diagrams representing intersection, union, and complement. The cross-hatched area of Figure 2.3(a) represents event A. The cross-hatched area on Figure 2.3(b) shows the intersection of events A and B. The union of events A and B is shown on part (c) of the diagram. The cross-hatched area of part (d) represents the complement of event A.

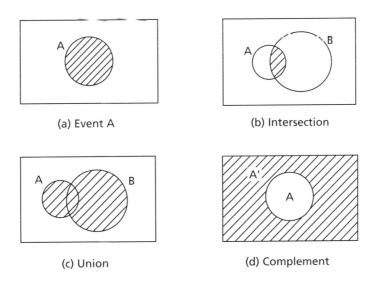

Figure 2.3: Set Relations on Venn Diagrams

If the events being considered are not mutually exclusive, and so there may be overlap between them, the Addition Rule becomes

$$\Pr [A \cup B) = \Pr [A] + \Pr [B] - \Pr [A \cap B] \tag{2.1}$$

In words, the probability of A or B or both is the sum of the probabilities of A and of B, less the probability of the overlap between A and B. The overlap is the intersection between A and B.

Chapter 2

Example 2.5

If one card is drawn from a well-shuffled bridge deck of 52 playing cards (13 of each suit), what is the probability that the card is a queen or a heart? Notice that a card can be both a queen and a heart. Then a queen of hearts (or queen∩heart) overlaps the two categories.

Answer: Pr [queen] = 4/52.
Pr [heart] = 13/52.
Pr [queen∩heart] = 1/52.

These quantities are shown on the Venn diagram of Figure 2.4:

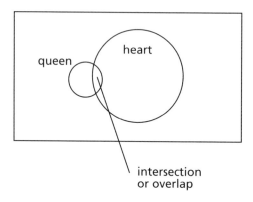

**Figure 2.4:
Venn Diagram for Queen of Hearts**

Then Pr [queen∪heart] = Pr [queen] + Pr [heart] − Pr [queen∩heart]

$$= \frac{4}{52} + \frac{13}{52} - \frac{1}{52} = \frac{16}{52}$$

The simple addition law, sometimes equation 2.1, and the definitions of intersections and unions can be used with Venn diagrams to solve problems involving three events with both single and double overlaps. This usually requires us to apply some form of the addition law several times. Often an appropriate approach is to find the frequency or probability corresponding to a series of simple areas on the diagram, each one representing either a part of only one event without overlap (such as $A \cap \overline{B} \cap \overline{C}$) or only a clearly defined overlap (such as $A \cap B \cap \overline{C}$).

Example 2.6

The class registrations of 120 students are analyzed. It is found that:

30 of the students do not take any of Applied Mechanics, Chemistry, or Computers.

15 of them take only Applied Mechanics.

25 of them take Chemistry and Computers but not Applied Mechanics.

20 of them take Applied Mechanics and Computers but not Chemistry.

Basic Probability

10 of them take all three of Applied Mechanics, Chemistry, and Computers.

A total of 45 of them take Chemistry.

5 of them take only Chemistry.

a) How many of the students take Applied Mechanics and Chemistry but not Computers?
b) How many of the students take only Computers?
c) What is the total number of students taking Computers?
d) If a student is chosen at random from those who take neither Chemistry nor Computers, what is the probability that he or she does not take Applied Mechanics either?
e) If one of the students who take at least two of the three courses is chosen at random, what is the probability that he or she takes all three courses?

Answer: Let's abbreviate the courses as AM, Chem, and Comp.

The number of items in the sample space, which is the total number of items under consideration, is often marked just above the upper right-hand corner of the rectangle. In this example that number is 120. Then the Venn diagram incorporating the given information for this problem is shown below. Two of the simple areas on the diagram correspond to unknown numbers. One of these is $(AM \cap Chem \cap \overline{Comp})$, which is taken by x students. The other is $(\overline{AM} \cap \overline{Chem} \cap Comp)$, so only Computers but not the other courses, and that is taken by y students.

In terms of quantities corresponding to simple areas on the Venn diagram, the given information that a total of 45 of the students take Chemistry requires that

$$x + 10 + 25 + 5 = 45$$

Then $x = 5$.

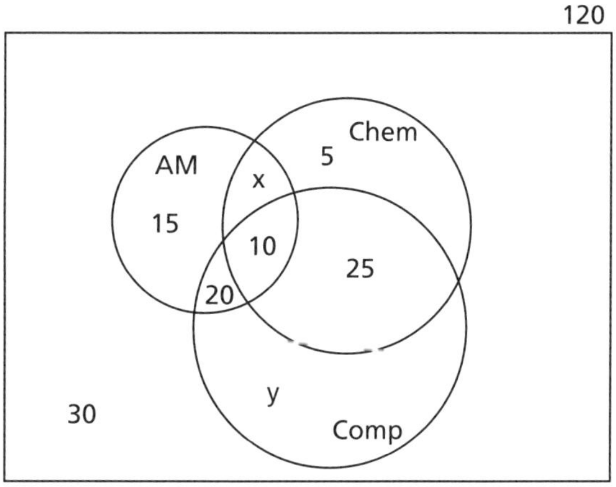

Figure 2.5: Venn Diagram for Class Registrations

15

Chapter 2

Let $n(...)$ be the number of students who take a specified course or combination of courses. Then from the total number of students and the number who do not take any of the three courses we have

$$n(AM \cup Chem \cup Comp) = 120 - 30 = 90$$

But from the Venn diagram and the knowledge of the total taking Chemistry we have

$$n(AM \cup Chem \cup Comp) = n(Chem) + n(AM \cap \overline{Chem} \cap \overline{Comp}) + n(AM \cap \overline{Chem} \cap Comp)$$
$$+ n(\overline{AM} \cap \overline{Chem} \cap Comp)$$

$$= 45 + 15 + 20 + y$$
$$= 80 + y$$

Then $y = 90 - 80 = 10$.

Now we can answer the specific questions.

a) The number of students who take Applied Mechanics and Chemistry but not Computers is 5.
b) The number of students who take only Computers is 10.
c) The total number of students taking Computers is $10 + 20 + 10 + 25 = 65$.
d) The number of students taking neither Chemistry nor Computers is $15 + 30 = 45$. Of these, the number who do not take Applied Mechanics is 30. Then if a student is chosen randomly from those who take neither Chemistry nor Computers, the probability that he or she does not take Applied Mechanics either is $\frac{30}{45} = \frac{2}{3}$.
e) The number of students who take at least two of the three courses is

$$n(AM \cap Chem \cap \overline{Comp}) + n(AM \cap \overline{Chem} \cap Comp) + n(\overline{AM} \cap Chem \cap Comp) + n(AM \cap Chem \cap Comp)$$

$$= 5 + 20 + 25 + 10$$
$$= 60$$

Of these, the number who take all three courses is 10. If a student taking at least two courses is chosen randomly, the probability that he or she takes all three courses is $\frac{10}{60} = \frac{1}{6}$.

2.2.2 Multiplication Rule

(a) The basic idea for calculating the number of choices can be described as follows: Say there are n_1 possible results from one operation. For each one of these, there are n_2 possible results from a second operation. Then there are ($n_1 \times n_2$) possible outcomes of the two operations together. In general, the numbers of possible results are given by *products* of the number of choices at each step. Probabilities can be found by taking ratios of possible results.

Basic Probability

Example 2.7

In one case a byte is defined as a sequence of 8 bits. Each bit can be either zero or one. How many different bytes are possible?

Answer: We have 2 choices for each bit and a sequence of 8 bits. Then the number of possible results is $(2)^8 = 256$.

(b) The simplest form of the *Multiplication Rule* for probabilities is as follows: If the events are *independent*, then the occurrence of one event does not affect the probability of occurrence of another event. In that case the probability of occurrence of more than one event together is the *product* of the probabilities of the separate events. (This is consistent with the basic idea of counting stated above.) If A and B are two separate events that are independent of one another, the probability of occurrence of both A and B together is given by:

$$\Pr[A \cap B] = \Pr[A] \times \Pr[B] \quad (2.2)$$

Example 2.8

If a player throws two fair dice, the probability of a double one (one on the first die and one on the second die) is $(1/6)(1/6) = 1/36$. These events are independent because the result from one die has no effect at all on the result from the other die. (Note that "die" is the singular word, and "dice" is plural.)

(c) If the events are *not independent*, one event affects the probability for the other event. In this case *conditional probability* must be used. The conditional probability of B given that A occurs, or *on condition* that A occurs, is written **Pr [B | A]**. This is read as the probability of B given A, or the probability of B on condition that A occurs. Conditional probability can be found by considering only those events which meet the condition, which in this case is that A occurs. Among these events, the probability that B occurs is given by the conditional probability, Pr [B | A]. In the reduced sample space consisting of outcomes for which A occurs, the probability of event B is Pr [B | A]. The probabilities calculated in parts (d) and (e) of Example 2.6 were conditional probabilities.

The multiplication rule for the occurrence of both A *and* B together when they are not independent is the product of the probability of one event and the conditional probability of the other:

$$\Pr[A \cap B] = \Pr[A] \times \Pr[B \mid A] = \Pr[B] \times \Pr[A \mid B] \quad (2.3)$$

This implies that conditional probability can be obtained by

$$\Pr[B \mid A] = \frac{\Pr[A \cap B]}{\Pr[A]} \quad (2.4)$$

or

$$\Pr[A \mid B] = \frac{\Pr[A \cap B]}{\Pr[B]} \quad (2.5)$$

These relations are often very useful.

Chapter 2

Example 2.9

Four of the light bulbs in a box of ten bulbs are burnt out or otherwise defective. If two bulbs are selected at random without replacement and tested, (i) what is the probability that exactly one defective bulb is found? (ii) What is the probability that exactly two defective bulbs are found?

Answer: A tree diagram is very useful in problems involving the multiplication rule. Let us use the symbols D_1 for a defective first bulb, D_2 for a defective second bulb, G_1 for a good first bulb, and G_2 for a good second bulb.

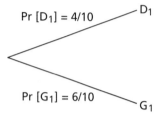

Figure 2.6: First Bulb

At the beginning the box contains four bulbs which are defective and six which are good. Then the probability that the first bulb will be defective is 4/10 and the probability that it will be good is 6/10. This is shown in the partial tree diagram at left.

Probabilities for the second bulb vary, depending on what was the result for the first bulb, and so are given by conditional probabilities. These relations for the second bulb are shown at right in Figure 2.7.

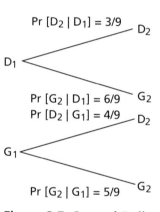

Figure 2.7: Second Bulb

If the first bulb was defective, the box will then contain three defective bulbs and six good ones, so the conditional probability of obtaining a defective bulb on the second draw is $\frac{3}{9}$, and the conditional probability of obtaining a good bulb is $\frac{6}{9}$.

If the first bulb was good the box will contain four defective bulbs and five good ones, so the conditional probability of obtaining a defective bulb on the second draw is $\frac{4}{9}$, and the conditional probability of obtaining a good bulb is $\frac{5}{9}$. Notice that these arguments hold only when the bulbs are selected "without replacement"; if the chosen bulbs had been replaced in the box and mixed well before another bulb was chosen, the relevant probabilities would be different.

Now let us combine the separate probabilities.

The probability of getting two defective bulbs must be $\left(\frac{4}{10}\right)\left(\frac{3}{9}\right) = \frac{12}{90}$, the probability of getting a defective bulb on the first draw and a good bulb on the second draw is $\left(\frac{4}{10}\right)\left(\frac{6}{9}\right) = \frac{24}{90}$, the probability of getting a good bulb on the first draw and a defective bulb on the second draw is $\left(\frac{6}{10}\right)\left(\frac{4}{9}\right) = \frac{24}{90}$, and the probability of getting two good

Basic Probability

bulbs is $\left(\frac{6}{10}\right)\left(\frac{5}{9}\right) = \frac{30}{90}$. In symbols we have:

$$\Pr[D_1 \cap D_2] = \Pr[D_1] \times \Pr[D_2|D_1] = \left(\frac{4}{10}\right)\left(\frac{3}{9}\right) = \frac{12}{90}$$

$$\Pr[D_1 \cap G_2] = \Pr[D_1] \times \Pr[G_2|D_1] = \left(\frac{4}{10}\right)\left(\frac{6}{9}\right) = \frac{24}{90}$$

$$\Pr[G_1 \cap D_2] = \Pr[G_1] \times \Pr[D_2|G_1] = \left(\frac{6}{10}\right)\left(\frac{4}{9}\right) = \frac{24}{90}$$

$$\Pr[G_1 \cap G_2] = \Pr[G_1] \times \Pr[G_2|G_1] = \left(\frac{6}{10}\right)\left(\frac{5}{9}\right) = \frac{30}{90}$$

Notice that both $D_1 \cap G_2$ and $G_1 \cap D_2$ correspond to obtaining 1 good bulb and 1 defective bulb.

The complete tree diagram is shown in Figure 2.8.

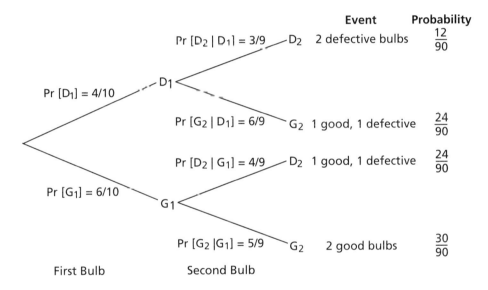

Figure 2.8: Complete Tree Diagram

Notice that all the probabilities of events add up to one, as they must:

$$\frac{12 + 24 + 24 + 30}{90} = 1$$

Now we have to answer the specific questions which were asked:

i) Pr [exactly one defective bulb is found] = $\Pr[D_1 \cap G_2] + \Pr[G_1 \cap D_2]$

$$= \frac{24 + 24}{90} = \frac{48}{90} = 0.533.$$

The first term corresponds to getting first a defective bulb and then a good bulb, and the second term corresponds to getting first a good bulb and then a defective bulb.

Chapter 2

ii) Pr [exactly two defective bulbs are found] = Pr $[D_1 \cap D_2]$ = $\frac{12}{90}$ = 0.133. There is only one path which will give this result.

Notice that testing could continue until either all 4 defective bulbs or all 6 good bulbs are found.

Example 2.10

A fair six-sided die is tossed twice. What is the probability that a five will occur at least once?

Answer: Note that this problem includes the possibility of obtaining two fives. On any one toss, the probability of a five is $\frac{1}{6}$, and the probability of no fives is $\frac{5}{6}$. This problem will be solved in several ways.

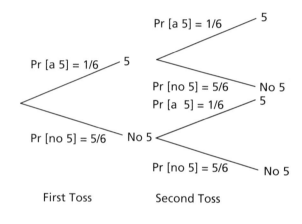

Figure 2.9: Tree Diagram for Two Tosses

First solution (considering all possibilities using a tree diagram):

Pr [5 on the first toss ∩ 5 on the second toss] = $\left(\frac{1}{6}\right)\left(\frac{1}{6}\right) = \frac{1}{36}$

Pr [5 on the first toss ∩ no 5 on the second toss] = $\left(\frac{1}{6}\right)\left(\frac{5}{6}\right) = \frac{5}{36}$

Pr [no 5 on the first toss ∩ 5 on the second toss] = $\left(\frac{5}{6}\right)\left(\frac{1}{6}\right) = \frac{5}{36}$

Pr [no 5 on the first toss ∩ no 5 on the second toss] = $\left(\frac{5}{6}\right)\left(\frac{5}{6}\right) = \frac{25}{36}$

Total of all probabilities (as a check) = 1

Then Pr [at least one five in two tosses] = $\frac{1}{36} + \frac{5}{36} + \frac{5}{36} = \frac{11}{36}$

Basic Probability

Second solution (using conditional probability):

The probability of at least one five is given by:

Pr [5 on the first toss] × Pr [at least one 5 in two tosses | 5 on the first toss]

   + Pr [no 5 on the first toss] × Pr [at least one 5 in two tosses | no 5 on the first toss].

But Pr [5 on the first toss] = Pr [5 on any one toss] = $\frac{1}{6}$

and Pr [at least one 5 in two tosses | 5 on the first toss] = 1 (a dead certainty!)

Also Pr [no 5 on the first toss] = Pr [no 5 on any toss] = $\frac{5}{6}$,

and Pr [at least one 5 in two tosses | no 5 on the first toss] = Pr [5 on the second toss] = $\frac{1}{6}$.

Then Pr [at least one 5 in two tosses] = $\left(\frac{1}{6}\right)(1) + \left(\frac{5}{6}\right)\left(\frac{1}{6}\right) = \frac{11}{36}$

Third solution (using the addition rule, eq. 2.1):

Pr [at least one 5 in two tosses]

= Pr [(5 on the first toss) ∪ (5 on the second toss)]

= Pr [5 on the first toss] + Pr [5 on the second toss]

   − Pr [(5 on the first toss) ∩ (5 on the second toss)]

= $\frac{1}{6} + \frac{1}{6} - \left(\frac{1}{6}\right)\left(\frac{1}{6}\right) = \frac{6}{36} + \frac{6}{36} - \frac{1}{36} = \frac{11}{36}$

Fourth solution: Look at the sample space (i.e., consider all possible outcomes). Let's use a matrix notation where each entry gives first the result of the first toss and then the result of the second toss, as follows:

1,1	1,2	1,3	1,4	1,5	1,6
2,1	2,2	2,3	2,4	2,5	2,6
3,1	3,2	3,3	3,4	3,5	3,6
4,1	4,2	4,3	4,4	4,5	4,6
5,1	5,2	5,3	5,4	5,5	5,6
6,1	6,2	6,3	6,4	6,5	6,6

Figure 2.10: Sample Space of Two Tosses

In the fifth row the result of the first toss is a 5, and in the fifth column the result of the second toss is a 5. This row and this column have been shaded and represent the part of the sample space which meets the requirements of the problem. This area contains 11 entries, whereas the whole sample space contains 36 entries, so Pr [at least one 5 in two tosses] = $\frac{11}{36}$.

Chapter 2

Fifth solution (and the fastest): The probability of no fives in two tosses is $\left(\frac{5}{6}\right)\left(\frac{5}{6}\right) = \frac{25}{36}$

Because the only alternative to no fives is at least one five,

$$\Pr[\text{at least one 5 in two tosses}] = 1 - \frac{25}{36} = \frac{11}{36}$$

Before we start to calculate we should consider whether another method may give a faster correct result!

Example 2.11

A class of engineering students consists of 45 people. What is the probability that no two students have birthdays on the same day, not considering the year of birth? To simplify the calculation, assume that there are 365 days in the year and that births are equally likely on all of them. Then what is the probability that some members of the class have birthdays on the same day?

Answer: The first person in the class states his birthday. The probability that the second person has a different birthday is $\frac{364}{365}$, and the probability that the third person has a different birthday than either of them is $\frac{363}{365}$. We can continue this calculation until the birthdays of all 45 people have been considered. Then the probability that no two students in the class have the same birthday is

$$(1)\left(\frac{364}{365}\right)\left(\frac{363}{365}\right)\left(\frac{362}{365}\right)\cdots\left(\frac{365-i+1}{365}\right)\cdots\left(\frac{365-45+1}{365}\right) = 0.059.$$ (The multiplication was done using a spreadsheet.) Then the probability that at least one pair of students have birthdays on the same day is $1 - 0.059 = 0.941$.

In fact, some days of the year have higher frequencies of births than others, so the probability that at least one pair of students would have birthdays on the same day is somewhat larger than 0.941.

The following example is a little more complex, but it involves the same approach. Because this case uses the multiplication rule, tree diagrams are very helpful.

Example 2.12

An oil company is bidding for the rights to drill a well in field A and a well in field B. The probability it will drill a well in field A is 40%. If it does, the probability the well will be successful is 45%. The probability it will drill a well in field B is 30%. If it does, the probability the well will be successful is 55%. Calculate each of the following probabilities:

  a) probability of a successful well in field A,
  b) probability of a successful well in field B,
  c) probability of both a successful well in field A and a successful well in field B,
  d) probability of at least one successful well in the two fields together,

Basic Probability

e) probability of no successful well in field A,
f) probability of no successful well in field B,
g) probability of no successful well in the two fields together (calculate by two methods),
h) probability of exactly one successful well in the two fields together.
Show a check involving the probability calculated in part h.

Answer:

For Field A:

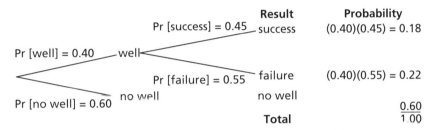

Figure 2.11: Tree Diagram for Field A

a) Then Pr [a successful well in field A] = Pr [a well in A] × Pr [success | well in A]
   = (0.40)(0.45)
   = 0.18   (using equation 2.3)

For Field B:

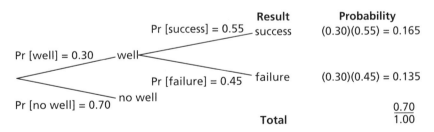

Figure 2.12: Tree Diagram for Field B

b) Then Pr [a successful well in field B] = Pr [a well in B] × Pr [success | well in B]
   = (0.30)(0.55)
   = 0.165 (using equation 2.3)

Chapter 2

c) Pr [both a successful well in field A and a successful well in field B]
    = Pr [a successful well in field A] × Pr [a successful well in field B]
    = (0.18)(0.165)
    = 0.0297 (using equation 2.2, since probability of success in
        one field is not affected by results in the other field)

d) Pr [at least one successful well in the two fields]
    = Pr [(successful well in field A)∪(successful well in field B)]
    = Pr [successful well in field A] + Pr [successful well in field B]
        − Pr [both successful]
    = 0.18 + 0.165 − 0.0297
    = 0.3153 or 0.315 (using equation 2.1)

e) Pr [no successful well in field A]
    = Pr [no well in field A] + Pr [unsuccessful well in field A]
    = Pr [no well in field A] + Pr [well in field A] × Pr [failure | well in A]
    = 0.60 + (0.40)(0.55)
    = 0.60 + 0.22
    = 0.82 (using equation 2.3 and the simple addition rule)

f) Pr [no successful well in field B]
    = Pr [no well in field B] + Pr [unsuccessful well in field B]
    = Pr [no well in field B] + Pr [well in field B] × Pr [failure | well in B]
    = 0.70 + (0.30)(0.45)
    = 0.70 + 0.135
    = 0.835 (using equation 2.3 and the simple addition rule)

g) Pr [no successful well in the two fields] can be calculated in two ways. One method uses the requirement that probabilities of all possible results must add up to 1. This gives:
    Pr [no successful well in the two fields] = 1 − Pr [at least one successful well in the two fields]
    = 1 − 0.3153
    = 0.6847 or 0.685
    The second method uses equation 2.2:
    Pr [no successful well in the two fields]
    = Pr [no successful well in field A] × Pr [no successful well in field B]
    = (0.82)(0.835)
    = 0.6847 or 0.685

h) Pr [exactly one successful well in the two fields]
    = Pr [(successful well in A) ∩ (no successful well in B)]
        + Pr [(no successful well in A) ∩ (successful well in B)]
    = (0.18)(0.835) + (0.82)(0.165)
    = 0.1503 + 0.1353
    = 0.2856 or 0.286 (using equation 2.2 and the simple addition rule)

Basic Probability

Check: For the two fields together,
Pr [two successful wells] = 0.0297 (from part c)
Pr [exactly one successful well] = 0.2856 (from part h)
Pr [no successful wells] = 0.6847 (from part g)
Total (check) = 1.0000

Problems

1. Past records show that 4 of 135 parts are defective in length, 3 of 141 are defective in width, and 2 of 347 are defective in both. Use these figures to estimate probabilities of the individual events assuming that defects occur independently in length and width.
   a) What is the probability that a part produced under the same conditions will be defective in length or width or both?
   b) What is the probability that a part will have neither defect?
   c) What are the fair odds against a defect (in length or width or both)?

2. In a group of 72 students, 14 take neither English nor chemistry, 42 take English and 38 take chemistry. What is the probability that a student chosen at random from this group takes:
   a) both English and chemistry?
   b) chemistry but not English?

3. A random sample of 250 students entering the university included 120 females, of whom 20 belonged to a minority group, 65 had averages over 80%, and 10 fit both categories. Among the 250 students, a total of 105 people in the sample had averages over 80%, and a total of 40 belonged to the minority group. Fifteen males in the minority group had averages over 80%.
   i) How many of those not in the minority group had averages over 80%?
   ii) Given a person was a male from the minority group, what is the probability he had an average over 80%?
   iii) What is the probability that a person selected at random was male, did not come from the minority group, and had an average less than 80%?

4. Two hundred students were sampled in the College of Arts and Science. It was found that: 137 take math, 50 take history, 124 take English, 33 take math and history, 29 take history and English, 92 take math and English, 18 take math, history and English. Find the probability that a student selected at random out of the 200 takes neither math nor history nor English.

5. Among a group of 60 engineering students, 24 take math and 29 take physics. Also 10 take both physics and statistics, 13 take both math and physics, 11 take math and statistics, and 8 take all three subjects, while 7 take none of the three.
   a) How many students take statistics?
   b) What is the probability that a student selected at random takes all three, given he takes statistics?

25

Chapter 2

6. Of 65 students, 10 take neither math nor physics, 50 take math, and 40 take physics. What are the fair odds that a student chosen at random from this group of 65 takes (i) both math and physics? (ii) math but not physics?

7. 16 parts are examined for defects. It is found that 10 are good, 4 have minor defects, and 2 have major defects. Two parts are chosen at random from the 16 without replacement, that is, the first part chosen is not returned to the mix before the second part is chosen. Notice, then, that there will be only 15 possible choices for the second part.
   a) What is the probability that both are good?
   b) What is the probability that exactly one part has a major defect?

8. There are two roads between towns A and B. There are three roads between towns B and C. John goes from town A to town C. How many different routes can he travel?

9. A hiker leaves point A shown in Figure 2.13 below, choosing at random one path from AB, AC, AD, and AE. At each subsequent junction she chooses another path at random, but she does not immediately return on the path she has just taken.
   a) What are the odds that she arrives at point X?
   b) You meet the hiker at point X. What is the probability that the hiker came via point C or E?

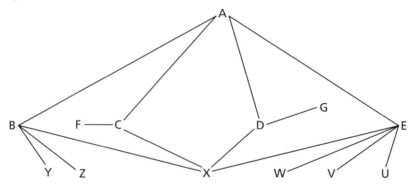

Figure 2.13: Paths for Hiker

10. The probability that a certain type of missile will hit the target on any one firing is 0.80. How many missiles should be fired so that there is at least 98% probability of hitting the target at lest once?

11. To win a daily double at a horse race you must pick the winning horses in the first two races. If the horses you pick have fair odds against of 3:2 and 5:1, what are the fair odds in favor of your winning the daily double?

12. A hockey team wins with a probability of 0.6 and loses with a probability of 0.3. The team plays three games over the weekend. Find the probability that the team:

Basic Probability

   a) wins all three games.
   b) wins at least twice and doesn't lose.
   c) wins one game, loses one, and ties one (in any order).

13. To encourage his son's promising tennis career, a father offers the son a prize if he wins (at least) two tennis sets in a row in a three-set series. The series is to be played with the father and the club champion alternately, so in the order father-champion-father or champion-father-champion. The champion is a better player than the father. Which series should the son choose if Pr [son beating the champion] = 0.4, and Pr [son beating his father] = 0.8? What is the probability of the son winning a prize for each of the two alternatives?

14. Three balls are drawn one after the other from a bag containing 6 red balls, 5 yellow balls and 3 green balls. What is the probability that all three balls are yellow if:
    a) the ball is replaced after each draw and the contents are well mixed?
    b) the ball is not replaced after each draw?

15. When buying a dozen eggs, Mrs. Murphy always inspects 3 eggs for cracks; if one or more of these eggs has a crack, she does not buy the carton. Assuming that each subset of 3 eggs has an equal probability of being selected, what is the probability that Mrs. Murphy will buy a carton which has 5 eggs with cracks?

16. Of 20 light bulbs, 3 are defective. Five bulbs are chosen at random. (a) Use the rules of probability to find the probability that none are defective. (b) What is the probability that at least one is defective?

17. Of flights from Saskatoon to Winnipeg, 89.5% leave on time and arrive on time, 3.5% leave on time and arrive late, 1.5% leave late and arrive on time, and 5.5% leave late and arrive late. What is the probability that, given a flight leaves on time, it will arrive late? What is the probability that, given a flight leaves late, it will arrive on time?

18. Eight engineering students are studying together. What is the probability that at least two students of this group have the same birthday, not considering the year of birth? Simplify the calculations by assuming that there are 365 days in the year and that all are equally likely to be birthdays.

19. The probabilities of the monthly snowfall exceeding 10 cm at a particular location in the months of December, January, and February are 0.2, 0.4, and 0.6, respectively. For a particular winter:
    a) What is the probability that snowfall will be less than 10 cm in all three of the months of December, January and February?
    b) What is the probability of receiving at least 10 cm snowfall in at least 2 of the 3 months?
    c) Given that the snowfall exceeded 10 cm in each of only two months, what is the probability that the two months were consecutive?

Chapter 2

20. A circuit consists of two components, A and B, connected as shown below.

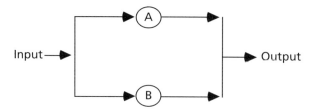

Figure 2.14: Circuit Diagram

Each component can fail (i) to an open circuit mode or
(ii) to a short circuit mode.

The probabilities of the components' failing to these modes in a year are:

	Probability of failing to	
	Open Circuit	Short Circuit
Component	Mode	Mode
A	0.100	0.150
B	0.200	0.100

The circuit fails to perform its intended function if (i) the component in at least one branch fails to the short circuit mode, or if (ii) both components fail to the open circuit mode.

Calculate the probability that the circuit will function adequately at the end of a two-year period.

21. Ten married couples are in a room.
    a) If two people are chosen at random find the probability that (i) one is male and one is female, (ii) they are married to each other.
    b) If 4 people are chosen at random, find the probability that 2 married couples are chosen.
    c) If the 20 people are randomly divided into ten pairs, find the probability that each pair is a married couple.

22. A box contains three coins, two of them fair and one two-headed. A coin is selected at random and tossed. If heads appears the coin is tossed again; if tails appears then another coin is selected from the two remaining coins and tossed.
    a) Find the probability that heads appears twice.
    b) Find the probability that tails appears twice.

23. The probability of precipitation tomorrow is 0.30, and the probability of precipitation the next day is 0.40.
    a) Use these figures to find the probability there will be no precipitation during the two days. State any assumption. What is the probability there will be some precipitation in the next two days?

Basic Probability

b) Why is this calculation not strictly correct? If figures were available, how could the probability of no precipitation during the next two days be calculated more accurately? Show this calculation in symbols.

2.3 Permutations and Combinations

Permutations and combinations give us quick, algebraic methods of counting. They are used in probability problems for two purposes: to count the number of equally likely possible results for the classical approach to probability, and to count the number of different arrangements of the same items to give a multiplying factor.

(a) Each separate arrangement of all or part of a set of items is called a *permutation*. The number of permutations is the number of different arrangements in which items can be placed. Notice that if the order of the items is changed, the arrangement is different, so we have a different permutation. Say we have a total of n items to be arranged, and we can choose r of those items at a time, where $r \leq n$. The number of permutations of n items chosen r at a time is written $_nP_r$. For permutations we consider both the identity of the items and their order.

Let us think for a minute about the number of choices we have at each step along the way. If there are n *distinguishable* items, we have n choices for the first item. Having made that choice, we have $(n-1)$ choices for the second item, then $(n-2)$ choices for the third item, and so on until we come to the r th item, for which we have $(n-r+1)$ choices. Then the total number of choices is given by the product $(n)(n-1)(n-2)(n-3)...(n-r+1)$. But remember that we have a short-hand notation for a related product, $(n)(n-1)(n-2)(n-3)...(3)(2)(1) = n!$, which is called n *factorial* or factorial n. Similarly, $r! = (r)(r-1)(r-2)(r-3)...(3)(2)(1)$, and $(n-r)! = (n-r)(n-r-1)((n-r-2)...(3)(2)(1)$. Then the total number of choices, which is called the *number of permutations of n items taken r at a time*, is

$$_nP_r = \frac{n!}{(n-r)!} = \frac{n(n-1)(n-2)...(2)(1)}{(n-r)(n-r-1)...(3)(2)(1)} \quad (2.6)$$

By definition, $0! = 1$. Then the number of choices of n items taken n at a time is $_nP_n = n!$.

Example 2.13

An engineer in technical sales must visit plants in Vancouver, Toronto, and Winnipeg. How many different sequences or orders of visiting these three plants are possible?

Answer: The number of different sequences is equal to $_3P_3 = 3! = 6$ different permutations. This can be verified by the following tree diagram:

Chapter 2

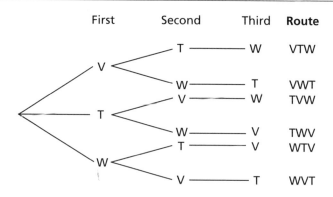

Figure 2.15: Tree Diagram for Visits to Plants

(b) The calculation of permutations is modified if some of the items cannot be distinguished from one another. We speak of this as calculation of the number of *permutations into classes*. We have already seen that if n items are all different, the number of permutations taken n at a time is $n!$. However, if some of them are indistinguishable from one another, the number of possible permutations is reduced. If n_1 items are the same, and the remaining $(n-n_1)$ items are the same of a different class, the number of permutations can be shown to be $\frac{n!}{n_1!(n-n_1)!}$. The numerator, $n!$, would be the number of permutations of n distinguishable items taken n at a time. But n_1 of these items are indistinguishable, so reducing the number of permutations by a factor $\frac{1}{n_1!}$, and another $(n-n_1)$ items are not distinguishable from one another, so reducing the number of permutations by another factor $\frac{1}{(n-n_1)!}$. If we have a total of n items, of which n_1 are the same of one class, n_2 are the same of a second class, and n_3 are the same of a third class, such that $n_1 + n_2 + n_3 = 1$, the number of permutations is $\frac{n!}{n_1!n_2!n_3!}$. This could be extended to further classes.

Example 2.14

A machinist produces 22 items during a shift. Three of the 22 items are defective and the rest are not defective. In how many different orders can the 22 items be arranged if all the defective items are considered identical and all the nondefective items are identical of a different class?

Answer: The number of ways of arranging 3 defective items and 19 nondefective items is $\frac{22!}{(3!)(19!)} = \frac{(22)(21)(20)}{(3)(2)(1)} = 1540$.

Basic Probability

Another modification of calculation of permutations gives *circular permutations*. If *n* items are arranged in a circle, the arrangement doesn't change if every item is moved by one place to the left or to the right. Therefore in this situation one item can be placed at random, and all the other items are placed in relation to the first item. Thus, the number of permutations of *n* distinct items arranged in a circle is $(n - 1)!$.

The principal use of permutations in probability is as a multiplying factor that gives the number of ways in which a given set of items can be arranged.

(c) *Combinations* are similar to permutations, but with the important difference that combinations take no account of order. Thus, AB and BA are different permutations but the same combination of letters. Then the number of permutations must be larger than the number of combinations, and the ratio between them must be the number of ways the chosen items can be arranged. Say on an examination we have to do any eight questions out of ten. The number of permutations of questions would be $_{10}P_8 = \frac{10!}{2!}$. Remember that the number of ways in which eight items can be arranged is $8!$, so the number of combinations must be reduced by the factor $\frac{1}{8!}$. Then the number of combinations of 10 distinguishable items taken 8 at a time is $\left(\frac{10!}{2!}\right)\left(\frac{1}{8!}\right)$. In general, the number of combinations of *n* items taken *r* at a time is

$$_nC_r = \frac{_nP_r}{r!} = \frac{n!}{(n-r)!r!} \tag{2.7}$$

$_nC_r$ gives the number of equally likely ways of choosing r items from a group of *n* distinguishable items. That can be used with the classical approach to probability.

Example 2.15

Four card players cut for the deal. That is, each player removes from the top of a well-shuffled 52-card deck as many cards as he or she chooses. He then turns them over to expose the bottom card of his "cut." He or she retains the cut card. The highest card will win, with the ace high. If the first player draws a nine, what is then his probability of winning without a recut for tie?

Answer: For the first player to win, each of the other players must draw an eight or lower. Then Pr [win] = Pr [other three players all get eight or lower].

There are $(4)(7) = 28$ cards left in the deck below nine after the first player's draw, and there are $52 - 1 = 51$ cards left in total. The number of combinations of three cards from 51 cards is $_{51}C_3$, all of which are equally likely. Of these, the number of combinations which will result in a win for the first player is the number of combinations of three items from 28 items, which is $_{28}C_3$.

Chapter 2

The probability that the first player will win is

$$\frac{_{28}C_3}{_{51}C_3} = \left(\frac{28!}{(25!)(3!)}\right)\left(\frac{(48!)(3!)}{51!}\right) = \left(\frac{(28)(27)(26)}{(3)(2)(1)}\right)\left(\frac{(3)(2)(1)}{(51)(50)(49)}\right) = \left(\frac{(28)(27)(26)}{(51)(50)(49)}\right) = \frac{19,656}{124,950} = 0.157$$

Like many other problems, this one can be done in more than one way. A solution by the multiplication rule using conditional probability is as follows:

Pr [player #2 gets eight or lower | player #1 drew a nine] = $\frac{28}{51}$

If that happens, Pr [player #3 gets eight or lower]

= Pr [third player gets eight or lower | first player drew a nine and second player drew eight or lower]

= $\frac{27}{50}$

If that happens, Pr [player #4 gets eight or lower]

= Pr [fourth player gets eight or lower | first player drew a nine and both second and third players drew eight or lower]

= $\frac{26}{49}$

The probability that the first player will win is

$$\left(\frac{28}{51}\right)\left(\frac{27}{50}\right)\left(\frac{26}{49}\right) = 0.157.$$

Problems

1. A bench can seat 4 people. How many seating arrangements can be made from a group of 10 people?
2. How many distinct permutations can be formed from all the letters of each of the following words: (a) them, (b) unusual?
3. A student is to answer 7 out of 9 questions on a midterm test.
   i) How many examination selections has he?
   ii) How many if the first 3 questions are compulsory?
   iii) How many if he must answer at least 4 of the first 5 questions?
4. Four light bulbs are selected at random without replacement from 16 bulbs, of which 7 are defective. Find the probability that
   a) none are defective.
   b) exactly one is defective.
   c) at least one is defective.
5. Of 20 light bulbs, 3 are defective. Five bulbs are chosen at random.
   a) Use permutations or combinations to find the probability that none are defective.
   b) What is the probability that at least one is defective?

(This is a modification of problem 15 of the previous set.)

Basic Probability

6. A box contains 18 light bulbs. Of these, four are defective. Five bulbs are chosen at random.
   a) Use permutations or combinations to find the probability that none are defective.
   b) What is the probability that exactly one of the chosen bulbs is defective?
   c) What is the probability that at least one of the chosen bulbs is defective?

7. How many different sums of money can be obtained by choosing two coins from a box containing a nickel, a dime, a quarter, a fifty-cent piece, and a dollar coin? Is this a problem in permutations or in combinations?

8. If three balls are drawn at random from a bag containing 6 red balls, 4 white balls, and 8 blue balls, what is the probability that all three are red? Use permutations or combinations.

9. In a poker hand consisting of five cards, what is the probability of holding:
   a) two aces and two kings?
   b) five spades?
   c) A, K, Q, J, 10 of the same suit?

10. In how many ways can a group of 7 persons arrange themselves
    a) in a row,
    b) around a circular table?

11. In how many ways can a committee of 3 people be selected from 8 people?

12. In playing poker, five cards are dealt to a player. What is the probability of being dealt (i) four-of-a-kind? (ii) a full house (three-of-a-kind and a pair)?

13. A hockey club has 7 forwards, 5 defensemen, and 3 goalies. Each can play only in his designated subgroup. A coach chooses a team of 3 forwards, 2 defense, and 1 goalie.
    a) How many different hockey teams can the coach assemble if position within the subgroup is not considered?
    b) Players A, B and C prefer to play left forward, center, and right defense, respectively. What is the probability that these three players will play on the same team in their preferred positions if the coach assembles the team at random?

14. A shipment of 17 radios includes 5 radios that are defective. The receiver samples 6 radios at random. What is the probability that exactly 3 of the radios selected are defective? Solve the problem
    a) using a probability tree diagram
    b) using permutations and combinations.

15. Three married couples have purchased theater tickets and are seated in a row consisting of just six seats. If they take their seats in a completely random fashion, what is the probability that
    a) Jim and Paula (husband and wife) sit in the two seats on the far left?
    b) Jim and Paula end up sitting next to one another.

Chapter 2

2.4 More Complex Problems: Bayes' Rule

More complex problems can be treated in much the same manner. You must read the question *very* carefully. If the problem involves the multiplication rule, a tree diagram is almost always very strongly recommended.

Example 2.16

A company produces machine components which pass through an automatic testing machine. 5% of the components entering the testing machine are defective. However, the machine is not entirely reliable. If a component is defective there is 4% probability that it will not be rejected. If a component is not defective there is 7% probability that it will be rejected.

a) What fraction of all the components are rejected?
b) What fraction of the components rejected are actually not defective?
c) What fraction of those not rejected are defective?

Answer: Let D represent a defective component, and G a good component.

Let R represent a rejected component, and A an accepted component.

Part (a) can be answered directly using a tree diagram.

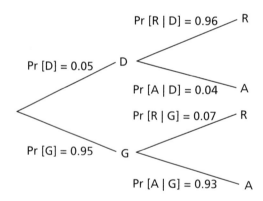

Figure 2.16: Testing Sequences

Now we can calculate the probabilities of the various combined events:

Pr [D ∩ R] =	Pr [D] × Pr [R | D] =	(0.05)(0.96) =	0.0480	Rejected
Pr [D ∩ A] =	Pr [D] × Pr [A | D] =	(0.05)(0.04) =	0.0020	Accepted
Pr [G ∩ R] =	Pr [G] × Pr [R | G] =	(0.95)(0.07) =	0.0665	Rejected
Pr [G ∩ A] =	Pr [G] × Pr [A | G] =	(0.95)(0.93) =	0.8835	Accepted
		Total =	1.0000	(Check)

Basic Probability

Because all possibilities have been considered and there is no overlap among them, we see that the "rejected" area is composed of only two possibilities, so the probability of rejection is the sum of the probabilities of two intersections. The same can be said of the "accepted" area.

Then $\Pr[R] = \Pr[D \cap R] + \Pr[G \cap R] = 0.0480 + 0.0665 = 0.1145$

and $\Pr[A] = \Pr[D \cap A] + \Pr[G \cap A] = 0.0020 + 0.8835 = 0.8855$

a) The answer to part (a) is that "in the long run" the fraction rejected will be the probability of rejection, 0.1145 or (with rounding) 0.114 or 11.4 %.

Now we can calculate the required quantities to answer parts (b) and (c) using conditional probabilities in the opposite order, so in a sense applying them backwards.

b) Fraction of components rejected which are not defective
   = probability that a component is good, given that it was rejected

$$= \Pr[G \mid R] = \frac{\Pr[G \cap R]}{\Pr[R]} = \frac{0.0665}{0.1145} = 0.58 \text{ or } 58\%.$$

c) Fraction of components passed which are actually defective
   = probability that a component is defective, given that it was passed

Using equation 2.4, this is $\Pr[D \mid A] = \dfrac{\Pr[D \cap A]}{\Pr[A]} = \dfrac{0.0020}{0.8855} = 0.0023$ or 0.23 %.

(Note that $\Pr[G \mid R] \neq \Pr[R \mid G]$, and $\Pr[D \mid A] \neq \Pr[A \mid D]$.)

Thus the fraction of defective components in the stream which is passed seems to be acceptably small, but the fraction of non-defective components in the stream which is rejected is unacceptably large. In practice, something would have to be done about that.

Note two points here about the calculation. First, to obtain answers to parts (b) and (c) of this problem we have applied conditional probability in two directions, first forward in the tree diagram, then backward. Both are legitimate applications of Equation 2.3 or 2.4. Second, we can go from the idea of the sample space, consisting of all possible results, to the reduced sample space, consisting of those outcomes which meet a particular condition. Here for $\Pr[D \mid A]$ the reduced sample space consists of all outcomes for which the component is not rejected. The conditional probability is the probability that an item in the reduced sample space will satisfy the requirement that the component is defective, or the long-run fraction of the items in the reduced sample space that satisfy the new requirement.

Bayes' Theorem or Rule is the name given to the use of conditional probabilities in both directions, with combination of all the intersections involving a particular

Chapter 2

event to give the probability of that event. The Bayesian approach can be summarized as follows:

- First, apply the multiplication rule with conditional probability forward along the tree diagram:
$$\Pr[A \cap B] = \Pr[A] \times \Pr[B|A] \tag{2.3 a}$$

- Second, apply the addition rule to reconstruct the probability of a particular event as a reduced sample space:
$$\Pr[B] = \Pr[A \cap B] + \Pr[\overline{A} \cap B] \tag{2.8}$$
where $\overline{A}$ represents "not A", the absence of A or complement of A.

- Third, apply the relation for conditional probability, in the opposite direction on the tree diagram from the first step, using this reduced sample space:
$$\Pr[A|B] = \frac{\Pr[A \cap B]}{\Pr[B]} \tag{2.5}$$

Bayes' Rule should always be used with a tree diagram. Thus, for Example 2.16 we have:

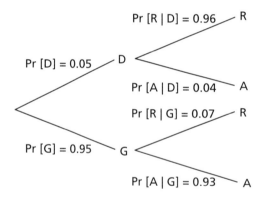

Figure 2.17:
Tree Diagram for Bayes' Rule

The steps corresponding to the reasoning behind Bayes' Rule for this tree diagram are:

First, $\Pr[D \cap R] = \Pr[D] \times \Pr[R|D]$, and so on, corresponding to equation 2.3 a.

Then, $\Pr[R] = \Pr[D \cap R] + \Pr[G \cap R]$, and similarly for $\Pr[A]$, corresponding to equation 2.8.

Then, $\Pr[G|R] = \dfrac{\Pr[G \cap R]}{\Pr[R]}$, and similarly for $\Pr[D|A]$, corresponding to equation 2.5.

An important use of Bayes' Rule is in modifying earlier estimates of probability with later observed data.

Basic Probability

Here is another example of the use of Bayes' Rule:

Example 2.17

A man has three identical jewelry boxes, each with two identical drawers. In the first box both drawers contain gold watches. In the second box both drawers contain silver watches. In the third box one drawer contains a gold watch, and the other drawer contains a silver watch. The man wants to wear a gold watch. If he selects a box at random, opens a drawer at random, and finds a silver watch, what is the probability that the other drawer in that box contains a gold watch?

Answer: (It is interesting at this point to guess what the right answer will be! Try it.)

If G stands for a gold watch and S stands for a silver watch, the three boxes and their contents can be shown as follows:

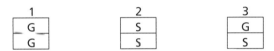

Figure 2.18: Jewelry Boxes

If the selected box contains both a silver watch and a gold watch, it must be Box 3.

Then we need to calculate the probability that the man chose Box 3 on condition that he found a silver watch, $\Pr[B_3|S]$, where B_3 stands for Box 3 and similar notations apply for other boxes. We start with a tree diagram and apply conditional probabilities along the tree.

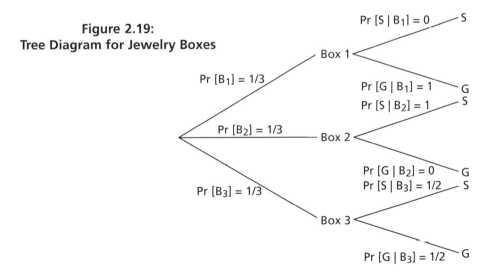

Figure 2.19: Tree Diagram for Jewelry Boxes

Using equation 2.5, $\Pr[S \cap B_i] = \Pr[B_i] \times \Pr[S|B_i]$, and similarly $\Pr[G \cap B_i] = \Pr[B_i] \times \Pr[G|B_i]$, so we have:

Chapter 2

i, Box No.	Pr $[S \cap B_i]$	Pr $[G \cap B_i]$
1	$\left(\frac{1}{3}\right)(0) = 0$	$\left(\frac{1}{3}\right)(1) = \frac{1}{3}$
2	$\left(\frac{1}{3}\right)(1) = \frac{1}{3}$	$\left(\frac{1}{3}\right)(0) = 0$
3	$\left(\frac{1}{3}\right)\left(\frac{1}{2}\right) = \frac{1}{6}$	$\left(\frac{1}{3}\right)\left(\frac{1}{2}\right) = \frac{1}{6}$
Total	$\frac{1}{2}$	$\frac{1}{2}$

Then $\Pr[S] = \sum_{i=1}^{3} \Pr[S \cap B_i] = 0 + \frac{1}{3} + \frac{1}{6} = \frac{1}{2}$

and $\Pr[G] = \sum_{i=1}^{3} \Pr[G \cap B_i] = \frac{1}{3} + 0 + \frac{1}{6} = \frac{1}{2}$

$$\text{Total} = 1 \ (\text{check})$$

Then we have $\Pr[B_3|S] = \dfrac{\Pr[B_3 \cap S]}{\Pr[S]} = \dfrac{1/6}{1/2} = \dfrac{1}{3}$

Then the probability that the other drawer contains a gold watch is $\frac{1}{3}$.

Other relatively complex problems will be encountered when the concepts of basic probability are combined with other ideas or distributions in later chapters.

Problems

1. Three different machines Ml, M2, and M3 are used to produce similar electronic components. Machines Ml, M2, and M3 produce 20%, 30% and 50% of the components respectively. It is known that the probabilities that the machines produce defective components are 1% for M1, 2% for M2, and 3% for M3. If a component is selected randomly from a large batch, and that component is defective, find the probability that it was produced: (a) by M2, and (b) by M3.

2. A flood forecaster issues a flood warning under two conditions only: (i) if fall rainfall exceeds 10 cm and winter snowfall is between 15 and 20 cm, or (ii) if winter snowfall exceeds 20 cm regardless of fall rainfall. The probability of fall rainfall exceeding 10 cm is 0.10, while the probabilities of winter snowfall exceeding 15 and 20 cm are 0.15 and 0.05 respectively.
   a) What is the probability that he will issue a warning any given spring?
   b) Given that he issues a warning, what is the probability that fall rainfall was greater than 10 cm?

Basic Probability

3. A certain company has two car assembly plants, A and B. Plant A produces twice as many cars as plant B. Plant A uses engines and transmissions from a subsidiary plant which produces 10% defective engines and 2% defective transmissions. Plant B uses engines and transmissions from another source where 8% of the engines and 4% of the transmissions are defective. Car transmissions and engines at each plant are installed independently.
   a) What is the probability that a car chosen at random will have a good engine?
   b) What is the probability that a car from plant A has a defective engine, or a defective transmission, or both?
   c) What is the probability that a car which has a good transmission and a defective engine was assembled at plant B?

4. It is known that of the articles produced by a factory, 20% come from Machine A, 30% from Machine B, and 50% from Machine C. The percentages of satisfactory articles among those produced are 95% for A, 85% for B and 90% for C. An article is chosen at random.
   a) What is the probability that it is satisfactory?
   b) Assuming that the article is satisfactory, what is the probability that it was produced by Machine A?

5. Of the feed material for a manufacturing plant, 85% is satisfactory, and the rest is not. If it is satisfactory, the probability it will pass Test A is 92%. If it is not satisfactory, the probability it will pass Test A is 9.5%. If it passes Test A it goes on to Test B; 99% will pass Test B if the material is satisfactory, and 16% will pass Test B if the material is not satisfactory. If it fails Test A it goes on to Test C; 82% will pass Test C if the material is satisfactory, but only 3% will pass Test C if the material is not satisfactory. Material is accepted if it passes both Test A and Test B. Material is rejected if it fails both Test A and Test C. Material is reprocessed if it fails Test B or passes Test C.
   a) What percentage of the feed material is accepted?
   b) What percentage of the feed material is reprocessed?
   c) What percentage of the material which is reprocessed was satisfactory?

6. In a small isolated town in Northern Saskatchewan, 90% of the Cola consumed by the townspeople is purchased from the General Store, while the rest is purchased from other vendors. Records show 60% of all the bottles sold are returned. According to a special study, a bottle purchased at the General Store is four times as likely to be returned as a bottle purchased elsewhere.
   a) Calculate the probability that a person buying a bottle of Cola from the General Store will return the empty bottle.
   b) If a Cola bottle is found lying in the street, what is the probability that it was not purchased at the General Store?

7. Three road construction firms, X, Y and Z, bid for a certain contract. From past experience, it is estimated that the probability that X will be awarded the contract is 0.40, while for Y and Z the probabilities are 0.35 and 0.25. If X does receive

the contract, the probability that the work will be satisfactorily completed on time is 0.75. For Y and Z these probabilities are 0.80 and 0.70.
   a) What is the probability that Y will be awarded the contract and complete the work satisfactorily?
   b) What is the probability that the work will be completed satisfactorily?
   c) It turns out that the work was done satisfactorily. What is the probability that Y was awarded the contract?

8. Two service stations compete with one another. The odds are 3 to 1 that a motorist will go to station A rather than station B. Given that a motorist goes to station B, the probability that he will be asked whether he wants his oil checked is 0.76. A survey indicates that of the motorists who are asked whether they want the oil checked, 79% went to station A. Given that a motorist goes to station A, what is the probability that he will be asked whether he wants his oil checked?

9. A machining process produces 98.6% good components. The rest are defective. Each component passes through a pneumatic gauging system. 96% of the defective components are rejected by the gauging system, but 5% of the good components are rejected also. All components rejected by the gauging system pass through a tester. The tester accepts 98% of the good components and 12% of the defective components which reach it. The components which are accepted by the tester go a second time through the gauging system, which now accepts 92% of the good components and 6% of the defective components which pass through it. The total reject stream consists of components rejected by the tester and components rejected by the second pass through the gauging system. The total accepted stream consists of components accepted by the gauging system in either pass.
   a) What percentage of all the components are rejected?
   b) What percentage of the total reject stream was accepted by the tester?
   c) What percentage of the total reject stream are not defective?

CHAPTER 3

Descriptive Statistics: Summary Numbers

Prerequisite: A good knowledge of algebra.

The purpose of descriptive statistics is to present a mass of data in a more understandable form. We may summarize the data in numbers as (a) some form of average, or in some cases a proportion, (b) some measure of variability or spread, and (c) quantities such as quartiles or percentiles, which divide the data so that certain percentages of the data are above or below these marks. Furthermore, we may choose to describe the data by various graphical displays or by the bar graphs called histograms, which show the distribution of data among various intervals of the varying quantity. It is often necessary or desirable to consider the data in groups and determine the frequency for each group. This chapter will be concerned with various summary numbers, and the next chapter will consider grouped frequency and graphical descriptions.

Use of a computer can make treatment of massive sets of data much easier, so computer calculations in this area will be considered in detail. However, it is necessary to have the fundamentals of descriptive statistics clearly in mind when using the computer, so the ideas and relations of descriptive statistics will be developed first for pencil-and-paper calculations with a pocket calculator. Then computer methods will be introduced and illustrated with examples.

First, consider describing a set of data by summary numbers. These will include measures of a central location, such as the arithmetic mean, markers such as quartiles or percentiles, and measures of variability or spread, such as the standard deviation.

3.1 Central Location

Various "averages" are used to indicate a central value of a set of data. Some of these are referred to as *means*.

(a) Arithmetic Mean

Of these "averages," the most common and familiar is the arithmetic mean, defined by

$$\bar{x} \text{ or } \mu = \frac{1}{N}\sum_{i=1}^{N} x_i \tag{3.1}$$

Chapter 3

If we refer to a quantity as a "mean" without any specific modifier, the arithmetic mean is implied. In equation 3.1 $\bar{x}$ is the mean of a sample, and μ is the mean of a population, but both means are calculated in the same way.

The arithmetic mean is affected by all of the data, not just any selection of it. This is a good characteristic in most cases, but it is undesirable if some of the data are grossly in error, such as "outliers" that are appreciably larger or smaller than they should be. The arithmetic mean is simple to calculate. It is usually the best single average to use, especially if the distribution is approximately symmetrical and contains no outliers.

If some results occur more than once, it is convenient to take frequencies into account. If f_i stands for the frequency of result x_i, equation 3.1 becomes

$$\bar{x} \text{ or } \mu = \frac{\sum x_i f_i}{\sum f_i} \tag{3.2}$$

This is in exactly the same form as the expression for the x-coordinate of the center of mass of a system of N particles:

$$\bar{x}_{\text{C of M}} = \frac{\sum x_i m_i}{\sum m_i} \tag{3.3}$$

Just as the mass of particle i, m_i, is used as the weighting factor in equation 3.3, the frequency, f_i, is used as the weighting factor in equation 3.2.

Notice that from equation 3.1

$$N\bar{x} - \sum_{i=1}^{N} x_i = 0$$

so

$$\sum_{i=1}^{N} (x_i - \bar{x}) = 0$$

In words, the sum of all the deviations from the mean is equal to zero.

We can also write equation 3.2 as

$$\bar{x} \text{ or } \mu = \sum_{j=1}^{N} x_j \left[\frac{f_j}{\sum_{\text{all } i} f_i} \right] \tag{3.2a}$$

The quantity μ in this expression is the mean of a population. The quantity $\dfrac{f_j}{\sum_{i=1}^{n} f_i}$ is the relative frequency of x_j.

Descriptive Statistics: Summary Numbers

To illustrate, suppose we toss two coins 15 times. The possible number of heads on each toss is 0, 1, or 2. Suppose we find no heads 3 times, one head 7 times, and two heads 5 times. Then the mean number of heads per trial using equation 3.2 is

$$\bar{x} = \frac{(0)(3)+(1)(7)+(2)(5)}{3+7+5} = \frac{17}{15} = 1.13$$

The same result can be obtained using equation 3.2a.

(b) Other Means

We must not think that the arithmetic mean is the only important mean. The *geometric mean, logarithmic mean,* and *harmonic mean* are all important in some areas of engineering. The geometric mean is defined as the *n*th root of the product of *n* observations:

$$\text{geometric mean} = \sqrt[n]{x_1 x_2 x_3 \ldots x_n} \qquad (3.4)$$

or, in terms of frequencies,

$$\text{geometric mean} = \sqrt[\Sigma f_i]{(x_1)^{f_1} (x_2)^{f_2} (x_3)^{f_3} \ldots (x_{n_1})^{f_{n_1}}}$$

Now taking logarithms of both sides,

$$\log(\text{geometric mean}) = \frac{\sum f_i \log x_i}{\sum f_i} \qquad (3.5)$$

The *logarithmic mean* of two numbers is given by the difference of the natural logarithms of the two numbers, divided by the difference between the numbers, or $\frac{\ln x_2 - \ln x_1}{x_2 - x_1}$. It is used particularly in heat transfer and mass transfer.

The *harmonic mean* involves inverses—i.e., one divided by each of the quantities. The harmonic mean is the inverse of the arithmetic mean of all the inverses, so

$$\frac{1}{\frac{1}{x_1} + \frac{1}{x_2} + \ldots}$$

In this book we will not be concerned further with logarithmic or harmonic means.

(c) Median

Another representative quantity, quite different from a mean, is the *median*. If all the items with which we are concerned are sorted in order of increasing magnitude (size), from the smallest to the largest, then the median is the middle item. Consider the five items: 12, 13, 21, 27, 31. Then 21 is the median. If the number of items is even, the median is given by the arithmetic mean of the two middle items. Consider the six items: 12, 13, 21, 27, 31, 33. The median is (21 + 27) / 2 = 24. If we interpret an

Chapter 3

item that is right at the median as being half above and half below, then in all cases the median is the value exceeded by 50% of the observations.

One desirable property of the median is that it is not much affected by outliers. If the first numerical example in the previous paragraph is modified by replacing 31 by 131, the median is unchanged, whereas the arithmetic mean is changed appreciably. But along with this advantage goes the disadvantage that changing the size of any item without changing its position in the order of magnitude often has no effect on the median, so some information is lost. If a distribution of items is very asymmetrical so that there are many more items larger than the arithmetic mean than smaller (or vice-versa), the median may be a more useful representative quantity than the arithmetic mean. Consider the seven items: 1, 1, 2, 3, 4, 9, 10. The median is 3, with as many items smaller than it as larger. The mean is 4.29, with five items smaller than it, but only two items larger.

(d) Mode

If the frequency varies from one item to another, the *mode* is the value which appears most frequently. As some of you may know, the word "mode" means "fashion" in French. Then we might think of the mode as the most "fashionable" item. In the case of continuous variables the frequency depends upon how many digits are quoted, so the mode is more usefully considered as the *midpoint of the class with the largest frequency* (see the grouped frequency approach in section 4.4). Using that interpretation, the mode is affected somewhat by the class width, but this influence is usually not very great.

3.2 Variability or Spread of the Data

The following groups all have the same mean, 4.25:

       Group A: 2, 3, 4, 8
       Group B: 1, 2, 4, 10
       Group C: 0, 1, 5, 11

These data are shown graphically in Figure 3.1.

It is clear that Group B is more variable (shows a larger spread in the numbers) than Group A, and Group C is more variable than Group B. But we need a quantitative measure of this variability.

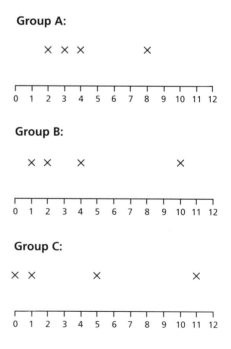

Figure 3.1: Comparison of Groups

Descriptive Statistics: Summary Numbers

(a) Sample Range

One simple measure of variability is the *sample range*, the difference between the smallest item and the largest item in each sample. For Group A the sample range is 6, for Group B it is 9, and for Group C it is 11. For small samples all of the same size, the sample range is a useful quantity. However, it is not a good indicator if the sample size varies, because the sample range tends to increase with increasing sample size. Its other major drawback is that it depends on only two items in each sample, the smallest and the largest, so it does not make use of all the data. This disadvantage becomes more serious as the sample size increases. Because of its simplicity, the sample range is used frequently in quality control when the sample size is constant; simplicity is particularly desirable in this case so that people do not need much education to apply the test.

(b) Interquartile Range

The *interquartile range* is the difference between the upper quartile and the lower quartile, which will be described in section 3.3. It is used fairly frequently as a measure of variability, particularly in the Box Plot, which will be described in the next chapter. It is used less than some alternatives because it is not related to any of the important theoretical distributions.

(c) Mean Deviation from the Mean

The *mean deviation from the mean*, defined as $\sum_{i=1}^{N}(x_i - \bar{x})/N$, where $\bar{x} = \sum x_i / N$, is *useless* because it is always zero. This follows from the discussion of the sum of deviations from the mean in section 3.1 (a).

(d) Mean Absolute Deviation from the Mean

However, the *mean absolute deviation from the mean*, defined as $\sum_{i=1}^{N}|x_i - \bar{x}|/N$ is used frequently by engineers to show the variability of their data, although it is usually not the best choice. Its advantage is that it is simpler to calculate than the main alternative, the standard deviation, which will be discussed below. For Groups A, B, and C the mean absolute deviation is as follows:

Group A: (2.25 + 1.25 + 0.25 + 3.75)/4 = 7.5/4 = 1.875.
Group B: (3.25 + 2.25 + 0.25 + 5.75)/4 = 11.5/4 = 2.875.
Group C: (4.25 + 3.25 + 0.75 + 6.75)/4 = 15/4 = 3.75.

Its disadvantage is that it is not simply related to the parameters of theoretical distributions. For that reason its routine use is not recommended.

(e) Variance

The *variance* is one of the most important descriptions of variability for engineers. It is defined as

Chapter 3

$$\sigma^2 = \frac{\sum_{i=1}^{N}(x_i - \mu)^2}{N} \quad (3.6)$$

In words it is the mean of the squares of the deviations of each measurement from the mean of the population. Since squares of both positive and negative real numbers are always positive, the variance is always positive. The symbol μ stands for the mean of the entire population, and σ^2 stands for the variance of the population. (Remember that in Chapter 1 we defined the population as a particular characteristic of all the items in which we are interested, such as the diameters of all the bolts produced under normal operating conditions.) Notice that variance is defined *in terms of the population mean*, μ. When we calculate the results from a sample (i.e., a part of the population) we do not usually know the population mean, so we must find a way to use the sample mean, which we can calculate. Notice also that the variance has units of the quantity squared, for example m^2 or s^2 if the original quantity was measured in meters or seconds, respectively. We will find later that the variance is an important parameter in probability distributions used widely in practice.

(f) Standard Deviation

The standard deviation is *extremely* important. It is defined as the square root of the variance:

$$\sigma = \sqrt{\frac{\sum_{i=1}^{N}(x_i - \mu)^2}{N}} \quad (3.7)$$

Thus, it has the same units as the original data and is a representative of the deviations from the mean. Because of the squaring, it gives more weight to larger deviations than to smaller ones. Since the variance is the mean square of the deviations from the population mean, the standard deviation is the root-mean-square deviation from the population mean. Root-mean-square quantities are also important in describing the alternating current of electricity. An analogy can be drawn between the standard deviation and the radius of gyration encountered in applied mechanics.

(g) Estimation of Variance and Standard Deviation from a Sample

The definitions of equations 3.6 and 3.7 can be applied directly if we have data for the complete population. But usually we have data for only a sample taken from the population. We want to *infer* from the data for the sample the parameters for the population. It can be shown that the sample mean, $\bar{x}$, is an unbiased estimate of the population mean, μ. This means that if very large random samples were taken from the population, the sample mean would be a good approximation of the population mean, with no systematic error but with a random error which tends to become smaller as the sample size increases.

Descriptive Statistics: Summary Numbers

However, if we simply substitute $\bar{x}$ for μ in equations 3.6 and 3.7, there will be a systematic error or bias. This procedure would *underestimate* the variance and standard deviation of the population. This is because the sum of squares of deviations from the sample mean, $\bar{x}$, is smaller than the sum of squares of deviations from *any* other constant value, including μ. $\bar{x}$ is an unbiased estimate of μ, but in general $\bar{x} \neq \mu$, so just substituting $\bar{x}$ for μ in equations 3.6 and 3.7 would tend to give estimates of variance and standard deviation that are too small. To illustrate this, consider the four numbers 11, 13, 10, and 14 as a sample. Their sample mean is 12. They might well come from a population of mean 13. Then the sum of squares of deviations from the population mean, $\sum_i (x_i - \mu)^2 = (11 - 13)^2 + (13 - 13)^2 + (10 - 13)^2 + (14 - 13)^2 = 2^2 + 0^2 + 3^2 + 1^2 = 14$, whereas $\sum_i (x_i - \bar{x})^2 = (11 - 12)^2 + (13 - 12)^2 + (10 - 12)^2 + (14 - 12)^2 = 1^2 + 1^2 + 2^2 + 2^2 = 10$. Thus, $\dfrac{\sum_i (x_i - \bar{x})^2}{N}$ would underestimate the variance.

The estimate of variance obtained using the sample mean in place of the population mean can be made unbiased by multiplying by the factor $\left(\dfrac{N}{N-1}\right)$. This is called Bessel's correction. The estimate of σ^2 is given the symbol s^2 and is called the variance estimated from a sample, or more briefly the *sample variance*. Sometimes this estimate will be high, sometimes it will be low, but in the long run it will show no bias if samples are taken randomly. The result of Bessel's correction is that we have

$$s^2 = \frac{\sum_{i=1}^{N}(x_i - \bar{x})^2}{N-1} \qquad (3.8)$$

The standard deviation is always the square root of the corresponding variance, so s is called the *sample standard deviation*. It is the estimate from a sample of the standard deviation of the population from which the sample came. The sample standard deviation is given by

$$s^2 = \sqrt{\frac{\sum_{i=1}^{N}(x_i - \bar{x})^2}{N-1}} \qquad (3.9)$$

Equations 3.8 and 3.9 (or their equivalents) should be used to calculate the variance and standard deviation from a sample unless the population mean is known. If the population mean is known, as when we know *all* the members of the population, we should use equations 3.6 and 3.7 directly. Notice that when N is very large, Bessel's correction becomes approximately 1, so then it might be neglected. How-

Chapter 3

ever, to avoid error we should always use equations 3.8 and 3.9 (or their equivalents) unless the population mean is known accurately.

(h) Method for Faster Calculation

A modification of equations 3.6 to 3.9 makes calculation of variance and standard deviation faster. In most cases in this book we have omitted derivations, but this case is an exception because the algebra is simple and may be helpful.

Equations 3.8 and 3.9 include the expression

$$\sum(x_i - \bar{x})^2 = \sum x_i^2 - 2\bar{x}\sum x_i + N\bar{x}^2$$

But by definition $\bar{x} = \dfrac{\sum x_i}{N}$

Then we have

$$\sum(x_i - \bar{x})^2 = \sum x_i^2 - \frac{2(\sum x_i)^2}{N} + \frac{N(\sum x_i)^2}{N^2}$$

$$= \sum x_i^2 - \frac{(\sum x_i)^2}{N} \quad (3.10)$$

Notice that $\sum x_i^2$ means we should square all the x's and then add them up. On the other hand, $(\sum x_i)^2$ means we should add up all the x's and square the result. They are not the same.

An alternative to equation 3.10 is

$$\sum(x_i - \bar{x})^2 = \sum x_i^2 - N(\bar{x})^2 \quad (3.10a)$$

Then we have

$$s^2 = \frac{\sum_{i=1}^{N} x_i^2 - \dfrac{\left(\sum_{i=1}^{N} x_i\right)^2}{N}}{N-1} = \frac{\sum_{i=1}^{N} x_i^2 - N(\bar{x})^2}{N-1} \quad (3.11)$$

It is often convenient to use equation 3.11 in the form for frequencies:

$$s^2 = \frac{\sum f_i x_i^2 - (\sum f_i x_i)^2 / (\sum f_i)}{(\sum f_i - 1)} \quad (3.12)$$

Descriptive Statistics: Summary Numbers

Equations 3.6 and 3.7 include $\sum_{i=1}^{N}(x_i - \mu)^2$, where for a complete population $\mu = \frac{1}{N}\sum_{i=1}^{N} x_i$. Then similar expressions to equations 3.10 to 3.12 (but dividing by N instead of $(N-1)$) apply for cases where the complete population is known.

The modified equations such as equation 3.11 or 3.12 should be used for calculation of variance (and the square root of one of them should be used for calculation of standard deviation) by hand or using a good pocket calculator because it involves fewer arithmetic operations and so is faster. However, some thought is required if a digital computer is used. That is because some computers carry relatively few significant figures in the calculation. Since in equation 3.11 the quantities $\sum_{i=1}^{N} x_i^2$ and $\frac{\left(\sum_{i=1}^{N} x_i\right)^2}{N}$ or $N(\bar{x})^2$ are of similar magnitudes, the differences in equation 3.11 may involve catastrophic loss of significance because of rounding of figures in the computation. Most present-day computers and calculators, however, carry enough significant figures so that this "loss of significance" is not usually a serious problem, but the possibility of such a difficulty should be considered. It can often be avoided by subtracting a constant quantity from each number, an operation which does not change the variance or standard deviation. For example, the variance of 3617.8, 3629.6, and 3624.9 is exactly the same as the variance of 17.8, 29.6, and 24.9. However, the number of figures in the squared terms is much smaller in the second case, so the possibility of loss of significance is greatly reduced. Then in general, fewer figures are required to calculate variance by subtracting the mean from each of the values, then squaring, adding, and dividing by the number of items (i.e., using equation 3.8 directly), but this adds to the number of arithmetic operations and so requires more time for calculations. If the calculating device carries enough significant figures to allow 3.11 or 3.12 to be used, that is the preferred method.

Microsoft Excel carries a precision of about 15 decimal digits in each numerical quantity. Statistical calculations seldom require greater precision in any final answer than four or five decimal digits, so "loss of significance" is very seldom a problem if Excel is being used. A comparison to verify that statement in a particular case will be included in Example 4.4.

(1) Illustration of Calculation

Now let us return to an example of calculations using the groups of numbers listed at the beginning of section 3.2.

Example 3.1

The numbers were as follows:

Chapter 3

> Group A: 2, 3, 4, 8
> Group B: 1, 2, 4, 10
> Group C: 0, 1, 5, 11

Find the sample variance and the sample standard deviation of each group of numbers. Use both equation 3.8 and equation 3.11 to check that they give the same result.

Answer: Since the mean of Group A (and also of the other groups) is 4.25, the sample variance of Group A using the basic definition, equation 3.8, is

$$[(2 - 4.25)^2 + (3 - 4.25)^2 + (4 - 4.25)^2 + (8 - 4.25)^2] / (4 - 1)$$
$$= [5.0625 + 1.5625 + 0.0625 + 14.0625] / 3 = 20.75 / 3 = 6.917,$$

so the sample standard deviation is $\sqrt{6.917} = 2.630$.

The variance of Group A calculated by equation 3.11 is

$$[2^2 + 3^2 + 4^2 + 8^2 - (4)(4.25)^2] / (4 - 1) = [4 + 9 + 16 + 64 - 72.25] / 3 = 6.917$$

(again). We can see that the advantage of equation 3.11 is greater when the mean is not a simple integer.

Using equation 3.11 on Group B gives

$$[1^2 + 2^2 + 4^2 + 10^2 - (4)(4.25)^2] / (4 - 1) = [1 + 4 + 16 + 100 - 72.25] / 3 = 48.75 / 3 = 16.25$$

for the sample variance, so the sample standard deviation is 4.031.

Using equation 3.11 on Group C gives

$$[0^2 + 1^2 + 5^2 + 11^2 - (4)(4.25)^2] / (4 - 1) = [0 + 1 + 25 + 121 - 72.25] / 3 = 74.75 / 3 = 24.917$$

for the variance, so the standard deviation is 4.992.

(j) Coefficient of Variation

A dimensionless quantity, the *coefficient of variation* is the ratio between the standard deviation and the mean for the same set of data, expressed as a percentage. This can be either (σ / μ) or $(s / \bar{x})$, whichever is appropriate, multiplied by 100%.

(k) Illustration: An Anecdote

A brief story may help the reader to see why variability is often important. Some years ago a company was producing nickel powder, which varied considerably in particle size. A metallurgical engineer in technical sales was given the task of developing new customers in the alloy steel industry for the powder. Some potential buyers said they would pay a premium price for a product that was more closely sized. After some discussion with the management of the plant, specifications for three new products were developed: fine powder, medium powder, and coarse powder. An order was obtained for fine powder. Although the specifications for this fine powder were within the size range of powder which had been produced in the past, the engineers in the plant found that very little of the powder produced at their best

Descriptive Statistics: Summary Numbers

guess of the optimum conditions would satisfy the specifications. Thus, the mean size of the specification was satisfactory, but the specified variability was not satisfactory from the point of view of production. To make production of fine powder more practical, it was necessary to change the specifications for "fine powder" to correspond to a larger standard deviation. When this was done, the plant could produce fine powder much more easily (but the customer was not willing to pay such a large premium for it!).

3.3 Quartiles, Deciles, Percentiles, and Quantiles

Quartiles, deciles, and percentiles divide a frequency distribution into a number of parts containing equal frequencies. The items are first put into order of increasing magnitude. *Quartiles* divide the range of values into four parts, each containing one quarter of the values. Again, if an item comes exactly on a dividing line, half of it is counted in the group above and half is counted below. Similarly, *deciles* divide into ten parts, each containing one tenth of the total frequency, and *percentiles* divide into a hundred parts, each containing one hundredth of the total frequency. If we think again about the median, it is the second or middle quartile, the fifth decile, and the fiftieth percentile. If a quartile, decile, or percentile falls between two items in order of size, for our purposes the value halfway between the two items will be used. Other conventions are also common, but the effect of different choices is usually not important. Remember that we are dealing with a quantity which varies randomly, so another sample would likely show a different quartile or decile or percentile.

For example, if the items after being put in order are 1, 2, 2, 3, 5, 6, 6, 7, 8, a total of nine items, the first or lower quartile is $(2 + 2)/2 = 2$, the median is 5, and the upper or third quartile is $(6 + 7)/2 = 6.5$.

Example 3.2

To start a program to improve the quality of production in a factory, all the products coming off a production line, under what we have reason to believe are normal operating conditions, are examined and classified as "good" products or "defective" products. The number of defective products in each successive group of six is counted. The results for 60 groups, so for 360 products, are shown in Table 3.1. Find the mean, median, mode, first quartile, third quartile, eighth decile, ninth decile, proportion defective in the sample, first estimate of probability that an item will be defective, sample variance, sample standard deviation, and coefficient of variation.

Table 3.1: Numbers of Defectives in Groups of Six Items

1	0	0	0	0	0	0	0	0	0	0	0
0	0	0	0	1	0	0	0	0	0	1	0
0	1	0	0	1	0	0	0	0	2	0	0
0	0	0	0	2	0	0	1	0	0	1	0
1	0	0	0	0	1	0	0	1	0	0	0

Chapter 3

Answer: The data in Table 3.1 can be summarized in terms of frequencies. If x_i represents the number of defectives in a group of six products and f_i represents the frequency of that occurrence, Table 3.2 is a summary of Table 3.1.

Table 3.2: Frequencies for Numbers of Defectives

Number of defectives, x_i	Frequency, f_i
0	48
1	10
2	2
>2	0

Then the mean number of defectives in a group of six products is
$$\frac{(48)(0)+(10)(1)+(2)(2)}{48+10+2} = \frac{14}{60} = 0.233$$
Notice that the mean is not necessarily a possible member of the set: in this case the mean is a fraction, whereas each number of defectives must be a whole number.

Among a total of 60 products, the median is the value between the 30th and 31st products in order of increasing magnitude, so (0 + 0) / 2 = 0.

The mode is the most frequent value, so 0.

The lower or first quartile is the value between the 15th and 16th products in order of size, thus between 0 and 0, so 0. The upper or third quartile is the value between the 45th and 46th products in order of size, thus between 0 and 0, so again 0. The eighth decile is the value larger than the 48th item and smaller than the 49th item, so between 0 and 1, or 0.5. The ninth decile is the value between the 54th and the 55th products, so between 1 and 1, so 1.

We have 14 defective products in a sample of 360 items, so the proportion defective in this sample is 14 / 360 = 0.0389 or 0.039. As we have seen from section 2.1, proportion or relative frequency gives an estimate of probability. Then we can estimate the probability that an item, chosen randomly from the population from which the sample came, will be defective. For this sample that first estimate of the probability that a randomly chosen item in the population will be defective is 0.039. This estimate is not very precise, but it would get better if the size of the sample were increased.

Now let us calculate the sample variance and standard deviation using equation 3.12:
$\sum f_i x_i^2 = (48)(0)^2 + (10)(1)^2 + (2)(2)^2 = 18$
$\sum f_i x_i = (48)(0) + (10)(1) + (2)(2) = 14$
$\sum f_i = 48 + 10 + 2 = 60$

Then from equation 3.12, $s^2 = \dfrac{\sum f_i x_i^2 - \left(\sum f_i x_i\right)^2 / \left(\sum f_i\right)}{\left(\sum f_i - 1\right)}$,

Descriptive Statistics: Summary Numbers

which gives $s^2 = \dfrac{18 - \left((14)^2/60\right)}{60-1} = 0.2497$,

so $s = 0.4997$ or 0.500.

The coefficient of variation is $\left(\dfrac{s}{\bar{x}}\right)(100\%) = \left(\dfrac{0.4997}{0.2333}\right)(100\%) = 214\%$.

The general term for a parameter which divides a frequency distribution into parts containing stated proportions of a distribution is a *quantile*. The symbol $Q(f)$ is used for the quantile, which is larger than a fraction f of a distribution. Then a lower quartile is $Q(0.25)$ or $Q(1/4)$, and an upper quartile is $Q(0.75)$.

In fact, if items are sorted in order of increasing magnitude, from the smallest to the largest, *each* item can be considered some sort of quantile, on a dividing line so that half of the item is above the line and half below. Then the *i*th item of a total of *n* items is a quantile larger than $(i - 0.5)$ items of the *n*, so the $\left(\dfrac{i-0.5}{n}\right)$ quantile or $Q\left(\dfrac{i-0.5}{n}\right)$. Say the sorted items are 1, 4, 5, 6, 7, 8, 9, a total of seven items. Think of each one as being exactly on a dividing line, so half above and half below the line. Then the second item, 4, is larger than one-and-a half items of the seven, so we can call it the $\dfrac{1.5}{7}$ quantile or $Q(0.21)$. Similarly, 5 is larger than two-and-a-half items of the seven, so it is the $\dfrac{2.5}{7}$ quantile or $Q(0.36)$. For purposes of illustration we are using small sets of numbers, but quantiles are useful in practice principally to characterize large sets of data.

Since proportion from a set of data gives an estimate of the corresponding probability, the quantile $Q\left(\dfrac{i-0.5}{n}\right)$ gives an estimate of the probability that a variable is smaller than the *i*th item in order of increasing magnitude. If an item is repeated, we have two separate estimates of this probability.

We can also use the general relation to find various quantiles. If we have a total of *n* items, then $Q\left(\dfrac{i-0.5}{n}\right)$ will be given by the *i*th item, even if *i* is not an integer. Consider again the seven items which are 1,4,5,6,7,8,9. The median, $Q\left(\dfrac{1}{2}\right)$, would be the item for which $\dfrac{i-0.5}{7} = \dfrac{1}{2}$, so $i = (7)\left(\dfrac{1}{2}\right) + 0.5 = 4$; that is, the fourth item, which is 6. That agrees with the definition given in section 3.1. Now, what is the first or lower quartile? This would be a value larger than one quarter of the items, or $Q(0.25)$. Then $\dfrac{i-0.5}{7} = \dfrac{1}{4}$, so $i = (7)\left(\dfrac{1}{4}\right) + 0.5 = 2.25$. Since this is a fraction, the

Chapter 3

first quartile would be between the second and third items in order of magnitude, so between 4 and 5. Then by our convention we would take the first quartile as 4.5. Similarly, for the third quartile, $Q(0.75)$, so we have $\frac{i-0.5}{7}=\frac{3}{4}$, $i = 5.75$, and the third quartile is between the 5th and 6th items in order of magnitude (7 and 8) and so is taken as $(7 + 8) / 2 = 7.5$.

Example 3.3

Consider the sample consisting of the following nine results :

2.3, 7.2, 3.7, 4.6, 5.0, 7.0, 3.7, 4.9, 4.2.

a) Find the median of this set of results by two different methods.
b) Find the lower quartile.
c) Find the upper quartile.
d) Estimate the probability that an item, from the population from which this sample came, would be less than 4.9.
e) Estimate the probability that an item from that population would be less than 3.7.

Answer: The first step is to sort the data in order of increasing magnitude, giving the following table:

i	1	2	3	4	5	6	7	8	9
$x(i)$	2.3	3.7	3.7	4.2	4.6	4.9	5	7	7.2

a) The basic definition of the median as the middle item after sorting in order of increasing magnitude gives $x(5) = 4.6$. Putting $\frac{i-0.5}{9} = 0.5$ gives $i = (9)(0.5) + 0.5 = 5$, so again the median is $x(5) = 4.6$.

b) The lower quartile is obtained by putting $\frac{i-0.5}{9} = 0.25$, which gives $i = (9)(0.25) + 0.5 = 2.75$. Since this is a fraction, the lower quartile is $\frac{x(2)+x(3)}{2} = \frac{3.7+3.7}{2} = 3.7$.

c) The upper quartile is obtained by putting $\frac{i-0.5}{9} = 0.75$, which gives $i = (9)(0.75) + 0.5 = 7.25$. Since this is again a fraction, the upper quartile is $\frac{x(7)+x(8)}{2} = \frac{5+7}{2} = 6$.

d) Probabilities of values smaller than the various items can be estimated as the corresponding fractions. 4.9 is the 6th item of the 9 items in order of increasing magnitude, and $\frac{6-0.5}{9} = 0.61$. Then the probability that an item, from the population from which this sample came, would be less than 4.9 is estimated to be 0.61.

e) 3.7 is the item of order both 2 and 3, so we have two estimates of the probability that an item from the same population would be less than 3.7. These are $\frac{2-0.5}{9}$ and $\frac{3-0.5}{9}$, or 0.17 and 0.28.

3.4 Using a Computer to Calculate Summary Numbers

A personal computer, either a PC or a Mac, is very frequently used with a spreadsheet to calculate the summary numbers we have been discussing. One of the spreadsheets used most frequently by engineers is Microsoft® Excel, which includes a good number of statistical functions. Excel will be used in the computer methods discussed in this book.

Using a computer can certainly reduce the labor of characterizing a large set of data. In this section we will illustrate using a computer to calculate useful summary numbers from sets of data which might come from engineering experiments or measurements. The instructions will assume the reader is already reasonably familiar with Microsoft Excel; if not, he or she should refer to a reference book on Excel; a number are available at most bookstores. Some of the main techniques useful in statistical calculations and recommended for use during the learning process are discussed briefly in Appendix B. Calculations involving formulas, functions, sorting, and summing are among the computer techniques most useful during both the learning process and subsequent applications, so they and simple techniques for producing graphs are discussed in that appendix. Furthermore, in Appendix C there is a brief listing of methods which are useful in practice for Excel once the concepts are thoroughly understood, but they should not be used during the learning process.

The *Help* feature on Excel is very useful and convenient. Access to it can be obtained in various ways, depending on the version of Excel which is being used. There is usually a Help menu, and sometimes there is a Help tool (marked by an arrow and a question mark, or just a question mark).

Further discussion and examples of the use of computers in statistical calculations will be found in section 4.5, Chapter 4. Some probability functions which can be evaluated using Excel will be discussed in later chapters.

Example 3.4

The numbers given at the beginning of section 3.2 were as follows:

Group A: 2, 3, 4, 8
Group B: 1, 2, 4, 10
Group C: 0, 1, 5, 11

Chapter 3

Find the sample variance and the sample standard deviation of each group of numbers. Use both equation 3.8 and equation 3.11 to check that they give the same result. This example is mostly the same as Example 3.1, but now it will be done using Excel.

Answer:

Table 3.3: Excel Worksheet for Example 3.4

	A	B	C	D	E
			Group A	Group B	Group C
1					
2	Entries		2	1	0
3			3	2	1
4			4	4	5
5			8	10	11
6	Sum		17	17	17
7	Arith. Mean	C6/4=, etc.	4.25	4.25	4.25
8	Deviations	C2-C7=, etc.	−2.25	−3.25	−4.25
9		C3-C7=, etc.	−1.25	−2.25	−3.25
10		C4-C7=, etc.	−0.25	−0.25	0.75
11		C5-C7=, etc.	3.75	5.75	6.75
12					
13	Deviations Sqd	C8^2=,etc	5.0625	10.5625	18.0625
14			1.5625	5.0625	10.5625
15			0.0625	0.0625	0.5625
16			14.0625	33.0625	45.5625
17	Sum Devn Sqd	Sums	20.75	48.75	74.75
18	Variance	C17/3=, etc.	**6.917**	**16.25**	**24.92**
19					
20	Entries Sqd	C2^2=, etc.	4	1	0
21			9	4	1
22			16	16	25
23			64	100	121
24	Sum Entries Sqd	Sums	93	121	147
25	Correction	4*C7^2=, etc.	72.25	72.25	72.25
26	Corrected Sum	C24-C25=, etc.	20.75	48.75	74.75
27	Variance	C26/3=, etc.	**6.917**	**16.25**	**24.92**
28	Std Dev, s	SQRT(C27)=, etc.	**2.630**	**4.031**	**4.992**

Descriptive Statistics: Summary Numbers

The worksheet is shown in Table 3.3. The letters A, B, C, etc. across the top are the column references, and the numbers 1, 2, 3, etc. on the left-hand side are the row references. The headings for Groups A, B, and C were placed in columns C, D, and E of row 1. Names of quantities were placed in column A. Statements of formulas are given in column B. The individual entries or values were placed in cells C2:E5, that is, rows 2 to 5 of columns C to E. Cell C6 was selected, and the AutoSum tool (see section (d) of Appendix B) was used to find the sum of the entries in Group A. The sums of the entries in the other two groups were found similarly. Note that the AutoSum tool may not choose the right set of cells to be summed in cell E6. Cell C7 was selected, and the formula =C6/4 was typed into it and entered, giving the result 4.25. Then the formula in cell C7 was copied, then pasted into cell D7 (to appear as =D6/4 because relative references were used) and entered; the same content was pasted into cell E7 as =E7/4 and entered. Again both results were 4.25.

According to equation 3.8 the sample variance is given by $s^2 = \dfrac{\sum_{i=1}^{N}(x_i - \bar{x})^2}{N-1}$.

Deviations from the arithmetic means were calculated in rows 8 to 11. Cell C8 was selected, and the formula =C2–C7 was typed into it and entered, giving the result –2.25. Notice that now, although the reference C2 is relative, the reference C7 is absolute. Then when the formula in cell C8 was copied, then pasted into cell C9, the formula became = C3 – C7; the formula was entered, giving the result –1.25. Pasting the formula into cells C10 and C11 and entering gave the results –0.25 and (+)3.75. Similarly, the formula = D2 – D7 was entered in cell D8 and copied to cells D9, D10, D11 and entered in each case. A similar formula was entered in cell E8, copied separately to cells E9, E10, E11, and entered in each.

Deviations were squared in rows 13 to 16. The formula = C8^2 in cell C13 was copied to cells D13 and E13, and similar operations were carried out in cells C14:E14, C15:E15, and C16:E16. Deviations were summed using the AutoSum tool in cells C17:E17, but we have to be careful again with the sum in cell E17. Then variances are the quantities in cells C17:E17 divided in each case by 4 – 1 = 3. Therefore the formula C17/3 was entered in cell C18, then copied to cell D18 and modified to D17/3 before being entered, and similarly for cell E18. As the quantities in cells C18:E18 were answers to specific questions, they were put in bold type by choosing the Bold tool (marked with B) on the standard tool bar. Furthermore, they were put in a format with three decimal places by choosing the Format menu, the Number format, Number, then writing in the code 0.000 before choosing OK or Return. This gave the answers according to equation 3.8.

According to equation 3.11 the sample variance is given by $s^2 = \dfrac{\sum_{i=1}^{N} x_i^2 - N(\bar{x})^2}{N-1}$.

Chapter 3

Squares of entries were placed in cells C20:E23 by entering =C2^2 in cell C20, copying, then pasting in cells D20 and E20, and repeating with modifications in C21:E21, C22:E22, and C23:E23. The squares of entries were summed using the AutoSum tool in cells C24, D24, and E24. Four times the squares of the arithmetic means, 4*C7^2, 4*D7^2, and 4*E7^2, were entered in cells C25, D25, and E25 respectively. These quantities were subtracted from the sums of squares of entries by entering =C24-C25 in cell C26, and corresponding quantities in cells D26 and E26. Then values of variance according to equation 3.11 were found in cells C27, D27, and E27. These also were put in bold type and formatted for three decimal places. Finally, standard deviations were found in cells C28, D28, and E28 by taking the square roots of the variances in cells C27, D27, and E27. As answers, these also were put in bold type and formatted for three decimals.

The results verify that equations 3.8 and 3.11 give the same results, but equation 3.11 generally involves fewer arithmetic operations.

Using Excel on a computer can save a good deal of time if the data set is large, but if as here the data set is small, hand calculations are probably quicker. Results of experimental studies often give very big data sets, so computer calculations are very often advantageous.

Example 3.5

To start a program to improve the quality of production in a factory, all the items coming off a production line, under what we have reason to believe are normal operating conditions, are examined and classified as "good" items or "defective" items. The number of defective items in each successive group of six is counted. The results for 60 groups, 360 items, are shown in Table 3.4. Find the mean, median, mode, first quartile, third quartile, eighth decile, ninth decile, proportion defective in the sample, first estimate of probability that an item will be defective, sample variance, sample standard deviation, and coefficient of variation.

Table 3.4: Numbers of Defectives in Groups of Six Items

1	0	0	0	0	0	0	0	0	0	0	0
0	0	0	0	1	0	0	0	0	0	1	0
0	1	0	0	1	0	0	0	0	2	0	0
0	0	0	0	2	0	0	1	0	0	1	0
1	0	0	0	0	1	0	0	1	0	0	0

This is the same as Example 3.2, but now we will use Excel.

Answer: The data of Table 3.4 were entered in column A of an Excel work sheet; extracts are shown in Table 3.5. These data were copied to column B, then sorted in ascending order as described in section (c) of Appendix B. The order numbers were

Descriptive Statistics: Summary Numbers

obtained in column C using the AutoFill feature with the fill handle, as also described in that section of Appendix B. Rows 3 to 62 show part of the discrete data of Example 3.2 after sorting and numbering on Microsoft Excel.

Table 3.5: Extracts of Work Sheet for Example 3.5

	A	B	C	D	E
1	Numbers of Defective Items				
2	Unsorted	Sorted	Order No.		
3	1	0	1		
4	0	0	2		
5	0	0	3		
	..	..	..	..	..
49	1	0	47		
50	0	0	48		
51	1	1	49		
52	0	1	50		
	..	..	..	..	..
60	0	1	58		
61	0	2	59		
62	0	2	60		

	A	B	C	D	E
64	Number	Frequency			
65	xi	fi	xi*fi	xi^2*fi	
66			A67*B67=, etc.	A67^2*B67=, etc.	
67	0	48	0	0	
68	1	10	10	10	
69	2	2	4	8	
70	Total=SUM	60	14	18	
71					
72	xbar=	C70/B70=			0.233
73	s^2=	(D70-(C70^2/B70))/(B70-1)=			0.250
74	s=	SQRT(E73)=			0.500
75	Coeff. of var.=	E74/E72=			214%

Chapter 3

With the sorted data in column B of Table 3.7 and the order numbers in column C, it is easy to pick off the frequencies of various numbers of defectives. Thus, the number of groups containing zero defectives is 48, the number containing one defective is 58 − 48 = 10, and the number containing two defectives is 60 − 58 = 2. The resulting numbers of defectives and the frequency of each were marked in cells A64:B69. The mode is the number of defectives with the largest frequency, so it is 0 in this example. Products x_i*f_i and $x_i^2*f_i$ were found in cells C65:D69. The formulas were entered in the form for relative references in cells C67 and D67, so copying them one and two lines below gave appropriate products. Then the Autosum tool (marked Σ) on the standard toolbar was used to sum the columns for each of f_i, x_if_i, and $x_i^2f_i$ and enter the results in row 70. The sum of the calculated frequencies should check with the total number of groups, which is 60 in this case. Then from equation 3.2, $\bar{x} = \dfrac{\sum f_i x_i}{\sum f_i} = \dfrac{14}{60} = 0.233$ in cell E72. From equation 3.12,

$$s^2 = \frac{\sum f_i x_i^2 - \left(\sum f_i x_i\right)^2 / \sum f_i}{\sum f_i - 1} = \frac{18 - (14)^2/60}{60-1} = 0.250$$ in cell E73, and the sample

standard deviation, s, is found in cell E74, with a result of 0.500. The coefficient of variation is given in cell E75 as 214%. Of course, all quantities must be clearly labeled on the spreadsheet. Labels are shown in rows 1, 2, 64, 65, 70, and 72 to 75, and explanations are given in rows 66 and 72 to 75.

Problems

1. The same dimension was measured on each of six successive parts as they came off a production line. The results were 21.14 mm, 21.87 mm, 21.53 mm, 21.37 mm, 21.61 mm and 21.93 mm. Calculate the mean and median.
2. For the measurements given in problem 1 above, find the variance, standard deviation, and coefficient of variation
   a) considering this set of values as a complete population, and
   b) considering this set of values as a sample of all possible measurements of this dimension.
3. Four items in a sequence were measured as 50, 160, 100, and 400 mm. Find their arithmetic mean, geometric mean, and median.
4. The temperature in a chemical reactor was measured every half hour under the same conditions. The results were 78.1°C, 79.2°C, 78.9°C, 80.2°C, 78.3°C, 78.8°C, 79.4°C. Calculate the mean, median, lower quartile, and upper quartile.
5. For the temperatures of problem 4, calculate the variance, standard deviation, and coefficient of variation
   a) considering this set of values as a complete population, and
   b) considering this set of values as a sample of all possible measurements of the temperature under these conditions.

Descriptive Statistics: Summary Numbers

6. The times to perform a particular step in a production process were measured repeatedly. The times were 20.3 s, 19.2 s, 21.5 s, 20.7 s, 22.1 s, 19.9 s, 21.2 s, 20.6 s. Calculate the arithmetic mean, geometric mean, median, lower quartile, and upper quartile.

7. For the times of problem 6, calculate the variance, standard deviation, and coefficient of variation
   a) considering this set of values as a complete population, and
   b) considering this set of values as a sample of all possible measurements of the times for this step in the process.

8. The numbers of defective items in successive groups of fifteen items were counted as they came off a production line. The results can be summarized as follows:

No. of Defectives	Frequency
0	57
1	57
2	18
3	5
4	3
>4	0

   a) Calculate the mean number of defectives in a group of fifteen items.
   b) Calculate the variance and standard deviation of the number of defectives in a group. Take the given data as a sample.
   c) Find the median, lower quartile, upper quartile, ninth decile, and 95th percentile.
   d) On the basis of these data estimate the probability that the next item produced will be defective.

9. Electrical components were examined as they came off a production line. The number of defective items in each group of eighteen components was recorded. The results can be summarized as follows:

No. of Defectives	Frequency
0	94
1	52
2	19
3	3
>3	0

   a) Calculate the mean number of defectives in a group of 18 components.
   b) Taking the given data as a sample, calculate the variance and standard deviation of the number of defectives in a group.
   c) Find the median, lower quartile, upper quartile, and 95th percentile.
   e) On the basis of these data, estimate the probability that the next component produced will be defective.

Chapter 3

Computer Problems

Use MS Excel in solving the following problems:

C10. The numbers of defective items in successive groups of fifteen items were counted as they came off a production line. The results can be summarized as follows:

No. of Defectives	Frequency
0	57
1	57
2	18
3	5
4	3
>4	0

a) Calculate the mean number of defectives in a group of fifteen items.
b) Calculate the variance and standard deviation of the number of defectives in a group. Take the given data as a sample.
c) Find the median, lower quartile, upper quartile, ninth decile, and 95th percentile.
d) On the basis of these data estimate the probability that the next item produced will be defective.

This is the same as Problem 8, but now it is to be solved using Excel.

C11. Electrical components were examined as they came off a production line. The number of defective items in each group of eighteen components was recorded. The results can be summarized as follows:

No. of Defectives	Frequency
0	94
1	52
2	19
3	3
>3	0

a) Calculate the mean number of defectives in a group of 18 components.
b) Taking the given data as a sample, calculate the variance and standard deviation of the number of defectives in a group.
c) Find the median, lower quartile, upper quartile, and 95th percentile.
e) On the basis of these data, estimate the probability that the next component produced will be defective.

This is the same as Problem 9, but now it is to be solved using Excel.

CHAPTER 4

Grouped Frequencies and Graphical Descriptions

Prerequisite: A good knowledge of algebra.

Like Chapter 3, this chapter considers some aspects of descriptive statistics. In this chapter we will be concerned with stem-and-leaf displays, box plots, graphs for simple sets of discrete data, grouped frequency distributions, and histograms and cumulative distribution diagrams.

4.1 Stem-and-Leaf Displays

These simple displays are particularly suitable for exploratory analysis of fairly small sets of data. The basic ideas will be developed with an example.

Example 4.1

Data have been obtained on the lives of batteries of a particular type in an industrial application. Table 4.1 shows the lives of 36 batteries recorded to the nearest tenth of a year.

Table 4.1: Battery Lives, years

4.1	5.2	2.8	4.9	5.6	4.0	4.1	4.3	5.4
4.5	6.1	3.7	2.3	4.5	4.9	5.6	4.3	3.9
3.2	5.0	4.8	3.7	4.6	5.5	1.8	5.1	4.2
6.3	3.3	5.8	4.4	4.8	3.0	4.3	4.7	5.1

For these data we choose "stems" which are the main magnitudes. In this case the digit before the decimal point is a reasonable choice: 1,2,3,4,5,6. Now we go through the data and put each "leaf," in this case the digit after the decimal point, on its corresponding stem. The decimal point is not usually shown. The result can be seen in Table 4.2. The number of stems on each leaf can be counted and shown under the heading of Frequency.

Table 4.2: Stem-and-Leaf Display

Stem	Leaf	Frequency
1	8	1
2	8 3	2
3	7 9 2 7 3 0	6
4	1 9 0 1 3 5 5 9 3 8 6 2 4 8 3 7	16
5	2 6 4 6 0 5 1 8 1	9
6	1 3	2

Chapter 4

From the list of leaves on each stem we have an immediate visual indication of the relative numbers. We can see whether or not the distribution is approximately symmetrical, and we may get a preliminary indication of whether any particular theoretical distribution may fit the data. We will see some theoretical distributions later in this book, and we will find that some of the distributions we encounter in this chapter can be represented well by theoretical distributions.

We may want to sort the leaves on each stem in order of magnitude to give more detail and facilitate finding parameters which depend on the order. The result of sorting by magnitude is shown in Table 4.3.

Table 4.3: Sorted Stem-and-Leaf Display

Stem	Leaf	Frequency
1	8	1
2	3 8	2
3	0 2 3 7 7 9	6
4	0 1 1 2 3 3 3 4 5 5 6 7 8 8 9 9	16
5	0 1 1 2 4 5 6 6 8	9
6	1 3	2

Another possibility is to double the number of stems (or multiply them further), especially if the number of data is large in relation to the initial number of stems. Stem "a" might have leaves from 0 to 4, and stem "b" might have leaves from 5 to 9. The result without sorting is shown in Table 4.4.

Table 4.4: Stem-and-Leaf Plot with Double Leaf

Stem	Leaf	Frequency
1b	8	1
2a	3	1
2b	8	1
3a	2 3 0	3
3b	7 9 7	3
4a	1 0 1 3 3 2 4 3	8
4b	9 5 5 9 8 6 8 7	8
5a	2 4 0 1 1	5
5b	6 6 5 8	4
6a	1 3	2

Of course, we might both double the number of stems and sort the leaves on each stem. In other cases it might be more appropriate to show two significant figures on each leaf, with appropriate separation between leaves. There are many possible variations.

4.2 Box Plots

A box plot, or box-and-whisker plot, is a graphical device for displaying certain characteristics of a frequency distribution. A narrow box extends from the lower quartile to the upper quartile. Thus the length of the box represents the interquartile range, a measure of variability. The median is marked by a line extending across the box. The smallest value in the distribution and the largest value are marked, and each is joined to the box by a straight line, the whisker. Thus, the whiskers represent the full range of the data.

Figure 4.1 is a box plot for the data of Table 4.1 on the life of batteries under industrial conditions. The labels, "smallest", "largest", "median", and "quartiles", are usually omitted.

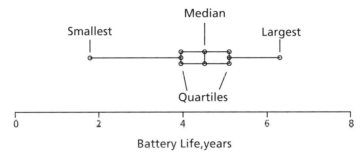

Figure 4.1: Box Plot for Life of Battery

Box plots are particularly suitable for comparing sets of data, such as before and after modifications were made in the production process. Figure 4.2 shows a comparison of the box plot of Figure 4.1 with a box plot for similar data under modified production conditions, both for the same sample size. Although the median has not changed very much, we can see that the sample range and the interquartile range for modified conditions are considerably smaller.

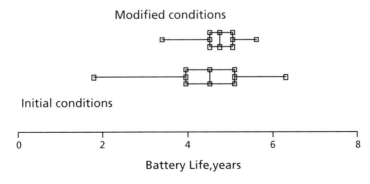

Figure 4.2: Comparison of Box Plots

Chapter 4

4.3 Frequency Graphs of Discrete Data

Example 3.2 concerned the number of defective items in successive samples of six items each. The data were summarized in Table 3.2, which is reproduced below.

Table 3.2: Frequencies for Numbers of Defectives

Number of defectives, x_i	Frequency, f_i
0	48
1	10
2	2
>2	0

These data can be shown graphically in a very simple form because they involve discrete data, as opposed to continuous data, and only a few different values. The variate is discrete in the sense that only certain values are possible: in this case the number of defective items in a group of six must be an integer rather than a fraction. The number of defective items in each group of this example is only 0, 1, or 2. The frequencies of these numbers are shown above. The corresponding frequency graph is shown in Figure 4.3. The isolated spikes correspond to the discrete character of the variate.

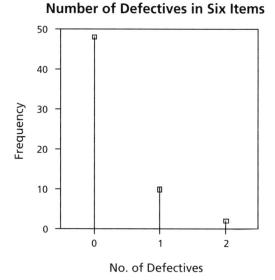

Figure 4.3: Distribution of Numbers of Defectives in Groups of Six Items

If the number of different values is very large, it may be desirable to use the grouped frequency approach, as discussed below for continuous data.

4.4 Continuous Data: Grouped Frequency

If the variate is continuous, any value at all in an appropriate range is possible. Between any two possible values, there are an infinite number of other possible values, although measuring devices are not able to distinguish some of them from one another. Measurements will be recorded to only a certain number of significant figures. Even to this number of figures, there will usually be a large number of possible values. If the number of possible values of the variate is large, too many occur on a table or graph for easy comprehension. We can make the data easier to comprehend by dividing the variate into intervals or classes and counting the frequency of occurrence for each class. This is called the *grouped frequency approach*.

Thus, frequency grouping is used to make the distribution more easily understood. The width of each class (the difference between its lower boundary and its upper boundary) should be constant from one class to another (there are exceptions to this statement, but we will omit them from this book). The number of classes should be from seven to twenty, depending chiefly on the size of the population or sample being represented. If the number of classes is too large, the result is too detailed and it is hard to see an underlying pattern. If the number of classes is too small, there is appreciable loss of information, and the pattern may be obscured. An empirical relation which gives an approximate value of the appropriate number of classes is Sturges' Rule:

$$\text{number of class intervals} \approx 1 + 3.3 \log_{10} N \tag{4.1}$$

where N is the total number of observations in the sample or population.

The procedure is to start with the range, the difference between the largest and the smallest items in the set of observations. Then the constant class width is given approximately by dividing the range by the approximate number of class intervals from equation 4.1. Round off the class width to a convenient number (remember that there is nothing sacred or exact about Sturges' Rule!).

The class boundaries must be clear with no gaps and no overlaps. For problems in this book choose the class boundaries halfway between possible magnitudes. This gives a definite and fair boundary. For example, if the observations are recorded to one decimal place, the boundaries should end in five in the second decimal place. If 2.4 and 2.5 are possible observations, a class boundary might be chosen as 2.45. The smallest class boundary should be chosen at a convenient value a little smaller than the smallest item in the set of observations.

Each class midpoint is halfway between the corresponding class boundaries.

Then the number of items in each class should be tallied and shown as class frequency in a table called a grouped frequency table. The *relative frequency* is the class frequency divided by the total of all the class frequencies, which should agree with the total number of items in the set of observations. The *cumulative* frequency is the total of all class frequencies smaller than a class *boundary*. The class boundary rather than class midpoint must be used for finding cumulative frequency because we can see from the table how many items are smaller than a class boundary, but we cannot know how many items are smaller than a class midpoint unless we go back to

Chapter 4

the original data. The *relative cumulative frequency* is the fraction (or percentage) of the total number of items smaller than the corresponding upper class boundary.

Let us consider an example.

Example 4.2

The thickness of a particular metal part of an optical instrument was measured on 121 successive items as they came off a production line under what was believed to be normal conditions. The results are shown in Table 4.5.

Table 4.5: Thicknesses of Metal Parts, mm

3.40	3.21	3.26	3.37	3.40	3.35	3.40	3.48	3.30	3.38	3.27
3.35	3.28	3.39	3.44	3.29	3.38	3.38	3.40	3.38	3.44	3.29
3.37	3.41	3.45	3.44	3.35	3.35	3.46	3.31	3.33	3.47	3.33
3.37	3.31	3.51	3.36	3.32	3.33	3.43	3.39	3.39	3.28	3.33
3.25	3.28	3.30	3.41	3.39	3.33	3.27	3.34	3.33	3.42	3.35
3.34	3.32	3.42	3.31	3.38	3.44	3.37	3.35	3.57	3.41	3.28
3.49	3.26	3.44	3.46	3.32	3.36	3.41	3.39	3.38	3.26	3.37
3.28	3.35	3.36	3.34	3.42	3.38	3.39	3.51	3.44	3.39	3.36
3.35	3.42	3.34	3.36	3.42	3.38	3.46	3.34	3.37	3.39	3.42
3.37	3.33	3.39	3.30	3.35	3.38	3.38	3.27	3.31	3.32	3.45
3.49	3.45	3.38	3.41	3.35	3.39	3.24	3.35	3.34	3.37	3.37

Thickness is a continuous variable, since any number at all in the appropriate range is a possible value. The data in Table 4.5 are given to two decimal places, but it would be possible to measure to greater or lesser precision. The number of possible results is infinite. The mass of numbers in Table 4.5 is very difficult to comprehend. Let us apply the methods of this section to this set of data.

Applying equation 3.1 to the numbers in Table 4.5 gives a mean of $\frac{407.59}{121} =$ 3.3685 or 3.369 mm. (We will see later that the mean of a large group of numbers is considerably more precise than the individual numbers, so quoting the mean to more significant figures is justified.) Since the data constitute a sample of all the thicknesses of parts coming off the production line under the same conditions, this is a sample mean, so $\bar{x} = 3.369$ mm. Then the appropriate relation to calculate the variance is equation 3.8:

$$s^2 = \frac{\sum_{i=1}^{N} x_i^2 - \frac{\left(\sum_{i=1}^{N} x_i\right)^2}{N}}{N-1}$$

Grouped Frequencies and Graphical Descriptions

$$s^2 = \frac{1373.4471 - (407.59)^2/121}{120}$$

$$= \frac{1373.4471 - 1372.971968}{120}$$

$$= \frac{0.475132}{120} = 0.003959 \text{ mm}^2$$

and the sample standard deviation is $\sqrt{0.003959} = 0.0629$ mm. The coefficient of variation is $\left(\dfrac{s}{\bar{x}}\right)(100\%) = (0.0629/3.369)(100\%) = 1.87\%$.

Note for Calculation: Avoiding Loss of Significance

Whenever calculations involve taking the difference of two quantities of similar magnitude, we must remember to make sure that enough significant figures are carried to give the desired accuracy in the result. In Example 4.2 above, the calculation of variance by equation 3.11 requires us to subtract 1372.971968 from 1373.4471, giving 0.475132. If the numbers being subtracted had been rounded to four figures as 1373.0 from 1373.4, the calculated result would have been 0.4. This would have been 16% in error.

To avoid such loss of significance, carry as many significant figures as possible in intermediate results. Do *not* round the numbers to a reasonable number of figures until a final result has been obtained. If a calculator is being used, leave intermediate results in the memory of the calculator. Similarly, if a spreadsheet is being used, do not reduce the number of figures, except perhaps for purposes of displaying a reasonable number of figures in a final result.

If the calculating device being used does not provide enough significant figures, it is often possible to reduce the number of required figures by subtracting a constant value from each figure. For instance, in Example 4.2 we could subtract 3 from each of the numbers in Table 4.5. This would not affect the final variance or standard deviation, but it would make the largest number 0.57 instead of 3.57, giving a square of 0.3249 instead of 12.7449, so requiring four figures instead of six at this point. The required number of figures in other quantities would be reduced similarly. However, most modern computing devices can easily retain enough figures so that this step is not required.

The median of the 121 numbers in Table 4.5 is the 61st number in order of magnitude. This is 3.37 mm. The fifth percentile is between the 6th and 7th items in order of magnitude, so (3.26 + 3.27) / 2 = 3.265 mm. The ninth decile is between the 108th and 109th numbers in increasing order of magnitude, so (3.44 + 3.45) / 2 = 3.445 mm.

Now let us apply the grouped frequency approach to the numbers in Table 4.5. The largest item in the table is 3.57, and the smallest is 3.21, so the range is 0.36. The number of class intervals according to Sturges' Rule should be approximately $1 + (3.3)(\log_{10} 121) = 7.87$. Then the class width should be approximately $0.36 / 7.87 = 0.0457$. Let us choose a convenient class width of 0.05. The thicknesses are stated to two decimal places, so the class boundaries should end in five in the third decimal. Let us choose the smallest class boundary, then, as 3.195. The resulting grouped frequency table is shown in Table 4.6.

Table 4.6: Grouped Frequency Table for Thicknesses

Lower Class Boundary, mm	Upper Class Boundary, mm	Class Midpoint, mm	Tally Marks	Class Frequency	Relative Frequency	Cumulative Frequency
3.195	3.245	3.220	||	2	0.017	2
3.245	3.295	3.270	||||| ||||| ||||	14	0.116	16
3.295	3.345	3.320	||||| ||||| ||||| ||||| ||||	24	0.198	40
3.345	3.395	3.370	||||| ||||| ||||| ||||| ||||| ||||| ||||| ||||| ||||| |	46	0.380	86
3.395	3.445	3.420	||||| ||||| ||||| ||||| ||	22	0.182	108
3.445	3.495	3.470	||||| |||||	10	0.083	118
3.495	3.545	3.520	||	2	0.017	120
3.545	3.595	3.570	|	1	0.008	121
			Total	121	1.000	

In this table the class frequency is obtained by counting the tally marks for each class. This becomes easier if we divide the tally marks into groups of five as shown in Table 4.6. The relative frequency is simply the class frequency divided by the total number of items in the table, i.e. the total frequency, which is 121 in this case. The cumulative frequency is obtained by adding together all the class frequencies for classes with values smaller than the current upper class boundary. Thus, in the third line of Table 4.6, the cumulative frequency of 40 is the sum of the class frequencies 2, 14 and 24. The corresponding relative cumulative frequency would be $\frac{40}{121} = 0.331$, or 33.1%. The cumulative frequency in the last line must be equal to the total frequency.

From Table 4.6 the mode is given by the class midpoint of the class with the largest class frequency, 3.370 mm. The mean, median and mode, 3.369, 3.37 and 3.370 mm, are in close agreement. This indicates that the distribution is approximately symmetrical.

Graphical representations of grouped frequency distributions are usually more readily understood than the corresponding tables. Some of the main characteristics of the data can be seen in *histograms* and cumulative frequency diagrams. A histogram is a bar graph in which the class frequency or relative class frequency is plotted against values of the quantity being studied, so the height of the bar indicates the class frequency or relative class frequency. Class midpoints are plotted along the horizontal axis. In principle, a histogram for continuous data should have the bars touching one another, and that should be done for problems in this book. However, the bars are often shown separated, and some computer software does not allow the bars to touch one another.

The histogram for the data of Table 4.5 is shown in Figure 4.4 for a class width of 0.05 mm as already calculated. Relative class frequency is shown on the right-hand scale.

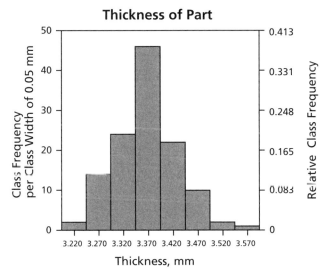

Figure 4.4: Histogram for Class Width of 0.05 mm

Histograms for class widths of 0.03 mm and 0.10 mm are shown in Figures 4.5 and 4.6 for comparison.

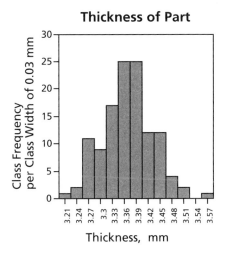

Figure 4.5: Histogram for Class Width of 0.03 mm

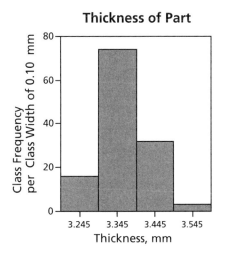

Figure 4.6: Histogram for Class Width of 0.10 mm

Chapter 4

Of these three, the class width of 0.05 mm in Figure 4.4 seems most satisfactory (in agreement with Sturges' Rule).

Cumulative frequencies are shown in the last column of Table 4.6. A cumulative frequency diagram is a plot of cumulative frequency vs. the upper class boundary, with successive points joined by straight lines. A cumulative frequency diagram for the thicknesses of Table 4.5 is shown in Figure 4.7.

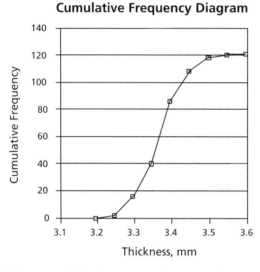

Figure 4.7: Cumulative Frequency Diagram for Thickness

The cumulative frequency diagram of Figure 4.7 could be changed into a relative cumulative frequency diagram by a change of scale for the ordinate.

Example 4.3

A sample of 120 electrical components was tested by operating each component continuously until it failed. The time to the nearest hour at which each component failed was recorded. The results are shown in Table 4.7.

Table 4.7: Times to Failure of Electrical Components, hours

1347	33	1544	1295	1541	14	2813	727	3385	2960
2075	215	346	153	735	1452	2422	1160	2297	594
2242	977	1096	965	315	209	1269	447	1550	317
3391	709	3416	151	2390	644	1585	3066	17	933
1945	844	1829	1279	1027	5	372	869	535	635
932	61	3253	47	4732	120	523	174	2366	323
1296	755	28	305	710	1075	74	1765	1274	180
1104	248	863	1908	2052	1036	359	202	1459	3
916	2344	581	1913	2230	1126	22	1562	219	166
678	1977	167	573	186	804	6	637	316	159
983	1490	877	152	2096	185	53	39	3997	310
1878	1952	5312	4042	4825	639	1989	132	432	1413

Grouped Frequencies and Graphical Descriptions

Once again, frequency grouping is needed to make sense of this mass of data. When the data are sorted in order of increasing magnitude, the largest value is found to be 5312 hours and the smallest is 3 hours. Then the range is 5312 − 3 = 5309 hours. There are 120 data points. Then applying Sturges' Rule, equation 4.1 indicates that the number of class intervals should be approximately $1 + 3.3 \log_{10} 120 = 7.86$. Then the class width should be approximately 5309 / 7.86 = 675 hours. A more convenient class width is 600 hours. Since times to failure are stated to the nearest hour, each class boundary should be a number ending in 0.5. The smallest class boundary must be somewhat less than the smallest value, 3. Then a convenient choice of the smallest class boundary is 0.5 hours. The resulting grouped frequency table is shown in Table 4.8. The corresponding histogram is Figure 4.8, and the cumulative frequency diagram (last column of Table 4.8 vs. upper class boundary) is Figure 4.9.

Table 4.8: Grouped Frequency Table for Failure Times

Lower Class Boundary, mm	Upper Class Boundary, mm	Class Midpoint, mm	Tally Marks	Class Frequency	Relative Frequency	Cumulative Frequency
0.5	600.5	300.5	||||| ||||| ||||| ||||| ||||| ||||| ||||| ||||| |||||		46	0.383	46
600.5	1200.5	900.5	||||| ||||| ||||| ||||| ||||| |||	28	0.233	74
1200.5	1800.5	1500.5	||||| ||||| |||||		16	0.133	90
1800.5	2400.5	2100.5	||||| ||||| ||||| ||	17	0.142	107
2400.5	3000.5	2700.5	|||	3	0.025	110
3000.5	3600.5	3300.5	|||||	5	0.042	115
3600.5	4200.5	3900.5	||	2	0.017	117
4200.5	4800.5	4500.5			1	0.008	118
4800.5	5400.5	5100.5	||	<u>2</u>	<u>0.017</u>	120
			Total	120	1.000	

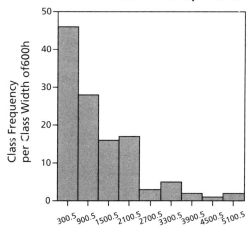

Figure 4.8: Histogram of Times to Failure for Electrical Components

Chapter 4

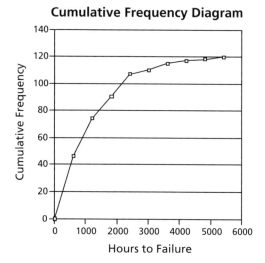

Figure 4.9:
Cumulative Frequency Diagram for Time to Failure

Figures 4.4 and 4.8 are both histograms for continuous data, but their shapes are quite different. Figure 4.4 is approximately symmetrical, whereas Figure 4.8 is strongly skewed to the right (i.e., the tail to the right is very long, whereas no tail to the left is evident in Figure 4.8). Correspondingly, the cumulative frequency diagram of Figure 4.7 is s-shaped, with its slope first increasing and then decreasing, whereas the cumulative frequency diagram of Figure 4.9 shows the slope generally decreasing over its full length.

Now the mean, median and mode for the data of Table 4.7 (corresponding to Figures 4.8 and 4.9) will be calculated and compared. The mean is $\sum x_i / N = 140746/120 = 1173$ hours. The median is the average of the two middle items in order of magnitude, 869 and 877, so 873 hours. The mode according to Table 4.8 is the midpoint of the class with the largest frequency, 300.5 hours, but of course the value would vary a little if the class width or starting class boundary were changed. Since Figure 4.8 shows that the distribution is very asymmetrical or skewed, it is not surprising that the mean, median and mode are so widely different.

The variance is given by equation 3.11,

$$s^2 = \frac{\sum_{i=1}^{N} x_i^2 - \frac{\left(\sum_{i=1}^{N} x_i\right)^2}{N}}{N-1}$$

$$= (317{,}335{,}200 - (140{,}746)^2/120) / 119$$
$$= (317{,}335{,}200 - 165{,}078{,}637.7) / 119$$
$$= 1{,}279{,}467 \text{ h}^2$$

Grouped Frequencies and Graphical Descriptions

and so the estimate of the standard deviation based on this sample is $s = \sqrt{1,279,467}$ = 1131 hours. The coefficient of variation is $\left(\dfrac{s}{\bar{x}}\right)(100\%) = 1131 / 1173 \times 100\% = 96.4\%$.

4.5 Use of Computers

In this section the techniques illustrated in section 3.4 will be applied to further examples. Further techniques, including production of graphs, will be shown. Once again, the reader is referred to brief discussions of some Excel techniques for statistical data in Appendix B.

Example 4.4

The thickness of a particular metal part of an optical instrument was measured on 121 successive items as they came off a production line under what was believed to be normal conditions. The results were shown in Table 4.5. Find the mean thickness, sample variance, sample standard deviation, coefficient of variation, median, fifth percentile, and ninth decile. Use Sturges' Rule in choosing a suitable class width for a grouped frequency distribution. Construct the resulting histogram and cumulative frequency diagram. Use the Excel spreadsheet in solving this problem, and check that rounding errors cause no appreciable loss of significance.

Answer: This is essentially the same problem as in Example 4.2, but now it will be solved using Microsoft Excel.

First the thicknesses were transferred from Table 4.5 to column B of a new work sheet. These data were sorted by increasing (ascending) thickness using the Sort command on the Data menu for later use in finding quantiles. Extracts of the work sheet are shown in Table 4.9. Notice again that each quantity must be clearly labeled.

Table 4.9: Extracts of Work Sheet for Example 4.4

	A	B	C	D	E	F
1	In column C	Thickness, xi mm	dev=xi-xbar	dev^2	xi*xi	Order no.
2	deviation =	3.21	-0.158512397	0.02512618	10.3041	1
3	B2:B122-B124	3.24	-0.128512397	0.01651544	10.4976	2
4		3.25	-0.118512397	0.01404519	10.5625	3
5		3.26	-0.108512397	0.01177494	10.6276	4
		..	..	..	..	..
119		3.49	0.121487603	0.01475924	12.1801	118
120		3.51	0.141487603	0.02001874	12.3201	119
121		3.51	0.141487603	0.02001874	12.3201	120
122		3.57	0.201487603	0.04059725	12.7449	121
123	Totals	407.59	6.66134E-14	0.47513223	1373.4471	
124	xbar, B123/121=	3.368512397	s^2=	D123/120=	0.003959	
125		s^2=	(E123-B123^2/121)/120=		0.003959	
126		diff =		E124-E125=	1.21E-15	

Chapter 4

	A	B	C	D	E	F
127			s=	SQRT(E125)=	0.062924	
128			s/xbar=	D127/B124=	1.87%	
129						
130						
131	A	B	C	D	E	F
132	Lower Class	Upper Class	Class	Class	Relative	Cumulative
133	Boundary	Boundary	Midpoint	Frequency	Class	Class
134	mm	mm	mm		Frequency	Frequency
135		3.195		0		0
136	3.195	3.245	3.22	2	0.017	2
137	3.245	3.295	3.27	14	0.116	16
138	3.295	3.345	3.32	24	0.198	40
139	3.345	3.395	3.37	46	0.380	86
140	3.395	3.445	3.42	22	0.182	108
141	3.445	3.495	3.47	10	0.083	118
142	3.495	3.545	3.52	2	0.017	120
143	3.545	3.595	3.57	1	0.0083	121
144	3.595	3.645	3.62	0		
145			Total	121		
146	In cells:					
147	A137:A144	B136:B144	C136:C144	D136:D144	E136:E144	F136:F144
148	The corresponding explanations are (same column):					
149	A136:A143+0.05=	A136:A144+0.05=	(A136:A144+B136:B144)/2=		D136:D144/D145=	
150				Frequency(B1:B122,B136:B144)=		
151						F135:F143+
152						D136:D144

Quantities in rows 2 to 122 were added using the Autosum tool; totals were placed in row 123. This gave a total thickness of 407.59 mm in cell B123 for the 121 items. Then the mean thickness, $\bar{x}$, was found in cell B124 to be 3.3685 mm. Next, deviations from the mean, $x_i - \bar{x}$, were found in column C using an array formula (which does a group of similar calculations together—see explanation in section (b) of Appendix B). The deviations calculated in this way were squared by the array formula =(C3:C123)^2, entered in cells D2:D122. (Remember that entering an array formula requires us to press more than one key simultaneously. See Appendix B.) Then the sample variance was found using equation 3.8 in cell E124 by dividing the sum of squares of deviations by 120. This gave 0.003959 mm². Notice that this method of calculation of variance requires more arithmetic steps than the alternative method, which will be used in the next paragraph. The first method is used in this example to provide a comparison giving a check on round-off errors, but the other method should be used unless such a comparison is required.

The squares of individual thicknesses, $(x_i)^2$, were found in cells E2:E122 by the array formula =B2^2. According to equation 3.11, the variance estimated from the sample is $s^2 = (\Sigma x_i^2 - (\Sigma x_i)^2 / N) / (N-1)$, where in this case N, the number of data points, is 121. Then in cell E125 the sample variance is calculated as 0.003959 mm^2, which agrees with the previous value. The sample standard deviation was found in cell D127, taking the square root of the variance. This gave 0.0629 mm. The coefficient of variation (from cell D128) is 1.87%, which was formulated as a percentage using the Format menu.

Now we can obtain some indications of error due to round-off in Microsoft Excel. In cell C123 the sum of all 121 deviations from the sample mean is shown as 6.66E – 14, whereas it should be zero. This is consistent with the statement that Excel stores values to a precision of about 15 decimal digits. The difference between the value of the sample variance in cell E124 and the value of the same quantity in cell E125 was calculated by the appropriate formula, =D125 – E125, and entered in cell E126. It is 1.21E – 15, again consistent with the statement regarding the precision of numbers calculated and stored in Excel. As these errors are very small in comparison to the quantities calculated, rounding errors are negligible.

The order numbers from 1 to 121 were entered in cells F3:F123. After the first two numbers were entered, the fill handle was dragged to produce the series. From the order numbers in cells F3:F123 and the thicknesses in cells B3:B123, numbers to calculate the median (order number 61, so in cell B63), fifth percentile (between order numbers 6 and 7, cells B8 and B9), and ninth decile (between order numbers 108 and 109, cells B110 and B111) were read. Then the median is 3.37 mm, the fifth percentile is (3.26 + 3.27) / 2 = 3.265 mm, and the ninth decile is (3.44 + 3.45) / 2 = 3.445 mm.

For the class width and the smallest class boundary for the grouped frequency table the reasoning is the same as in Example 4.3. The largest thickness, in cell B123, is 3.57 mm, and the smallest thickness, in cell B3, is 3.21 mm, so the range is 3.57 – 3.21 = 0.36 mm. Since there are 121 items, the number of class intervals according to Sturges' Rule should be approximately $1 + (3.3)(\log_{10} 121) = 7.87$. This calls for a class width of approximately 0.36 / 7.87 = 0.0457 mm, and we choose a convenient value of 0.05 mm. The smallest class boundary should be a little smaller than the smallest thickness and halfway between possible values of the thickness, which was measured to two decimal places. Then the smallest class boundary was chosen as 3.195 mm.

Column headings for the grouped frequency table were entered in cells A132:F134. The smallest class boundary, 3.195 mm, was entered in cell A136. To obtain an extra class of zero frequency for the cumulative frequency distribution, 3.195 was entered also in cell B135, and zero was entered in cell D135. For a class width of 0.05 mm the next lower class boundary of 3.245 was entered in cell A137, and the fill handle was dragged to 3.595 in cell A144. Upper class boundaries were entered in cells B136:B144 by the array function =A136:A144 + 0.05. Class midpoints were entered in cells C136:C144 by the array function =(A136:A144 + B136:B144)/2.

Chapter 4

A saving in time can be obtained at this point by using one of Excel's built-in functions (see section (e) of Appendix B). Class frequencies were entered in cells D135:D144 by the array formula =FREQUENCY(B2:B122,B135:B143), where the cells B2:B122 contain the data array (thickness in mm in this case) and the cells B135:B143 contain the corresponding upper class boundaries. For further information, from the Help menu select Microsoft Excel Help, and then the Frequency worksheet function. Note that the number of cells in D135:D144 is nine, one more than the number of cells in B135:B143. The last item in column D (cell D144) is 0 and represents the frequency above the largest effective upper class boundary, 3.595 mm. The class frequencies in cells D135:D144 agree with the values given in Table 4.6. The total frequency was found in cell D145 using the Autosum tool. It is 121, as before. Relative class frequencies in cells E136:E143 were found using the array formula =D136:D143/121. Again the results agree with previous results. The first cumulative frequency in cell F135 is the same as the corresponding class frequency, so it is given by =D135. Cumulative class frequencies in cells F136:F143 were found by the array formula =F135:F142+D136:D143. They can be checked by comparison with the largest order numbers in the upper part of Table 4.9 corresponding to a thickness less than an upper class boundary. For example, the largest order number corresponding to a thickness less than the upper class boundary 3.495 is 118. Minor changes, such as centering, were made in formatting cells A132:F145. Instead of the function Frequency, the function Histogram can be used if it is available.

To produce the histogram, the class midpoints (cells D133:D141) and the class frequencies (cells E133:E141) were selected; from the Insert menu, Chart was selected. The "Chart Wizard" guided choices for the chart. A simple column chart was chosen with data series in columns, x-axis titled "Thickness, mm", y-axis titled "Class frequency", and no legend. The chart was opened as a new sheet titled "Example 4.4."

The chart was modified by selecting it and opening the Chart menu. One modification was of the font size for the titles of axes. The x-axis title was chosen, and from the Format menu the Selected Axis Title was chosen, then the font size was changed from 10 point to 12 point. The y-axis title was modified similarly. To make the bars of the histogram touch one another without gaps, a bar was clicked and from the Format menu the Selected Data Series was chosen; the Option tab was clicked, and then the gap width was reduced to zero. This left the histogram in solid black. To remedy this, the bars were double-clicked: the screen for Format Data Point appeared with the Patterns tab, and the Fill Effects bar was clicked. A suitable diagonal pattern was selected for the fill of each bar, with the diagonals sloping in different directions on adjacent bars. The final histogram is very similar to Figure 4.4, differing from it mainly as a result of using different software, CA-Cricket Graph III vs. Excel.

Grouped Frequencies and Graphical Descriptions

To obtain the cumulative frequency diagram, first the upper class boundaries, cells B135:B144, were selected. Then the corresponding cumulative class frequencies, cells F135:F144, were selected while holding down Crtl in Excel for Windows or Command in Excel for the Macintosh, because this is a nonadjacent selection to be added to the selection of class boundaries. Then from the Insert menu, Chart was clicked. A simple line chart was chosen with horizontal grids. The data series are in columns, the first column contains x-axis labels, and the first row gives the first data point. A choice was made to have no legend. The chart title was chosen to be "Cumulative Frequency Diagram." The title for the x-axis was chosen to be "Thickness, mm." The title for the y-axis was chosen to be "Cumulative Frequency." The result is essentially the same as Figure 4.7.

Example 4.5

A sample of 120 electrical components was tested by operating each component continuously until it failed. The time to the nearest hour at which each component failed was recorded. The results were shown in Table 4.7. Calculate the mean, median, mode, variance, standard deviation, and coefficient of variation for these data. Prepare a grouped frequency table from which a histogram and cumulative frequency diagram could be prepared. Calculate using Excel.

Answer: This is a repeat of most of Example 4.3, but using Excel.

The times to failure, t_i hours, were entered in column B, rows 3 to 122, of a new work sheet. They were sorted from the smallest to the largest using the Sort command on the Data menu. The work sheet must include headings, labels, and explanations. Extracts of the work sheet are shown in Table 4.10. This is similar to the work sheet of Example 4.4, which was shown in Table 4.9.

Table 4.10: Extracts from Work Sheet, Example 4.5

	A	B	C	D	E	F
1		Time, t_i h	t_i^2	Order No.		
2			(B3:B122)^2=			
3		3	9	1		
4		5	25	2		
..	..	..	..	..		
61		863	744769	59		
62		869	755161	60		
63		877	769129	61		
64		916	839056	62		
..	..	..	..	..		
120		4732	22391824	118		
121		4825	23280625	119		
122		5312	28217344	120		
123	Sums	140742	317324464			
124	Mean, tbar=					

Chapter 4

125	B123/120=	1172.85				
126	s^2=	(C123-B123*B123/120)/(120-1)=			1.28E6	
127	s=	SQRT(E126)=			1130	
128	c.v.= s/xbar=	E127/B125=			96%	
129						

	Lower Class Boundary h	Upper Class Boundary h	Class Midpoint h	Class Frequency	Relative Class Frequency	Cumulative Class Frequency
130						
131						
132						
133						
134		0.5		0		0
135	0.5	600.5	300.5	46	0.383333	46
136	600.5	1200.5	900.5	28	0.233333	74
137	1200.5	1800.5	1500.5	16	0.133333	90
138	1800.5	2400.5	2100.5	17	0.141667	107
139	2400.5	3000.5	2700.5	3	0.025	110
140	3000.5	3600.5	3300.5	5	0.041667	115
141	3600.5	4200.5	3900.5	2	0.016667	117
142	4200.5	4800.5	4500.5	1	0.008333	118
143	4800.5	5400.5	5100.5	2	0.016667	120
144			Total	120		
145	In cells:					
146	A136:A143	B135:B143	C135:C143	D134:D143	E135:E143	F135:F143
147	the corresponding explanations are (same			column):		
148	A135:A142+600=		(A135:A143+B135:B143)/2=			D134:D142+ D135:D143=
149		A135:A143+600=		Frequency(B3:B122,B135:B143)=		

In cells E135:E143 the explanation is D135:D143/D144.

Appendix C lists some functions which should not be used during the learning process but are useful shortcuts once the reader has learned the fundamentals thoroughly.

Concluding Comment

In this chapter and the one before, we have seen several types of frequency distributions from numerical data. In the next few chapters we will encounter theoretical probability distributions, and some of these will be found to represent satisfactorily some of the frequency distributions of these chapters.

Problems

1. The daily emissions of sulfur dioxide from an industrial plant in tonnes/day were as follows:

4.2	6.7	5.4	5.7	4.9	4.6	5.8	5.2	4.1	6.2
5.5	4.9	5.1	5.6	5.9	6.8	5.8	4.8	5.3	5.7

Grouped Frequencies and Graphical Descriptions

    a) Prepare a stem-and leaf display for these data.
    b) Prepare a box plot for these data.

2. A semi-commercial test plant produced the following daily outputs in tonnes/day:

1.3	2.5	1.8	1.4	3.2	1.9	1.3	2.8	1.1	1.7
1.4	3.0	1.6	1.2	2.3	2.9	1.1	1.7	2.0	1.4

    a) Prepare a stem-and leaf display for these data.
    b) Prepare a box plot for these data.

3. Over a period of 60 days the percentage relative humidity in a vegetable storage building was measured. Mean daily values were recorded as shown below:

60	63	64	71	67	73	79	80	83	81
86	90	96	98	98	99	89	80	77	78
71	79	74	84	85	82	90	78	79	79
78	80	82	83	86	81	80	76	66	74
81	86	84	72	79	72	84	79	76	79
74	66	84	78	91	81	64	76	78	82

    a) Make a stem-and-leaf display with at least five stems for these data. Show the leaves sorted in order of increasing magnitude on each stem.
    b) Make a frequency table for the data, with a maximum bound of 100.5% relative humidity (since no relative humidity can be more than 100%). Use Sturges' rule to approximate the number of classes.
    c) Draw a frequency histogram for these data.
    d) Draw a relative cumulative frequency diagram.
    e) Find the median, lower quartile, and upper quartile.
    f) Find the arithmetic mean of these data.
    g) Find the mode of these data from the grouped frequency distribution.
    h) Draw a box plot for these data.
    i) Estimate from these data the probability that the mean daily relative humidity under these conditions is less than 85%.

4. A random sample was taken of the thickness of insulation in transformer windings, and the following thicknesses (in millimeters) were recorded:

18	21	22	29	25	31	37	38	41	39
44	48	54	56	56	57	47	38	35	36
29	37	32	42	43	40	48	36	37	37
36	38	40	41	44	39	38	34	24	32
39	44	42	30	37	30	42	37	34	37
32	24	42	36	49	39	23	34	36	40

    a) Make a stem-and-leaf display for these data. Show at least five stems. Sort the data on each stem in order of increasing magnitude.
    b) Estimate from these data the percentage of all the windings that received more than 30 mm of insulation but less than 50 mm.

c) Find the median, lower quartile, and ninth decile of these data.
d) Make a frequency table for the data. Use Sturges' rule.
e) Draw a frequency histogram.
f) Add and label an axis for relative frequency.
g) Draw a cumulative frequency graph.
h) Find the mode.
i) Show a box plot of these data.

5. The following scores represent the final examination grades for an elementary statistics course:

23	60	79	32	57	74	52	70	82	36
80	77	81	95	41	65	92	85	55	76
52	10	64	75	78	25	80	98	81	67
41	71	83	54	64	72	88	62	74	43
60	78	89	76	84	48	84	90	15	79
34	67	17	82	69	74	63	80	85	61

a) Make a stem-and-leaf display for these data. Show at least five stems. Sort the data on each stem in order of increasing magnitude.
b) Find the median, lower quartile, and upper quartile of these data.
c) What fraction of the class received scores which were less than 65?
d) Make a frequency table, starting the first class interval at a lower class boundary of 9.5. Use Sturges' Rule.
e) Draw a frequency histogram.
f) Draw a relative frequency histogram on the same x-axis.
g) Draw a cumulative frequency diagram.
h) Find the mode.
i) Show a box plot of these data.

Computer Problems

Use MS Excel in solving the following problems:

C6. For the data given in Problem 3:
a) Sort the given data and find the largest and smallest values.
b) Make a frequency table, starting the first class interval at a lower bound of 59.5% relative humidity. Use Sturges' rule to approximate the number of classes.
c) Find the median, lower quartile, eighth decile, and 95th percentile.
d) Find the arithmetic mean and the mode.
e) Find the variance and standard deviation of these data taken as a complete population, using both a basic definition and a method for faster calculation.
f) From the calculations of part (e) check or verify in two ways the statement that Excel stores numbers to a precision of about fifteen decimal places.

Grouped Frequencies and Graphical Descriptions

C7. For the data given in Problem 4, perform the same calculations and determinations as in Problem C6. Choose a reasonable lower boundary for the smallest class.

C8. For the data given in Problem 5:
   a) Sort the data and find the largest and smallest values.
   b) Find the median, upper quartile, ninth decile, and 90th percentile.
   c) Make a frequency table. Use Sturges' rule to approximate the number of classes.
   d) Find the arithmetic mean and mode.
   e) Find the variance of the data taken as a sample.

CHAPTER 5

Probability Distributions of Discrete Variables

For this chapter the reader should have a solid understanding of sections 2.1, 2.2, 3.1, and 3.2.

We saw in Chapters 3 and 4 some frequency distributions for discrete and continuous variates. Examples included frequencies of various numbers of defective items in samples taken from production lines, and frequencies of various classes of thicknesses of items produced industrially.

Now we want to look at the *probabilities* of various possible results. If we know enough about the probability distributions, we can calculate the probability of each result. For instance, we can calculate the probability of each possible number of defective items in a sample of fixed size. From that we might calculate the probability of finding (for example) three or more defective items in a sample of 18 items. That might be useful in assessing the implications for quality control of finding three defectives in such a sample. Similarly, if we know enough about the probability distribution we can calculate the probabilities of parts which are thicker than appropriate limits.

The number of defective items in a sample of 18 items is a real number expressing a result determined by chance. We can't predict the number of defective items in the next sample, but we may be able to calculate some probabilities. The probability of any particular number of defective items would be a function of the parameters of the problem. A quantity such as this is called a *random variable*.

The distinction between a discrete and a continuous random variable is the same as the distinction between a discrete and a continuous frequency distribution: only certain results are possible for a discrete random variable, but any of an infinite number of results within a certain range are possible for a continuous random variable. The random variable describing the number of defective items in a sample of 18 parts is discrete because the number of defective items in this case must be either zero or a positive whole number no more than 18, and not any other number between zero and 18. Another example of a discrete random variable is the number of failures in an electronic device in its first five years of operation. On the other hand, the time between successive failures of an electronic device is a continuous random variable because there are an infinite number of possible results between any two possible results that we may choose (even though practical measurement devices may not be able to distinguish some of them from one another because they report results to a finite number of figures). Another example of a continuous random variable is a measurement of the diameter of a part as it comes from a production line. We cannot predict any particular value of the random variable but, with sufficient data of the type discussed in Chapter 4, we may be able to find the probability of a result in a particular interval.

This chapter is concerned with discrete variables, and the next chapter is concerned with cases where the variable is continuous. Both types of variables are fundamental to some of the applications discussed in later chapters. In this chapter we will start with a general discussion of discrete random variables and their probability and distribution functions. Then we will look at the idea of mathematical expectation, or the mean of a probability distribution, and the concept of the variance of a probability distribution. After that, we will look in detail at two important discrete probability distributions, the Binomial Distribution and the Poisson Distribution.

5.1 Probability Functions and Distribution Functions

(a) Probability Functions

Say the possible values of a discrete random variable, X, are $x_0, x_1, x_2, \ldots x_k$, and the corresponding probabilities are $p(x_0), p(x_1), p(x_2) \ldots p(x_k)$. Then for any choice of i, $p(x_i) \geq 0$, and $\sum_{i=0}^{k} p(x_i) = 1$, where k is the maximum possible value of i. Then $p(x_i)$ is a *probability function*, also called a probability mass function. An alternative notation is that the probability function of X is written $\Pr[X = x_i]$. In many cases $p(x_i)$ (or $\Pr[X = x_i]$) and x_i are related by an algebraic function, but in other cases the relation is shown in the form of a table. The relation can be represented by isolated spikes on a bar graph, as shown for example in Figure 5.1. By convention the random variable is represented by a capital letter (for example, X), and particular values are represented by lower-case letters (for example, x, x_i, x_0).

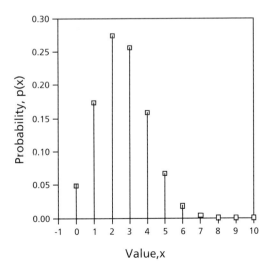

Figure 5.1: Example of a Probability Function for a Discrete Random Variable

Chapter 5

(b) Cumulative Distribution Functions

Cumulative probabilities, Pr $[X \leq x]$, where X still represents the random variable and x now represents an upper limit, are found by adding individual probabilities.

$$\Pr[X \leq x] = \sum_{x_i \leq x} p(x_i) \tag{5.1}$$

where $p(x_i)$ is an individual probability function. For example, if x_i can be only zero or a positive integer,

$$\Pr[X \leq 3] = p(0) + p(1) + p(2) + p(3)$$

The functional relationship between the cumulative probability and the upper limit, x, is called the *cumulative distribution function*, or the probability distribution function.

Note that since $\Pr[X \leq 2] = p(0) + p(1) + p(2)$,

we have $\quad p(3) = \Pr[X \leq 3] - \Pr[X \leq 2]$.

In general,

$$p(x_i) = \Pr[X \leq x_i] - \Pr[X \leq x_{i-1}] \tag{5.2}$$

As an illustration, consider the random variable that represents the number of heads obtained on tossing five fair coins. The probability of obtaining heads on any one coin is $\frac{1}{2}$. The probability function and cumulative distribution are given by the binomial distribution, which will be considered in detail in section 5.3. The probability function of possible results is shown in Table 5.1 and Figure 5.2.

Table 5.1: Probability Function for Tossing Coins

r, no. of heads	Probability, p(r)
0	$\frac{1}{32}$
1	$\frac{5}{32}$
2	$\frac{10}{32}$
3	$\frac{10}{32}$
4	$\frac{5}{32}$
5	$\frac{1}{32}$
Total	1

Probability Distributions of Discrete Variables

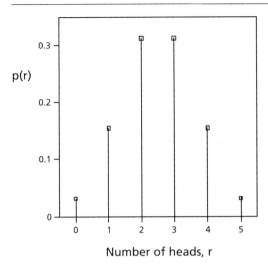

Figure 5.2: Probability Function for Results of Tossing Five Fair Coins

The corresponding cumulative distribution function is shown in Figure 5.3. The graph of the cumulative distribution function for a discrete random variable is a stepped function because there can be no change in the cumulative probability between possible values of the variable.

Using this cumulative distribution function with equation 5.2,

$$p(3) = \Pr[R \leq 3] - \Pr[R \leq 2] = \frac{26}{32} - \frac{16}{32} = \frac{10}{32} = 0.3125.$$

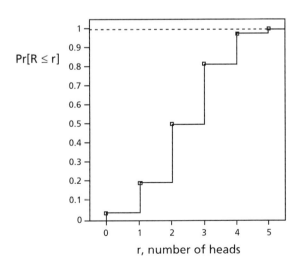

Figure 5.3: Cumulative Distribution for Tossing Five Fair Coins

Chapter 5

5.2 Expectation and Variance

(a) Expectation of a Random Variable

The mathematical expectation or expected value of a random variable is an arithmetic mean that we can expect to closely approximate the mean result from a very long series of trials, if a particular probability function is followed. The expected value is the mean of all possible results for an *infinite* number of trials. We must know the complete probability function in order to calculate the expectation. The expectation of a random variable X is denoted by $E(X)$ or μ_x or μ. The last two symbols indicate that the expectation or expected value is the mean value of the distribution of the random variable.

Let us go back to the empirical approach to probability. The probability of a particular result would be given to a good approximation by the relative frequency of that result from an extremely large number of trials:

$$\Pr[x_i] \approx \frac{f(x_i)}{\sum_{\text{all } i} f(x_i)} \tag{5.3}$$

If the number of trials became infinite, this relation would become exact.

We also have from equation 3.2a that

$$\bar{x} = \sum_{j=1}^{N} x_j \left[\frac{f(x_j)}{\sum_{\text{all } i} f(x_i)} \right] \tag{5.4}$$

The factor within square brackets in equation 5.4 is the relative frequency for factor j. Then for an infinite number of trials we have, using equation 5.3, that

$$E(X) = \mu_X = \sum_{\text{all } x_i} (x_i) \Pr[x_i] \tag{5.5}$$

In words, the expectation or the mean value of the random variable X is given by the sum, for all possible outcomes, of the products given by multiplying each outcome by its probability. If we repeated an experiment a very large number of times, the arithmetic mean of the results would closely approximate the expected value if the stated probability distribution was followed. These relations apply, as written, to discrete random variables, but a similar relation will be found in section 6.2 for a continuous random variable. Equation 5.5 will be used from this point on to calculate *expectation* of a discrete random variable.

The relation for the expected value can be illustrated for the random variable, R, which was shown in Figures 5.2 and 5.3. It is the number of heads obtained on tossing five fair coins.

$$\mu_R = E(R) = (0)\left(\frac{1}{32}\right) + (1)\left(\frac{5}{32}\right) + (2)\left(\frac{10}{32}\right) + (3)\left(\frac{10}{32}\right) + (4)\left(\frac{5}{32}\right) + (5)\left(\frac{1}{32}\right)$$

$$= 2.500$$

Probability Distributions of Discrete Variables

Notice that, like the arithmetic mean, the expected value is not necessarily a possible result from a single trial.

Example 5.1

The probability that a thirty-year-old man will survive a fixed length of time is 0.995. The probability that he will die during this time is therefore 1 – 0.995 = 0.005. An insurance company will sell him a $20,000 life insurance policy for this length of time for a premium of $200.00. What is the expected gain for the insurance company?

Answer: If the man lives through the fixed length of time, the company's gain will be $200.00. The probability of this is 0.995. On the other hand, if the man dies during this time, the company's gain will be +$200.00 – $20,000.00 = – $19,800.00. The probability of this is 0.005.

Using the working expression, equation 5.5, the expected gain for the company is

$$E(X) = (\$200.00)(0.995) + (-\$19,800.00)(0.005)$$
$$= \$199.00 - \$99.00 = \$100.00$$

The idea of fair odds was introduced in section 2.1(f) as an alternative expression giving the same information as probability. It is easy to show from expectation that the relations given in that section are correct. If the probability of "success" in a particular trial is p and the only possible results are "success" and "failure," the probability of "failure" must be $1 - p$. If the process is completely fair, the expectation of gain for any individual must be zero. If the wager for "success" is $1, and the wager against "success" is $A, the individual's gain in the case of "success" is $A and his gain in the case of "loss" is – $1. Then we must have

$$(p)(\$A) + (1-p)(-\$1) = 0$$
$$(p)(\$A) = (1-p)(\$1)$$
$$\frac{\$A}{\$1} = \frac{1-p}{p}$$

The ratio of one wager to the other is called the odds. Then the fair odds *against* "success" must be $\frac{1-p}{p}$ to 1. Similarly, the fair odds *for* "success" must be $p/(1-p)$ to 1.

(b) Variance of a Discrete Random Variable

The variance was defined for the frequency distribution of a population by equation 3.6 as $\sum_{i=1}^{N}(x_i - \mu)^2 / N$ —that is, the mean value of $(x_i - \mu)^2$. Since the quantity corresponding to the mean for a probability distribution is the expectation, the variance of a discrete random variable must be

Chapter 5

$$\sigma_X^2 = E(x-\mu_x)^2$$
$$= \sum_i (x_i - \mu_X)^2 \Pr[x_i] \tag{5.6}$$

An alternative form, like the one found in equation 3.10a, is faster to calculate. It is obtained as follows:

$$E[(X-\mu_X)^2] = E[X^2 - 2(\mu_X)(X) + \mu_x^2]$$
$$= E[X^2] - 2\mu_x E[X] + \mu_x^2$$

But $E[X] = \mu_X$. Then

$$\sigma_X^2 = E\left[(X-\mu_X)^2\right] = E\left[X^2\right] - 2\mu_x^2 + \mu_x^2$$

or
$$\sigma_X^2 = E\left[X^2\right] - \mu_x^2 \tag{5.7}$$

where
$$E\left[X^2\right] = \sum_{\text{all } i} x_i^2 \Pr(x_i) \tag{5.8}$$

The *standard deviation* is always simply the square root of the corresponding variance. Then

$$\sigma_x = \sqrt{E\left[(X-\mu_X)^2\right]}$$
$$= \sqrt{E(X^2) - [E(X)]^2}$$

Let us continue with the previous illustration for the random variable, R, given by the number of heads obtained on tossing five fair coins.

$$E(R^2) = \sum_{\text{all } i} r_i^2 \Pr[r_i]$$

$$= (0)^2 \left(\frac{1}{32}\right) + (1)^2 \left(\frac{5}{32}\right) + (2)^2 \left(\frac{10}{32}\right) + (3)^2 \left(\frac{10}{32}\right) + (4)^2 \left(\frac{5}{32}\right) + (5)^2 \left(\frac{1}{32}\right)$$

$$= 7.500$$

From the previous calculation, $E(R) = \mu_R = 2.500$

Then $\sigma_R^2 = E(R^2) - \mu_R^2$

$$= 7.500 - (2.500)^2$$
$$= 1.25$$

and $\sigma_R = \sqrt{1.25} = 1.118$

Probability Distributions of Discrete Variables

Example 5.2

A probability function is given by $p(0) = 0.3164$, $p(1) = 0.4219$, $p(2) = 0.2109$, $p(3) = 0.0469$, and $p(4) = 0.0039$. Find its mean and variance.

Answer: The mean or expected value is

$(0)(0.3164) + (1)(0.4219) + (2)(0.2109) + (3)(0.0469) + (4)(0.0039) = 1.000$.

The variance is

$(0)^2(0.3164) + (1)^2(0.4219) + (2)^2(0.2109) + (3)^2(0.0469) + (4)^2(0.0039) - (1.000)^2 = 1.750 - 1.000 = 0.750$.

Problems

1. The probabilities of various numbers of failures in a mechanical test are as follows:

    Pr[0 failures] = 0.21, Pr[1 failure] = 0.43, Pr[2 failures] = 0.28, Pr[3 failures] = 0.08, Pr[more than 3 failures] = 0.
    (a) Show this probability function as a graph.
    (a) Sketch a graph of the corresponding cumulative distribution function.
    (b) What is the expected number of failures—that is, the mathematical expectation of the number of failures?

2. Three items are selected at random without replacement from a box containing ten items, of which four are defective. Calculate the probability distribution for the number of defectives in the sample. What is the expected number of defectives in the sample?

3. An experiment was conducted wherein three balls were drawn at random from a barrel containing two blue balls, three red balls, and five green balls.
    a) Find the mean and variance of the probability distribution of the number of green balls chosen.
    b) What is the probability that all the balls will have the same color?

4. A modified version of the game of Yahtzee has been developed and consists of throwing three dice once. The points associated with the possible results are as follows:

Result	Points
Three of a kind	500
A pair	100
All different	50

    a) Find the probability distribution of the number of points.
    b) Find the expected value of the number of points.
    c) Find the standard deviation of the number of points.

Chapter 5

5. A discrete random variable, X, has three possible results with the following probabilities:
$$\Pr[X = 1] = 1/6$$
$$\Pr[X = 2] = 1/3$$
$$\Pr[X = 3] = 1/2$$
No other results can occur.
   (a) Sketch a graph of the probability function.
   (b) What is the mean or expected value of this random variable?
   (c) What are the variance and standard deviation of this random variable?

6. i) Find the probability that, when 5 fair six-sided dice are rolled, the result is:
   a) 5-of-a-kind (all 5 numbers the same);
   b) 4-of-a-kind (4 numbers the same and 1 different);
   c) a "full house" (3 of one number, 2 of another number);
   d) 3-of-a-kind (the other 2 numbers being different from one another);
   e) a single pair;
   f) two pairs;
   g) all 5 numbers different.

   Check that all above probabilities add to 1.

   ii) The players agree to take turns rolling the dice and to collect according to a payout scheme. If the payouts are $1000 for 5-of-a-kind, $40 for 4-of-a-kind, $20 for a full house, $5 for 3-of-a-kind, $2 for a pair and $4 for two pair, what is the expected value on a single roll of 5 dice?

7. A local body shop is run by four employees. However, with such a small staff, absenteeism creates many difficulties financially. If only one employee is absent, the day's total income is reduced by 50%, and if more than one is absent, the shop is closed for that day. When all four are working, an income of $1000 per day can be realized. The shop's expenses are $600 per day when opened and $400 per day when closed. If, on the average, one particular employee misses ten of 100 days and the remaining three miss five of 100 days each, what is the expected daily profit for the company? Assume all absences are independent.

8. A factory produces 3 diesel-generator sets per week. At the end of each week, the sets are tested. If the sets are acceptable, they are shipped to purchasers. The probability that a set proves to be acceptable is 0.70. The second possibility is that minor adjustments can be made so that a set will become acceptable for shipping; this has a probability of 0.20. The third possible outcome is that the set has to go to the diagnostic shop for major adjustment and be shipped at a later date; this has a probability of 0.10. Outcomes for different sets are independent of one another.
   (a) Find the probability of each possible number of sets, for one week's production, which are acceptable without any adjustment.

(b) What is the expected number of sets which are tested and found to be acceptable without adjustment?

(c) What is the cumulative probability distribution for the number of sets which are tested and found to be acceptable without adjustment? Sketch the corresponding graph.

9.

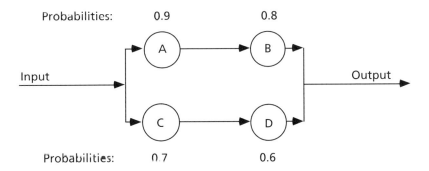

Figure 5.4: Series-Parallel System

A system consists of two branches in parallel, each branch having two components. The probabilities of successful operation of components A, B, C, and D are 0.9, 0.8, 0.7, and 0.6, as shown above. If a component fails, the output from its branch is zero. If only one branch operates, the output is 50%. Of course, if both branches operate, the output is 100%.

a) Find the probability of zero output.
b) Find the expected percentage output.

10. For constant rate of input, the rate of output of a system is determined by whether A, B, and C operate, as shown below.

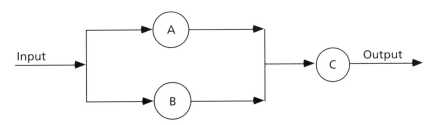

Figure 5.5: Parallel Components, then Series

The probabilities that components A, B, and C operate are as follows:
Pr $[A]$ = 0.70, Pr $[B]$ = 0.60, Pr $[C]$ = 0.90.

Chapter 5

If all of A, B, and C operate, the system output is 100. If both A and C operate but not B, or both B and C but not A, the system output is 80. If both A and B fail, the system output is 0. If C fails, the system output is 0.
a) Find the probability of each possible output.
b) Find the expected output.

(c) More Complex Problems

Now let us look at two more complex examples. To solve them we will need to use our knowledge of basic probability as well as knowledge of expected values. We will have to read each problem very carefully. In the great majority of cases, a tree diagram will be very desirable.

Example 5.3

A manufacturer has two expansion options available to him. The profits of the expansions depend on the cost of energy. The fair odds are 3:2 in favor of energy costs being greater than 8¢/kwh. The manufacturer is twice as likely to choose option 1 as option 2, regardless of circumstances.

If the cost of energy is less than 8¢/kwh, then expansion option 1 will yield returns of +$150,000, $0, and −$50,000 with probabilities of 60%, 20%, and 20%, respectively. Under those conditions, expansion option 2 will yield returns of +$100,000, +$20,000, and −$20,000 with probabilities of 70%, 10%, and 20%, respectively.

If the cost of energy is greater than 8¢/kwh, then option 1 will yield returns of +$100,000, $0, and −$50,000 with probabilities of 60%, 20%, and 20%, respectively, while option 2 will yield returns of +$80,000, $0, and −$50,000 with probabilities of 70%, 10%, and 20%, respectively.

a) What is the probability that option 2 will be pursued and that energy prices will exceed 8¢/kwh?
b) What is the manufacturer's expected return from expansion?
c) Given that several years later the expansion yielded a return greater than zero, what is the probability that option 2 was chosen?

Answer: The first step will be to draw a tree diagram. (See Figure 5.6.)

a) $\Pr[(\text{option 2}) \cap (\text{energy} > 8¢/\text{kwh})] =$
= $(\Pr[\text{energy} > 8¢/\text{kwh}]) \times (\Pr[(\text{option 2}) | (\text{energy} > 8¢/\text{kwh})])$
= $\left(\frac{3}{5}\right)\left(\frac{1}{3}\right) = \frac{1}{5}$ or 0.200.

b) Expected return = $\sum_{\text{all possibilities}} (\text{return for each possibility}) \times (\Pr[\text{that return}])$

= [(0.16)(150) + (0.05333)(0) + (0.05333)(−50) + (0.09333)(+100) +
+ (0.01333)(+20) + (0.02667)(−20) + (0.24)(+100) + (0.08)(0) +
+ (0.08)(−50) + (0.14)(+80) + (0.02)(0) + (0.04)(−50)] thousand dollars

= 59.6 thousand dollars
= $59,600.

Probability Distributions of Discrete Variables

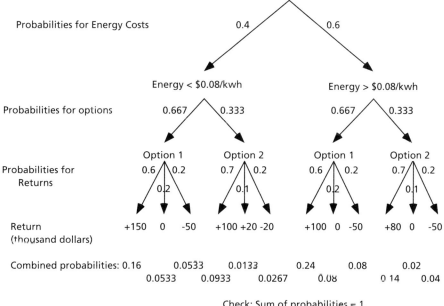

Figure 5.6: Expansion Options

c) $\Pr[\text{option 2} \mid (\text{return} > 0)] = \dfrac{\Pr[(\text{option 2}) \cap (\text{return} > 0)]}{\Pr(\text{return} > 0)}$

$= \dfrac{\Pr[(\text{option 2}) \cap (\text{return} > 0)]}{\Pr[(\text{option 2}) \cap (\text{return} > 0)] + \Pr[(\text{option 1}) \cap (\text{return} > 0)]}$

$= \dfrac{0.0933 + 0.01333 + 0.14}{(0.09333 + 0.01333 + 0.14) + (0.16 + 0.24)}$

$= \dfrac{0.2467}{0.2467 + 0.4000}$

$= 0.381 \text{ or } 38.1\%.$

(Note that part (c) involves Bayesian probability.)

Example 5.4

A flood forecaster issues a flood warning under two conditions only:
  i) Winter snowfall exceeds 20 cm regardless of fall rainfall; or
  ii) Fall rainfall exceeds 10 cm and winter snowfall is between 15 and 20 cm.

The probability of winter snowfall exceeding 20 cm is 0.05. The probability of winter snowfall between 15 and 20 cm is 0.10. The probability of fall rainfall exceeding 10 cm is 0.10.

Chapter 5

a) What is the probability that the forecaster will issue a warning any given spring?
b) Given that he issues a warning, what is the probability that winter snow fall was greater than 20 cm?
c) The probability of flooding is 0.75 for condition (i) above, 0.60 for condition (ii) above, and 0.05 for conditions where no flooding is anticipated. If the cost of a flood after a warning is $100,000, a flood with no warning is $1,000,000, no flood after a warning is $200,000, and zero for no warning and no flood, what is the expected cost in any given year?

Answer: Again, the first step is to draw a tree diagram using the given information.

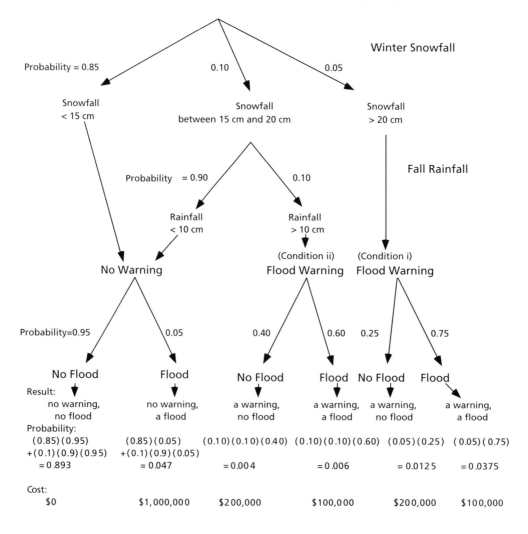

Figure 5.7: Flood Probabilities

Probability Distributions of Discrete Variables

If Pr [winter snowfall > 20 cm] = 0.05 and Pr [15 cm < winter snowfall < 20 cm] = 0.10, then Pr [winter snowfall < 15 cm] = 1 − 0.05 − 0.10 = 0.85.

If Pr [fall rainfall > 10 cm] = 0.10, then Pr [fall rainfall < 10 cm] = 1 − 0.10 = 0.90.

a) Using the tree diagram, Pr [warning] = 0.05 + (0.10)(0.10) = 0.05 + 0.01 = 0.06.

b) Pr [winter snowfall > 20 cm | warning] = $\dfrac{\Pr\left[(\text{winter snowfall} > 20\text{cm}) \cap \text{warning}\right]}{\Pr[\text{warning}]}$

$= \dfrac{\Pr\left[(\text{winter snowfall} > 20\text{cm})\right]}{\Pr[\text{warning}]} = \dfrac{0.05}{0.06} =$

$= 0.83$

(Notice that this calculation used Bayes' Rule.)

c) In order to calculate expected costs, we will need probabilities of each combination of warning or no warning and flood or no flood. These are shown in the second-last line of Figure 5.7. We should apply a check on these calculations: do the probabilities add up to 1?
0.893 + 0.047 + 0.004 + 0.006 + 0.0125 + 0.0375 = 1.000 (check).

Now using equation 4.5, the expected cost in any given year is

($100,000)(0.0375) + ($200,000)(0.0125) + ($100,000)(0.006) + ($200,000)(0.04) + ($1,000,000)(0.047) + ($0)(0.893) = $61,850

Problems

1. Every student in a certain program of studies takes all three of courses A, B, and C. The average enrollment in the program is 50 students.

    Past history shows that on the average:
    (1) 5 students in course A receive marks of at least 75%.
    (2) 7.5 students in course B receive marks of at least 75%.
    (3) 6 students in course C receive marks of at least 75%.
    (4) 80% of students who receive marks of at least 75% in course A also do so in course B.
    (5) 50% of students who receive marks of at least 75% in course B also do so in course C.
    (6) 60% of students who receive marks of at least 75% in course C also do so in course A.
    (7) 10 students receive marks of at least 75% in one or more of these classes.
    A sponsor gives a scholarship of $500 to anyone who receives a mark of at least 75% in all three courses. What can the sponsor expect to pay on average?

Chapter 5

2. A box contains a fair coin and a two-headed coin. A coin is selected at random and tossed. If heads appears, the other coin is tossed; if tails appears, the same coin is tossed.
   a) Find the probability that heads appears on the second toss.
   b) Find the expected number of heads from the two tosses.
   c) If heads appeared on the first toss, find the probability that it also appeared on the second toss.

3. A box contains two red and two green balls. A contestant in a game show selects a ball at random. If the ball is green, he receives no prize for the draw and puts the ball on one side. If the ball is red, he receives $1000 and puts the ball back in the box. The game is over when both green balls are drawn or after three draws, whichever comes first.
   a) What is the probability of the contestant receiving no prize at all?
   b) What is the expected prize?
   c) If the game lasts for three draws, what is the probability that a green ball was selected on the first draw?

4. The probabilities of the monthly snowfall exceeding 10 cm at a particular location in the months of December, January and February are 0.20, 0.40 and 0.60, respectively. For a particular winter:
   a) What is the probability of not receiving 10 cm of snowfall in *any* of the months of December, January and February in a particular winter?
   b) What is the probability of receiving at least 10 cm snowfall in a month, in at least two of the three months of that winter?
   c) Given that the snowfall exceeded 10 cm in each of only two months, what is the probability that the two months were consecutive?
   d) Find the expected number of months in which monthly snowfall does not exceed 10 cm.

5. The probability that Jim will hit a target on a certain range is 25% for any one shot, regardless of what happened on the previous shot or shots. He fires four shots.
   a) What is the probability that Jim will hit the target exactly twice?
   b) What is the probability that he will hit the target at least once?
   c) Find the expected number of hits on the target.
   d) If five persons who are equally good marksmen as Jim shoot at five targets, what is the probability that exactly two targets are hit at least once?

6. Three boxes containing red, white and blue balls are used in an experiment. Box #1 contains two red, three white and five blue balls; Box #2 contains one red and three white balls; and Box #3 contains three red, one white and three blue balls. The experiment consists of drawing a ball at random from Box #1 and placing it with the other balls in Box #2, then drawing a ball at random from Box #2 and placing it in Box #3.

a) Draw the probability distribution of the number of red balls in Box #3 at the end of the experiment.
b) What is the expected number of red balls in Box #3 at the end of the experiment?
c) Given that at the end of the experiment there are three red balls in Box #3, what is the probability that a white ball was picked from Box #1?
d) After the experiment is completed, a ball is drawn from Box #3. What is the probability that the ball is white?

7. Two octahedral dice with faces marked 1 through 8 are constructed to be out of balance so that the 8 is 1.5 times as probable as the 2 through 7, and the sum of the probabilities of the 1 and the 8 equals that of the other pairs on opposing faces, i.e. the 2 and 7, the 3 and 6, and the 4 and 5.
   a) Find the probability distribution and the mean and variance of the number that can show up on one roll of the two dice.
   b) Find the probabilities of getting between 5 and 9 (inclusive) on at least 3 out of 10 rolls of the two dice.
   c) Find the probability of getting one occurrence of between 2 and 4, five occurrences of between 5 and 9, and four occurrences of between 9 and 16, in 10 rolls of the two dice.
   All ranges of numbers are inclusive.

8. A panel of people is assembled to test the ability to correctly distinguish an "improved" product from an older product. The panelists are chosen from a population consisting of 20% rural and 80% urban people. Two-thirds of the population are younger than 30 years of age, while one-third are older. The probability that the urban panelists under 30 years of age will correctly identify the improved product is 12%, while for older urban panelists, the probability increases to 45%. Regardless of age, rural panelists are twice as likely as urban panelists to correctly identify the improved product.
   a) What is the probability that any one panelist chosen at random from this population will correctly identify the improved product?
   b) For a panel of 10 persons, what is the expected number of panelists who will correctly identify the improved product?
   c) If a panelist has correctly identified the improved product, what is the probability that the panelist is under 30 years of age?
   d) If a panelist is under 30 years of age, what is the probability that the panelist will correctly identify the improved product?

9. Certain devices are received at an assembly plant in batches of 50. The sampling scheme used to test all batches has been set up in the following way. One of the 50 devices is chosen randomly and tested. If it is defective, all the remaining 49 items in that batch are returned to the supplier for individual testing; if the tested device is not defective, another device is chosen randomly and tested. If the second item is not defective, the complete batch is accepted without any more testing; if the second device is defective, a third device is chosen randomly and

Chapter 5

tested. If the third device is not defective, the complete batch is accepted without any more testing, but the one defective device is replaced by the supplier. If the third device is defective, all remaining 47 items in that batch are returned to the supplier for individual testing.

The receiver pays for all initial single-item tests. However, whenever the remaining devices in a batch are returned to the supplier for individual tests, the costs of this extra testing are paid by the supplier. If a batch is returned to the supplier, the superintendent must ensure that the receiver is sent 50 items which have been tested and shown to be good. Assume that the superintendent accepts the results of the receiver's tests. Each device is worth $60.00 and the cost of testing is $10.00 per device.

Consider a batch which contains 12 defective items and 38 good items.
a) What is the probability that the batch will be accepted?
b) What is the expected cost to the supplier of the testing and of replacing defectives?
c) Of the 12 defective items in the batch, find the expected number which will be accepted.

10. An oil refinery has a problem with air pollution. In any one year the probability of escape of SO_2 is 23%, and probability of escape of a sticky oil is 16%. Escape of SO_2 and escape of the oil will not occur at the same time. If the wind direction is right, the SO_2 or oil will blow away from the city and no damage will result. The probability of this is 55%. Otherwise, an escape of SO_2 will result in damage claims of $80,000, an escape of oil will result in damage claims of $45,000, and there will be possibility of a fine. If the pollutant is SO_2, under these conditions there is 90% probability of a fine, which will be $150,000. If the pollutant is oil, the probability of a fine depends on whether the oil affects a prominent politician's house or not. If oil causes damage, the probability it will affect his house is 5%. If it affects his house, the probability of a fine is 96%. If it does not affect his house, the probability of a fine is 65%. If there is a fine for pollution by oil, it is $175,000. Answer the following questions for the next year.
a) What is the probability there will be damage claims for escape of SO_2?
b) What is the probability there will be damage claims for escape of oil?
c) What is the probability of a $150,000 fine?
d) What is the expected cost for damages and fines?

11. A mining company is planning strategy with respect to its operations. It has the option of developing 3 properties, but only in a given sequence of A, B, and C. The probability of A being successful and yielding a net profit of $1.5 million is 0.7, and the probability of its failing and causing a loss of $0.5 million is 0.3. If A is successful, B has 0.6 probability of being successful and producing a gain of $1.2 million, and 0.4 probability of being a failure and causing a loss of $.75 million. If A is a failure, B has 0.4 probability of being a success with a gain of $1 million, and 0.6 probability of being a failure with a loss of $1.8 million. If

both A and B are failures, then the company will not proceed with C. If both A and B are successes, C will be a success with probability of 0.9 and a gain of $2.5 million, or a failure with probability of 0.1 and a loss of $1.5 million. If either A or B is a failure (but not both) then C is attempted. In that case, the probability of success of C would be 0.3 but a gain of $5 million would result; failure of C, probability 0.7, would result in a loss of $0.8 million. The company decides to proceed with this strategy.

a) What is the expected gain or loss?
b) Given that A is a failure, what is the expected total gain from projects B and C?
c) Given that there is a net loss for all three (or two) projects taken together, what is the probability that B was a failure?

5.3 Binomial Distribution

This important distribution applies in some cases to repeated trials where there are only two possible outcomes: heads or tails, success or failure, defective item or good item, or many other possible pairs. The probability of each outcome can be calculated using the multiplication rule, perhaps with a tree diagram, but it is usually much faster and more convenient to use a general formula.

The *requirements* for using the binomial distribution are as follows:

- The outcome is determined completely by chance.
- There are only two possible outcomes.
- All trials have the *same* probability for a particular outcome in a single trial. That is, the probability in a subsequent trial is *independent* of the outcome of a previous trial. Let this constant probability for a single trial be p.
- The number of trials, n, must be fixed, regardless of the outcome of each trial.

(a) Illustration of the Binomial Distribution

All items from a production line are tested as they are produced. Each item is classified as either defective (D) or good (G). There are no other possible outcomes. $\Pr[D] = 0.100$, $\Pr[G] = 1 - \Pr[D] = 0.900$. Let us consider all the possible results for a sample consisting of three items, calculating their probabilities from basic principles using the multiplication rule of section 2.2.2.

Outcome	Probability of that Outcome	
G G G	$(0.900)^3$	= 0.729
D G G	$(0.100)(0.900)^2$	= 0.081
G D G	$(0.900)(0.100)(0.900)$	= 0.081
G G D	$(0.900)^2(0.100)$	= 0.081
D D G	$(0.100)^2(0.900)$	= 0.009
G D D	$(0.900)(0.100)^2$	= 0.009
D G D	$(0.100)(0.900)(0.100)$	= 0.009
D D D	$(0.100)^3$	= <u>0.001</u>
	Total	= 1.000 (Check)

Notice that the outcome containing three good items appeared once, and so did the outcome containing three defective items. The outcome containing two good items and one defective appeared three times, which is the number of permutations of two items of one class and one item of another class. The outcome containing one good item and two defectives also appears three times (as D D G, G D D, and D G D); again, this is the number of permutations of one item of one class and two items of another class.

(b) Generalization of Results

Now we'll develop more general results. Let the probability that an item is defective be p. Let the probability that an item is good be q, such that $q = 1 - p$. Notice that the definitions of p and q can be interchanged, and other terms such as "success" and "failure" can be used instead (and often are). Let the fixed number of trials be n. The probability that all n items are defective is p^n. The probability that exactly r items are defective and $(n-r)$ items are good, in any one sequence, is $p^r q^{(n-r)}$. But r defective items and $(n-r)$ good items can be arranged in various ways. How many different orders are possible? This is the number of permutations into two classes, consisting of r defective items and $(n-r)$ good items, respectively. From section 2.2.3 this number of permutations is given by $\dfrac{n!}{r!(n-r)!}$. But this is exactly the expression for the number of combinations of n items taken r at a time, $_nC_r$. Then the general expression for the probability of exactly r defective items (or successes, heads, etc.) in any order in n trials must be $p^r q^{(n-r)}$ multiplied by $_nC_r$, or

$$\Pr[R = r] = {}_nC_r\, p^r q^{(n-r)} \tag{5.9}$$

The lefthand side of this equation should be read as the probability that exactly r items are defective (or successes, heads, etc.).

The name given to this discrete probability distribution is the binomial distribution. This name arises because the expression for probability in equation 5.9 is the same as the $(n+1)$th term in the binomial expansion of $(q+p)^n$.

Tables of cumulative binomial probabilities are found in many reference books. Individual binomial probabilities, like those given in equation 5.9, are found from cumulative binomial probabilities by subtraction using equation 5.2. Both individual and cumulative probabilities can be calculated also using computer software such as Excel. That will be discussed briefly in section 5.3(f).

(c) Application of the Binomial Distribution

The binomial distribution is often used in quality control of items manufactured by a production line when each item is classified as either defective or nondefective. To meet the requirements of the binomial distribution the probability that an item is defective must be constant. This condition is *not* met by sampling *without replacement* from a small batch because, as we have seen from Example 2.7, in that case the probability that the second item drawn will be defective depends on whether the first item drawn was defective or not, and so on. The condition of constant probability is met to an acceptable approximation if the total number of trials is much less than the batch size, so for a *sufficiently small sample* from a *large enough batch*. Then the probability of a defect (or "success" etc.) on a single trial will be approximately constant.

The condition is met for sampling item by item from *continuous* production under constant conditions. It is also met for sampling from a small batch if each item which is removed as a specimen is returned to the batch and mixed thoroughly with the other items, once it has been examined and classified as defective or good. This, however, is not often a practical procedure: if we know that an item is defective, we should not mix it with other items of production. Indeed, sometimes we can't, because the test procedure may destroy the sample.

Example 5.5

On the basis of past experience, the probability that a certain electrical component will be satisfactory is 0.98. The components are sampled item by item from continuous production. In a sample of five components, what are the probabilities of finding (a) zero, (b) exactly one, (c) exactly two, (d) two or more defectives?

Answer: The requirements of the binomial distribution are met.

$n = 5$, $p = 0.98$, $q = 0.02$, where p is taken to be the probability that an item will be satisfactory, and so q is the probability that an item will be defective.

(a) Pr [0 defectives] = $(0.98)^5$ = 0.9039 or 0.904.
(b) Pr [1 defective] = $_5C_1 (0.98)^4 (0.02)^1$
$= (5)(0.98)^4(0.02)^1 = 0.0922$ or 0.092.
(c) Pr [2 defectives] = $_5C_2 (0.98)^3(0.02)^2$
$= \frac{(5)(4)}{2} (0.98)^3(0.02)^2 = 0.0038.$
(d) Pr [2 or more defectives] = 1 − Pr [0 def.] − Pr [1 def.]
= 1 − 0.9039 − 0.0922
= 0.0038.

Example 5.6

A company is considering drilling four oil wells. The probability of success for each well is 0.40, independent of the results for any other well. The cost of each well is $200,000. Each well that is successful will be worth $600,000.

 a) What is the probability that one or more wells will be successful?
 b) What is the expected number of successes?
 c) What is the expected gain?
 d) What will be the gain if only one well is successful?
 e) Considering all possible results, what is the probability of a loss rather than a gain?
 f) What is the standard deviation of the number of successes?

Chapter 5

Answer: The binomial distribution applies. Let us start by calculating the probability of each possible result. We use $n = 4$, $p = 0.40$, $q = 0.60$.

No. of Successes	Probability	
0	$(1)(0.40)^0(0.60)^4$	$= 0.1296$
1	$(4)(0.40)^1(0.60)^3$	$= 0.3456$
2	$\dfrac{(4)(3)}{2}(0.40)^2(0.60)^2$	$= 0.3456$
3	$(4)(0.40)^3(0.60)^1$	$= 0.1536$
4	$(1)(0.40)^4(0.60)^0$	$= \underline{0.0256}$
	Total	$= 1.000$ (check)

(Notice that $_nC_r = {_nC_{(n-r)}}$)

Now we can answer the specific questions.

a) Pr [one or more successful wells] = 1− Pr [no successful wells]
   $= 1 - 0.1296$
   $= 0.8704$ or 0.870.

b) Expected number of successes $= (1)(0.3456) + (2)(0.3456) + (3)(0.1536) + (4)(0.0256)$
   $= 1.600$.

c) Expected gain $= (1.6)(\$600{,}000) - (4)(\$200{,}000) = \$160{,}000$.

d) If only one well is successful, gain $= (1)(\$600{,}000) - (4)(\$200{,}000)$
   $= -\$200{,}000$ (so a loss).

e) There will be a loss if 0 or 1 well is successful, so the probability of a loss is $(0.1296 + 0.3456) = 0.4752$ or 0.475.

f) Using equation 4.3, $\sigma_x^2 = E(X^2) - \mu_x^2$,
where $E(X^2) = (0.3456)(1)^2 + (0.3456)(2)^2 + (0.1536)(3)^2 + (0.0256)(4)^2 = 3.5200$,
so $\sigma^2 = 3.5200 - (1.600)^2 = 0.9600$.
The standard deviation of the number of successes is $\sqrt{0.9600} = 0.980$.

(d) Shape of the Binomial Distribution

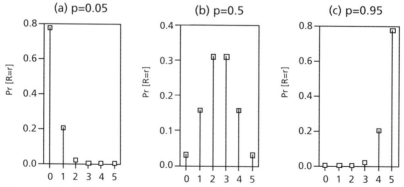

Figure 5.8: Effect of Varying Probability of Success in a Single Trial when the Number of Trials is 5

Probability Distributions of Discrete Variables

Figure 5.8 compares the shapes of the distributions for p equal to 0.05, 0.50, and 0.95, all for n equal to 5. When p is close to zero or one, the distribution is very skewed, and the distribution for p equal to p_1 is the mirror image of the distribution for p equal to $(1-p_1)$. When p is equal to 0.500, the distribution is symmetrical.

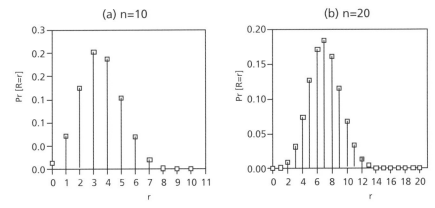

Figure 5.9: Effect of Varying Number of Trials when the Probability of Success Is 0.35

Figure 5.9 compares the shape of the distributions for n equal to 10 and 20, both for p equal to 0.35. At this intermediate value of p, the distribution is rather skewed for small numbers of trials, but it becomes more symmetrical and bell-shaped as n increases.

(e) Expected Mean and Standard Deviation

For any discrete random variable, equation 5.5 gives that the expected mean is $E(R) = \mu\,(or\,\mu_R) = \Sigma$ (number of "successes")(probability of that number of "successes") for all possible results.

For the binomial distribution, from equation 5.9 the probability of r "successes" in n trials is given by

$$\Pr[R = r] = {}_nC_r\,(1-p)^{n-r}p^r$$

Then

$$\mu = \sum_{r=0}^{n}(r)\Pr[R=r] = \sum_{r=0}^{n}(r)({}_nC_r)(1-p)^{(n-r)}p^r$$

If the algebra is followed through, the result is

$$\mu = np \tag{5.10}$$

Thus, the mean value of the binomial distribution is the product of the number of trials and the probability of "success" in a single trial. This seems to be intuitively correct.

From equation 5.6, for any discrete probability distribution,

Chapter 5

$$\sigma^2 = E(r-\mu)^2 = \sum_{r=0}^{n}(r-\mu)^2 \Pr[R=r]$$

Substituting for the probability for the binomial distribution and following through the algebra gives

$$\sigma^2 = np(1-p)$$

or

$$\sigma^2 = npq \qquad (5.11)$$

The standard deviation is always given by the square root of the corresponding variance, so the standard deviation for the binomial distribution is

$$\sigma = \sqrt{npq} \qquad (5.12)$$

Example 5.7

Calculate the expected number of successes and the standard deviation of the number of successes for Example 5.6 and compare with the results of parts b and f of that example.

Answer: Binomial distribution with $n = 4$, $p = 0.4$, $q = 0.6$.

Then the expected number of successes from equation 5.6 is $np = (4)(0.400) = 1.60$. This agrees with the results of part b of Example 5.6.

The standard deviation of the number of successes from equation 5.8 is $\sqrt{(4)(0.400)(0.600)} = \sqrt{0.960} = 0.980$. This agrees with the results of part f of Example 5.6.

Example 5.8

Twelve doughnuts sampled from a manufacturing process are weighed each day. The probability that a sample will have no doughnuts weighing less than the design weight is 6.872%.

a) What is the probability that a sample of twelve doughnuts contains exactly three doughnuts weighing less than the design weight?
b) What is the probability that the sample contains more than three doughnuts weighing less than the design weight?
c) In a sample of twelve doughnuts, what is the expected number of doughnuts weighing less than the design weight?

Answer: In 12 doughnuts Pr [0 doughnuts < design weight] = 0.06872.

Assuming that Pr [a single doughnut < design weight] is the same for all doughnuts and that weights of doughnuts vary randomly, the binomial distribution will apply. Let this probability that a single doughnut will weigh less than the design weight be p.

Probability Distributions of Discrete Variables

Then $(1-p)^{12} = 0.06872$.

$1 - p = 0.8000$

Then Pr [a doughnut < design weight] = 1 − 0.8000 = 0.2000. Then $p = 0.2$, and $n = 12$.

a) Pr [exactly 3 doughnuts in 12 are below design weight] = $_{12}C_3(1-p)^9 p^3$

$$= \frac{(12)(11)(10)}{(3)(2)}(0.8000)^9 (0.2000)^3$$

$$= 0.2362 \text{ or } 23.6\%.$$

b)
Number less than design weight	Probability	
0	$(0.8)^{12}$	= 0.0687
1	$_{12}C_1(0.8)^{11}(0.2)^1$	= 0.2062
2	$_{12}C_2(0.8)^{10}(0.2)^2$	= 0.2835
3	$_{12}C_3(0.8)^9(0.2)^3$	= 0.2362
Sum		0.7946

Therefore, Pr [more than three doughnuts are below design weight] =
= 1 − (Pr [R = 0] + Pr [R = 1] + Pr [R = 2] + Pr [R = 3])
= 1 − 0.7946
= 0.2054 or 0.205 = 20.5%.

c) Expected number of doughnuts below the design weight is (n)(p) = (12)(0.200) = 2.4.

(f) Use of Computers

If a computer with suitable software is available, calculations for the binomial distribution can be done easily. If Excel is available, the function BINOMDIST will be found to be very useful. There is not usually a great advantage to use of a computer if only individual terms of the distribution are required, as equation 5.9 is convenient for that purpose. But if cumulative expressions are required, such as the probability of six or fewer occurrences, the computer can greatly reduce the amount of labor required.

The parameters required by the Excel function BINOMDIST are r, n, p, and an indication of whether a cumulative expression or an individual term is required. As in the earlier part of this section, r is the number of "successes" in a total of n trials, and p is the probability of "success" in each trial. The fourth parameter should be entered as TRUE if the cumulative distribution function is required, giving the probability of *at most r* "successes"; the fourth parameter should be entered as FALSE if the required quantity is the individual probability, the probability of *exactly r* "successes." For example, if we want the probability of six or fewer "successes" in a total of 12 trials when the probability of "success" in a single trial is 0.245, the parameters for Excel in the function BINOMDIST are 6, 12, 0.245, TRUE. The function returns the corresponding probability, which is 0.9873.

Chapter 5

(g) Relation of Proportion to the Binomial Distribution

Assuming that the only alternative to a rejected item is an accepted item, the sample size is fixed and independent of the results, and the probability of rejection is constant and independent of other factors such as previous results, we have seen that the number of rejects in a sample of size n is governed by the binomial distribution. If the probability that an item will be rejected is p, the probability that there will be exactly x rejects in the sample is ${}_nC_x \, p^x \, (1-p)^{(n-x)}$. The mean number of rejects will be np, and the variance of the number of rejects will be $np(1-p)$.

We can look at the sample from a somewhat different viewpoint, focusing on the *proportion* of rejects rather than their number. The ratio $\dfrac{x}{n}$ is an unbiased estimate of p, the proportion of rejects in the population, and we use the symbol $\hat{p}$ for this estimate. The probability that the estimate of proportion from the sample will be $\hat{p} = \dfrac{x}{n}$ is the same as the probability that there will be exactly x rejected items in a sample of size n, and that is ${}_nC_x \, p^x \, (1-p)^{(n-x)}$. If we associate the number 1 with each rejected item and the number 0 with each item which is not rejected, then x, the number of rejected items, can be interpreted as the sum of the zeros and ones for a sample of size n. Then $\hat{p} = \dfrac{x}{n}$ is a sample mean. Since n is a constant, in the whole population the mean proportion rejected is

$$\mu_{\hat{p}} = E(\hat{P}) = E\left(\frac{X}{n}\right) = \frac{np}{n} = p \tag{5.13}$$

This seems reasonable.

Similarly, using the relations for variance of a variable multiplied or divided by a constant that will be discussed in section 8.2, we find that the variance of the proportion rejected is

$$\sigma_{\hat{p}}^2 = \sigma_{X/n}^2 = \frac{\sigma_X^2}{n^2} = \frac{np(1-p)}{n^2} = \frac{p(1-p)}{n} \tag{5.14}$$

Example 5.9

The true proportion of defective items in a continuous stream is 0.0100. A random sample of size 400 is taken.

(a) Calculate the probabilities that the sample will give sample estimates of the proportion defective of $\dfrac{0}{400}, \dfrac{1}{400}, \dfrac{2}{400}, \dfrac{3}{400}, \dfrac{4}{400}$, and $\dfrac{5}{100}$, respectively.

(b) Calculate the standard deviation of the proportion defective.

Probability Distributions of Discrete Variables

Answer:

(a) $p = 0.01$, $n = 400$

$\Pr[\hat{p} = 0] = \Pr[0 \text{ defective items}] = {}_{400}C_0 \, (0.01)^0 (0.99)^{400}$

$\qquad\qquad\qquad\qquad = (1)(1)(0.01795) \qquad\qquad\qquad\qquad = 0.0180$

$\Pr[\hat{p} = \dfrac{1}{400} = 0.00250] = {}_{400}C_1 \, (0.01)^1 (0.99)^{399} =$

$\qquad\qquad\qquad\qquad = (400)(0.01)(0.01813) \qquad\qquad\qquad = 0.0725$

$\Pr[\hat{p} = \dfrac{2}{400} = 0.00500] = {}_{400}C_2 \, (0.01)^2 (0.99)^{398} =$

$\qquad\qquad\qquad\qquad = \dfrac{(400)(399)}{2} (0.01)^2 (0.99)^{398} \qquad = 0.1462$

$\Pr[\hat{p} = \dfrac{3}{400} = 0.00750] = {}_{400}C_3 \, (0.01)^3 (0.99)^{397} =$

$\qquad\qquad\qquad\qquad = \dfrac{(400)(399)(398)}{(3)(2)} (0.01)^3 (0.99)^{397} \doteq 0.1959$

$\Pr[\hat{p} = \dfrac{4}{400} = 0.01000] = {}_{400}C_4 \, (0.01)^4 (0.99)^{396} =$

$\qquad\qquad\qquad\qquad = \dfrac{(400)(399)(398)(397)}{(4)(3)(2)} (0.01)^4 (0.99)^{396} = 0.1964$

$\Pr[\hat{p} = \dfrac{5}{100} = 0.01250] = {}_{400}C_5 \, (0.01)^5 (0.99)^{395} =$

$\qquad\qquad\qquad\qquad = \dfrac{(400)(399)(398)(397)(396)}{(5)(4)(3)(2)} (0.01)^5 (0.99)^{395} = 0.1571$

Thus, the probability that the sample will give an estimate of the proportion defective that agrees exactly with the true proportion (0.01) is less than 20%, and the probability of getting any one of the three estimates, 0.0075 or 0.01 or 0.0125, is less than 55%.

Calculations of probabilities of sample estimates can be continued. The results are shown in Figure 5.10.

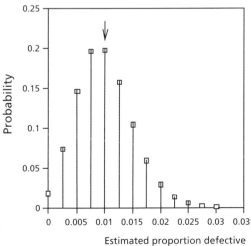

Figure 5.10: Probabilities of Estimates When True Proportion Is 0.0100

Chapter 5

We see that there can be a wide range of estimates from a sample, even when the sample size is as large as 400.

(b) The standard deviation of the proportion defective is given, according to equation 5.14, by $\sqrt{\frac{(p)(1-p)}{n}} = \sqrt{\frac{(0.01)(0.99)}{400}} = 0.004975$.

The standard deviation is nearly half of the true proportion defective. Again, this indicates that an estimate from a sample of this size will not be very reliable.

(h) Nested Binomial Distributions

These are situations in which one binomial distribution is enclosed within another binomial distribution.

Example 5.10

A boiler containing eight welds is manufactured in a small shop. When the boiler is completed, each weld is checked by an inspector. If more than one weld is defective on a single boiler, the person who made that boiler is reported to the foreman.

a) If 9.0% of all welds made by Joe Smith are defective, what percentage of all boilers made by him will have more than one defective weld?
b) Over a long period of time how many times will Joe Smith be reported to the foreman for each 15 boilers he makes?
c) If Joe makes 15 boilers in a shift, what is the probability that he will be reported for more than two of these 15 boilers?

Answer: a) The probabilities of various numbers of defective welds on a single boiler are given by the binomial distribution with $n = 8$, $p = 0.090$, $q = 1 - 0.090 = 0.910$.

The probability of exactly r defective welds on a boiler is given by
$\Pr[R = r] = {}_8C_r (0.910)^{(8-r)}(0.090)^r$.

More than one defective weld corresponds to all results except zero defective welds and one defective weld.

$\Pr[R = 0] = (1) (0.910)^8 (0.090)^0 = 0.4703$
$\Pr[R = 1] = (8) (0.910)^7 (0.090)^1 = 0.3721$

(Four figures are being carried in intermediate results, and final answers will be shown to three figures.)

$\Pr[\text{more than one defective weld in a single boiler}] = 1 - 0.4703 - 0.3721 = 0.1577$.
Then 15.8% of boilers made by Joe will have more than one defective weld.

b) Now the problem shifts to the outer Binomial problem for the number of times Joe will be reported to the foreman for each 15 boilers he makes. Then $n = 15$,

Probability Distributions of Discrete Variables

$p = $ Pr [being reported for 1 boiler] $= 0.1577$, and $q = 1 - p = 0.8423$. (Notice that the value of p, the probability of too many defects in a single boiler in the outer binomial distribution, is given by the result of calculations for the inner binomial distribution.)

Under these conditions the expected number of times Joe will be reported to the foreman is $\mu = np = (15)(0.1577) = 2.37$.

c) As in part b, this corresponds to a binomial problem with $n = 15$, $p = 0.1577$, $q = 0.8423$.

In general, $\quad$ Pr $[R = r] = {}_{15}C_r (0.8423)^{(15-r)}(0.1577)^r$

Then specifically, Pr $[R = 0] = (1)(0.8423)^{15}(0.1577)^0 = 0.0762$

$$\text{Pr } [R = 1] = (15)(0.8423)^{14}(0.1577)^1 = 0.2141$$

$$\text{Pr } [R = 2] = \frac{(15)(14)}{2}(0.8423)^{13}(0.1577)^2 = 0.2805$$

The probability that Joe will be reported to the foreman for more than two of the 15 boilers he makes in a shift is $1 - 0.0762 - 0.2141 - 0.2805 = 0.429$ or 42.9%.

(i) Extension: Multinomial Distribution

The multinomial distribution is similar to the binomial distribution except that there are more than two possible results from each trial. The details of the multinomial distribution are given in various references, including the book by Walpole and Myers (see the List of Selected References in section 15.2). For example, mechanical components coming off a production line might be classified on the basis of a particular dimension as undersize, acceptable, or oversize (three possible outcomes). If the outcome of any one trial is determined completely by chance, all trials are independent and have the same set of probabilities for the various possible outcomes, and the number of trials is fixed, the multinomial distribution would apply.

Notice that if we consider separately just one result and lump together all other results from each trial, the multinomial distribution becomes a binomial distribution. Thus, in the example of mechanical components just cited, if undersized and oversized are lumped together as unacceptable, the distribution becomes binomial.

Problems

1. Under normal operating conditions 1.5% of the transistors produced in a factory are defective. An inspector takes a random sample of forty transistors and finds that two are defective.
   a) What is the probability that exactly two transistors will be defective from a random sample of forty under normal operating conditions?
   b) What is the probability that more than two transistors will be defective from a random sample of forty if conditions are normal?
2. A control system is set up so that when production conditions are normal, only 6% of items from the production line gives readings beyond a particular limit. If

Chapter 5

more than two of six successive items are beyond the limit, production is stopped and all machine settings are examined. What is the probability that production will be stopped in this way when production conditions are normal?

3. A company supplying transistors claims that they produce no more than 2% defectives. A purchaser picks 50 at random from an order of 5000 and tests the 50. If he finds more than 1 defective, he rejects the order. If the supplier's claim is true and 2% of the transistors are defective, what is the probability that the order will be rejected?

4. An experiment was conducted wherein three balls were drawn at random from a barrel containing two blue balls, three red balls, and five green balls. We want to find the mean and variance of the probability distribution of the number of green balls chosen. Explain why this problem involving three colours can not be handled using a binomial distribution. Suppose we consider both the blue balls and the red balls together as not-green. Now find the required mean and variance.

5. A binomial distribution is known to have the following cumulative probability distribution: $\Pr[X \leq 0] = 1/729$, $\Pr[X \leq 1] = 13/729$, $\Pr[X \leq 2] = 73/729$, $\Pr[X \leq 3] = 233/729$, $\Pr[X \leq 4] = 473/729$, $\Pr[X \leq 5] = 665/729$, $\Pr[X \leq 6] = 1.0000$.
   a) What is n, the number of trials?
   b) Find p and q, the probabilities of success and failure.
   c) Verify that with these values of n, p and q the cumulative probabilities are as stated.
   d) What is the probability that the number of successes, r, lies within one standard deviation of the mean?
   e) What is the coefficient of variation?

6. Ten judges are asked to pick the best tasting orange juice from two samples labeled A and B. If, in fact, A and B are the same orange juice, what is the probability that eight or more of the judges will declare the same sample to be the best? Assume that no judge says that they are equal.

7. A sample of eleven electric bulbs is drawn every day from those manufactured at a plant. The eleven bulbs are tested before shipment to the customer. An analysis of the test data collected over a number of years reveals that the probability of finding no defective bulb in a sample of eleven bulbs is 0.5688. Probabilities of defective bulbs are random and independent of previous results.
   a) What is the probability of finding exactly three defective bulbs in a sample?
   b) What is the probability of finding three or more defective bulbs in a sample?

8. There are ten multiple choice questions on an examination. If there are five choices per question, what is the probability that a student will answer at least five questions correctly just by picking one answer at random from the possibilities for each question? State any assumptions.

9. Among a group of five people selected at random from a particular population it is known that the probability that no one will be 30 or over is 0.01024.
   a) What is the probability that exactly one person in the group is under 30?
   b) Calculate the mean and variance of the probability distribution of the number of persons over 30 and compare to the formula values for this type of distribution.
   c) Given three such groups, what is the probability that two out of three groups have no more than two persons 30 or over?
   d) State any assumptions.

10. A fraction 0.014 of the output from a production line is defective. A sample of 95 items is taken. Assume defective items occur randomly and independently.
    a) What is the standard deviation of the proportion defective in a sample of this size?
    b) What is the probability that the proportion of defective items in the sample will be within two standard deviations of the fraction defective in the whole population?

11. Surveys have indicated that in a given region 75% of car occupants use seat belts regardless of where they sit in the car. Use of seat belts in the region is random and shows no regular pattern. The surveys have shown also that in 40% of cars the driver is the sole occupant, in 25% there are two occupants, in 20% three occupants, in 10% four occupants, and in 5% five occupants.
    a) What is the probability that a car picked at random will have exactly three persons not using their seat belts? Remember to consider all possible number of occupants.
    b) What is the probability that of three cars chosen at random, exactly two have all occupants wearing belts?

12. A small hotel has rooms on only four floors, with four smoke detectors on each floor. Because of improper maintenance, the probability that any one detector is functioning is only 0.55. The probabilities that smoke detectors are functioning are randomly and independently distributed.
    a) What is the probability that exactly one smoke detector is working on the top floor?
    b) What is the probability that there is exactly one detector working on each of two floors and there are two detectors working on each of the other two floors?
    c) What is the probability that there will be no functioning smoke detectors on one particular floor? What is the probability that there will be at least one functioning smoke detector on that floor?
    d) What is the probability that on at least one of the four floors there will be no functioning smoke detectors?
    e) What is the probability that there will be at least 15 functioning smoke detectors in the hotel at any one time?

13. The FIXIT company is to bring in seven new products in a sales line for which the probability that each new product will be successful is 0.15. Probabilities of success for the various products are random and independent. The cost of bringing in a new product is $75,000. If each product is successful, the expected revenue from sales for it will be $800,000 .
    a) What is the expected net profit from the seven products?
    b) What is the probability that the total net profit will be at least $1,000,000?
    c) What is the probability that none of the products will be successful?
    d) If the number of successful products is three or more, the sales engineer will be promoted. What is the probability that this will happen?

14. The probability that a certain type of IC chip will fail after installation is 0.06. A memory board for a computer contains twelve such chips. The operation will be satisfactory if ten or more of the chips on the board do not fail.
    a) What is the probability that a memory board operates satisfactorily?
    b) If there are five such memory boards in a given computer, what is the probability that at least four of them operate satisfactorily?
    c) State any assumptions.

15. 5% of a large lot of electrical components are defective. Six batches of four components each are drawn from this lot at random.
    a) What is the probability that any one batch contains fewer than two defectives?
    b) What is the probability that at least five of the six batches contain fewer than two defectives each?
    c) State any assumptions.

16. 20% of a large lot of mechanical components are found to be faulty. Five batches of five components each are drawn from this lot. What is the probability that at least four of these batches contain fewer than two defectives? State any assumptions.

17. A consultant collected data on bolt failures in an anchor assembly used in tower construction. A large number of anchor assemblies, each containing the same number of bolts, were examined and each bolt was graded either a success or a failure. The probability distribution of the number of satisfactory bolts in an assembly had a mean value of 3.5 and a variance of 1.05. Satisfactory and unsatisfactory bolts occur randomly and independently. Calculate the probabilities associated with the possible numbers of satisfactory bolts in an assembly. If an assembly is considered to be adequate if there are three or fewer bolt failures, what is the probability that an assembly chosen at random will be inadequate?

18. Each automobile leaving a certain motor company's plant is equipped with five tires of a particular brand. Tires are assigned to cars randomly and independently. The tires on each of 100 such automobiles were examined for major defects with the following results.

No. of Tires with Defects	0	1	2	3	4	5
No. of Automobiles (occurrences)	75	18	4	2	0	1

a) Estimate the probability that a randomly selected tire from this manufacturer will contain a major defect.
b) Suppose you buy an automobile of this make. From the results of (a) calculate the probability that it will have at least one tire with a major defect.
c) What is the probability that, in a fleet purchase of six of these cars, at least half the cars have no defective tires?
d) What is the expected number of defective tires in the fleet purchase of six cars?
e) If the replacement cost of a defective tire is $120, what is the total expected replacement cost for this fleet purchase?

19. Thirteen electronic components from a manufacturing process are tested every day. Components for testing are chosen randomly and independently. It was found over a long period of time that 51.33% of such samples have no defectives.
    a) What is the probability of a sample containing exactly two defective components?
    b) What is the probability of finding three or more defective components in a sample?
    c) The assembly line has a weekly bonus system as follows: Each man receives a bonus of $500 if none of the five daily samples that week contained a defective. The bonus is $250 if only one sample out of the five contained a defective, and none of the others contained any. What is the expected bonus per man per week?

20. Truck tires are tested over rough terrain. 25% of the trucks fail to complete the test run without a blowout. Of the next fifteen trucks through the test, find the probability that:
    a) exactly three have one or more blowouts each;
    b) fewer than four have blowouts;
    c) more than two have blowouts.
    d) What would be the expected number of trucks with blowouts of the next fifteen tested?
    e) What would be the standard deviation of the number of trucks with blowouts of the next fifteen tested?
    f) If fifteen trucks are tested on each of three days, what is the probability that more than two trucks have blowouts on exactly two of the three days?
    g) State any assumptions.

21. An elevator arrives empty at the main floor and picks up five passengers. It can stop at any of seven floors on its way up. What is the probability that no two passengers get off at the same floor? Assume that the passengers act independently and that a passenger is equally likely to get off at any one of the floors.

Chapter 5

22. In a particular computer chip 8 bits form a byte, and the chip contains 112 bytes. The probability of a bad bit, one which contains a defect, is 1.2 E-04.
    a) What is the probability of a bad byte, i.e. a byte which contains a defect?
    b) The chip is designed so that it will function satisfactorily if at least 108 of its 112 bytes are good. What is the probability that the chip will not function satisfactorily?

23. In a particular computer chip 8 bits form a byte, and the chip contains 112 bytes. The probability of a bad bit, one which contains a defect, is 2.7 E-04.
    a) What is the probability of a bad byte, i.e. a byte which contains a defect?
    b) The chip is designed so that it will function satisfactorily if at least 108 of its 112 bytes are good. What is the probability that the chip will not function satisfactorily?

Computer Problems

C24. Under normal operating conditions the probability that a mechanical component will be defective when it comes off the production line is 0.035. A sample of 40 components is taken. In one case, four of the components are found to be defective. If the operating conditions are still correct, what is the probability that that many or more components will be defective in a sample of size 40?

C25. A computer chip is organized into bits, bytes, and cells. Each byte contains 8 bits, and each cell contains 112 bytes. The probability that any one bit will be bad (or corrupted) is 1.E–11 (i.e. 10^{-11}).
   a) What is the probability that any one byte will contain a bad bit and so will be bad and give an error in a calculation? Note that you can neglect the probability that a byte will contain more than one bad bit.
   b) What is the probability there will be no bad bytes in a cell?
   c) What is the probability there will be exactly one bad byte in a cell?
   d) What is the probability there will be exactly two bad bytes (and so also exactly 110 good bytes) in a cell?
   e) What is the probability there will be exactly three bad bytes (and so also exactly 109 good bytes) in a cell?
   f) What is the probability there will be two or more bad bytes in a cell? Calculate this in three ways: i) Use the results of some of parts (a) (b) (c).
      ii) Use the results of parts (d) and (e).
      iii) Use a cumulative probability.
      Do they give the same answer? If not, explain why not.

C26. In order to estimate the fraction defective among electrical components as they are produced under normal conditions, a sample containing 1000 components is taken and each component is classified as defective or non-defective. Nine components are found to be defective in this sample.

Probability Distributions of Discrete Variables

a) What is the best estimate from this sample of the proportion defective in the population?
b) Assuming that that estimate is exactly correct, what is the standard deviation of the proportion defective? Then what are the limits of the interval from the best estimate minus two standard deviations to the best estimate plus two standard deviations? What is the probability of a result outside this interval?
c) Assuming the estimate in part (a) is exactly correct, what is the probability that more than three defective components will be found in a sample of 100 components?

C27. A sample containing 400 items is taken from the output of a production line. A fraction 0.016 of the items produced by the line are defective. Assume defective items occur randomly and independently.

a) What is the probability that the proportion defective in the sample will be no more than 0.0250?
b) What is the standard deviation of the proportion defective in a sample of this size?
c) What sample proportion defective would be two standard deviations less than the proportion defective in the whole population?

5.4 Poisson Distribution

This is a discrete distribution that is used in two situations. It is used, when certain conditions are met, as a probability distribution in its own right, and it is also used as a convenient approximation to the binomial distribution in some circumstances. The distribution is named for S.D. Poisson, a French mathematician of the nineteenth century.

The Poisson distribution applies in its own right where the *possible* number of discrete occurrences is much larger than the *average* number of occurrences in a given interval of time or space. The number of possible occurrences is often not known exactly. The outcomes must occur randomly, that is, completely by chance, and the probability of occurrence must not be affected by whether or not the outcomes occurred previously, so the occurrences are independent. In many cases, although we can count the occurrences, such as of a thunderstorm, we cannot count the corresponding nonoccurrences. (We can't count "non-storms"!)

Examples of occurrences to which the Poisson distribution often applies include counts from a Geiger counter, collisions of cars at a specific intersection under specific conditions, flaws in a casting, and telephone calls to a particular telephone or office under particular conditions. For the Poisson distribution to apply to these outcomes, they must occur randomly.

Chapter 5

(a) Calculation of Poisson Probabilities

The probability of exactly r occurrences in a fixed interval of time or space under particular conditions is given by

$$\Pr[R = r] = \frac{(\lambda t)^r e^{-\lambda t}}{r!} \tag{5.13}$$

where t (in units of time, length, area or volume) is an interval of time or space in which the events occur, and λ is the mean rate of occurrence per unit time or space (so that the product λt is dimensionless). As usual, e is the base of natural logarithms, approximately 2.71828. Then the probability of no occurrences, $r = 0$, is $e^{-\lambda t}$, the probability of exactly one occurrence, $r = 1$, is $\lambda t\, e^{-\lambda t}$, the probability of exactly two occurrences, $r = 2$, is $\dfrac{(\lambda t)^2 e^{-\lambda t}}{2!}$, and so on. Once one of these probabilities is calculated it is often more convenient to calculate other members of the sequence from the following recurrence formula:

$$\Pr[R = r + 1] = \left(\frac{\lambda t}{r+1}\right) \Pr[R = r] \tag{5.14}$$

The basic relation for the Poisson distribution, equation 5.13, can be derived from a differential equation or as a limiting expression from the binomial distribution.

Cumulative Poisson probabilities can be found in many reference books. Once again, Poisson probabilities for single events can be found by subtraction using equation 5.2: the probability of x_i is just the difference between the cumulative probability that $X \leq x_i$ and the cumulative probability that $X \leq x_{i-1}$.

Example 5.11

From tables for the cumulative Poisson distribution to three decimal points, for $\lambda t = 10.5$,

$$\Pr[X \leq 12] = \sum_{k=0}^{12} \frac{(e^{-\lambda t})(\lambda t)^k}{k!} \text{ is equal to } 0.742,$$

$$\Pr[X \leq 11] = \sum_{k=0}^{11} \frac{(e^{-\lambda t})(\lambda t)^k}{k!} \text{ is equal to } 0.639, \text{ and}$$

$$\Pr[X \leq 10] = \sum_{k=0}^{10} \frac{(e^{-\lambda t})(\lambda t)^k}{k!} \text{ is equal to } 0.521.$$

Then for $\lambda t = 10.5$, we have $\Pr[R=12] = 0.742 - 0.639 = 0.103$, compared with 0.1032 from equation 5.13, and

$$\Pr[R = 11 \text{ or } 12] = \sum_{k=0}^{12} \frac{(e^{-\lambda t})(\lambda t)^k}{k!} - \sum_{k=0}^{10} \frac{(e^{-\lambda t})(\lambda t)^k}{k!} = 0.742 - 0.521 = 0.221,$$

compared with $\Pr[R = 11] + \Pr[R = 12] = \dfrac{(e^{-10.5})(10.5)^{11}}{11!} + \dfrac{(e^{-10.5})(10.5)^{12}}{12!}$

$$= 0.1180 + 0.1032 = 0.2212.$$

These figures check (to three decimal points).

The shape of the probability function for the Poisson distribution is usually skewed, particularly for small values of (λt). Figure 5.11 shows the probability function for $\lambda t = 0.5$. Its mode is for zero occurrences, and probabilities decrease very rapidly as

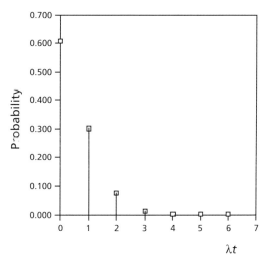

Figure 5.11: Probability Function for Poisson Distribution, $\lambda t = 0.5$

the number of occurrences becomes larger. For comparison, Figure 5.12 shows the probability function for $\lambda t = 5.0$. It is considerably more symmetrical.

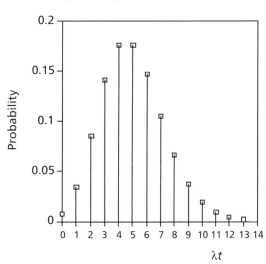

Figure 5.12: Poisson Probability Function for $\lambda t = 5.0$

Chapter 5

Example 5.12

The number of meteors found by a radar system in any 30-second interval under specified conditions averages 1.81. Assume the meteors appear randomly and independently.

    a) What is the probability that no meteors are found in a one-minute interval?
    b) What is the probability of observing at least five but not more than eight meteors in two minutes of observation?

Answer: a) $\lambda = (1.81) / (0.50 \text{ minute}) = 3.62 / \text{minute}$.

For a one-minute interval, $\mu = \lambda t = 3.62$.

Pr [none in one minute] $= e^{-\lambda t} = e^{-3.62} = 0.0268$.

b) For two minutes, $\mu = \lambda t = (3.62)(2) = 7.24$.

$$\Pr[R=r] = \frac{(\lambda t)^r e^{-\lambda t}}{r!}.$$

Then $\Pr[R=5] = \dfrac{(7.24)^5 e^{-7.24}}{5!} = 0.1189$.

From equation 5.14, $\Pr[R=r+1] = \left(\dfrac{\lambda t}{r+1}\right) \Pr[R=r]$

so $\Pr[R=6] = \left(\dfrac{7.24}{6}\right)(0.1189) = 0.1435$,

$\Pr[R=7] = \left(\dfrac{7.24}{7}\right)(0.1435) = 0.1484$,

and $\Pr[R=8] = \left(\dfrac{7.24}{8}\right)(0.1484) = 0.1343$.

Then Pr [at least five but not more than eight meteors in two minutes]
    = Pr [5 or 6 or 7 or 8 meteors in two minutes]
    = 0.1189+0.1435+0.1484+0.1343
    = 0.545

Example 5.13

The average number of collisions occurring in a week during the summer months at a particular intersection is 2.00. Assume that the requirements of the Poisson distribution are satisfied.

    a) What is the probability of no collisions in any particular week?
    b) What is the probability that there will be exactly one collision in a week?
    c) What is the probability of exactly two collisions in a week?
    d) What is the probability of finding not more than two collisions in a week?

Probability Distributions of Discrete Variables

e) What is the probability of finding more than two collisions in a week?
f) What is the probability of exactly two collisions in a particular two-week interval?

Answer: $\lambda = 2.00$/week, $t = 1$ week, so $\lambda t = 2.00$.

a) $\Pr[R = 0] = e^{-\lambda t} = e^{-2.00} = 0.135$

b) Pr [exactly one collision in a week]
$= \Pr[R = 1] = (\lambda t)e^{-\lambda t} = 2.00 e^{-2.00}$
$= 0.271$

c) Pr [exactly two collisions in a week]
$= \Pr[R = 2] = \dfrac{(\lambda t)^2 e^{-\lambda t}}{2!} = \dfrac{(2.00)^2 e^{-2.00}}{2!}$
$= 0.271$

d) Pr [not more than two collisions in a week]
$= \Pr[R \leq 2]$
$= \Pr[R = 0] + \Pr[R = 1] + \Pr[R = 2]$
$= 0.135 + 0.271 + 0.271$
$= 0.677$

e) Pr [more than two collisions in a week]
$= \Pr[R > 2]$
$= 1 - \Pr[R \leq 2]$
$= 1 - 0.677$
$= 0.323$

f) Now we still have $\lambda = 2.00$/week, but $t = 2$ weeks, so $\lambda t = 4.00$
Then Pr [exactly two collisions in a two-week interval]
$= \dfrac{(\lambda t)^2 e^{-\lambda t}}{2!} = \dfrac{(4.00)^2 e^{-4.00}}{2!}$
$= 0.147$

Example 5.14

The demand for a particular type of pump at an isolated mine is random and independent of previous occurrences, but the average demand in a week (7 days) is for 2.8 pumps. Further supplies are ordered each Tuesday morning and arrive on the weekly plane on Friday morning. Last Tuesday morning only one pump was in stock, so the storesman ordered six more to come in Friday morning.

a) Find the probability that one pump will still be in stock on Friday morning when new stock arrives.

Chapter 5

b) Find the probability that stock will be exhausted and there will be unsatisfied demand for at least one pump by Friday morning.
c) Find the probability that one pump will still be in stock this Friday morning and at least five will be in stock next Tuesday morning.

Answer: First we have to recognize that the Poisson distribution will apply.

$$\lambda = \frac{2.8}{7 \text{ days}} = 0.4 \text{ / day}.$$

a) From Tuesday morning to Friday morning is three days.
Then $\lambda t = (0.4 \text{ / day})(3 \text{ days}) = 1.2$.
Pr [no demand in three days] $= e^{-\lambda t} = e^{-1.2} = 0.3012$.
Then Pr [one pump will still be in stock Friday morning when new stock arrives] $= 0.301$.

b) Pr [demand for two or more pumps in three days] =
$= 1 - $ Pr [demand for zero or one pump in three days]
$= 1 - $ Pr [demand for no pumps in three days] $-$ Pr [demand for one pump in three days]
$= 1 - 0.3012 - \frac{(0.3012)(1.2)}{1}$ (using equation 5.14)
$= 0.3374$.

Then Pr [unsatisfied demand for at least one pump by Friday morning] $= 0.337$.

c) From part (a), Pr [one pump will still be in stock this Friday morning] $= 0.3012$.

From Friday morning to Tuesday morning is four days, so $(\lambda t) = (0.4 \text{ /day})(4 \text{ days}) = 1.6$.

After the new stock arrives we will have $1 + 6 = 7$ pumps in stock Friday morning. If we have at least five in stock Tuesday morning, the demand in four days is ≤ 2 pumps.

Pr [demand for 0 pumps in 4 days] $= e^{-1.6} = $ 0.2019.

Pr [demand for 1 pump in 4 days] $= \frac{(e^{-1.6})(1.6)}{(1)} = $ 0.3230.

Pr [demand for 2 pumps in 4 days] $= \frac{(e^{-1.6})(1.6)^2}{2} = $ 0.2584.

Then Pr [demand for 2 or fewer pumps in 4 days] $= 0.7834$.

Then Pr [at least 5 will be in stock next Tuesday morning | one pump in stock Friday morning] $= 0.7834$. Note that this is a conditional probability.

Probability Distributions of Discrete Variables

Then Pr [(one in stock Friday morning) ∩ (at least five in stock on Tuesday morning)] =

= Pr [one in stock Friday morning] × Pr[at least 5 in stock Tuesday A.M. | one in stock Friday A.M.]

= (0.3012)(0.7834)

= 0.236.

(b) Mean and Variance for the Poisson Distribution

Since the Poisson distribution is discrete, the mean and variance can be found from the previous general relations. Equation 5.5 gives

$$\mu = E(R) = \sum_{\text{all } r}(r)(\Pr[R=r])$$

When the probability function of equation 5.13 is substituted in this expression and the algebra is worked through, the result is that the mean or expectation of the number of occurrences according to the Poisson Probability Distribution is

$$\mu = \lambda t \tag{5.15}$$

Therefore an alternative form of the probability function for the Poisson distribution is

$$\Pr[R=r] = \frac{\mu^r e^{-\mu}}{r!} \tag{5.16}$$

Similarly, from equation 5.6,

$$\sigma^2 = E(r-\mu)^2 = \sum_{\text{all } r}(r-\mu)^2 \left(\Pr[R=r]\right)$$

Again, the probability function 5.13 can be substituted. The result of this derivation for the Poisson Distribution is that

$$\sigma^2 = \lambda t \tag{5.17}$$

Thus, the variance of the number of occurrences for the Poisson distribution is equal to the mean number of occurrences, μ.

(c) Approximation to the Binomial Distribution

Let us compare the results from the binomial distribution for $\mu = 1.2$, from various combinations of values of n and p, with the results from the Poisson distribution for $\mu = \lambda t = 1.2$. In each case let us calculate Pr [R=0] and Pr [R=1]. The results are shown in Table 5.1.

Table 5.1: Comparison of Binomial and Poisson Distributions

For the **Binomial Distribution:**

n	p	μ	Pr [R=0]		Pr [R=1]	
4	0.3	1.2	$(1)(0.3)^0(0.7)^4$	= 0.240	$(4)(0.3)^1(0.7)^3$	= 0.412
8	0.15	1.2	$(1)(0.15)^0(0.85)^8$	= 0.272	$(8)(0.15)^1(0.85)^7$	= 0.385
20	0.06	1.2	$(1)(0.06)^0(0.94)^{20}$	= 0.290	$(20)(0.06)^1(0.94)^{19}$	= 0.370
100	0.012	1.2	$(1)(0.012)^0(0.988)^{100}$	= 0.299	$(100)(0.012)^1(0.988)^{99}$	= 0.363
200	0.006	1.2	$(1)(0.006)^0(0.994)^{200}$	= 0.300	$(200)(0.006)^1(0.994)^{199}$	= 0.362

Chapter 5

For the **Poisson Distribution**:

n	p	μ	Pr [R=0]		Pr [R=1]	
—	—	1.2	$(1.2)^0(e^{-1.2}) =$	0.301	$(1.2)^1(e^{-1.2}) =$	0.361

In the part of Table 5.1 for the binomial distribution, n is gradually increased and p is correspondingly decreased so that the product ($np = \mu$) stays constant. The results are compared to the corresponding probabilities according to the Poisson distribution for this value of μ. At least in this instance we find that as n increases and p decreases so that μ stays constant, the resulting probabilities for the binomial distribution approach the probabilities for the Poisson distribution. In fact, this relationship between the binomial and Poisson distributions is general. One way of deriving the Poisson distribution is to take the limit of the binomial distribution as n increases and p decreases such that the product np (equal to μ) remains constant.

Thus the Poisson distribution is a good approximation to the binomial distribution if n is sufficiently large and p is sufficiently small. The usual rule of thumb (that is, a somewhat arbitrary rule) is that if $n \geq 20$ *and* $p \leq 0.05$, the approximation is reasonably good. That rule should be used for problems in this book. The error at the limit of the approximation according to this rule depends on the parameters, but some indication can be seen if we look at the case where $\mu = 1.2$, $p = 0.05$, and so $n = \frac{1.2}{0.05} = 24$. At this point Pr [R=0] by the Poisson distribution is 3.2% higher than Pr [R = 0] by the binomial distribution, and Pr [R = 1] by the Poisson distribution is 2.0% lower than Pr [R=1] by the binomial distribution.

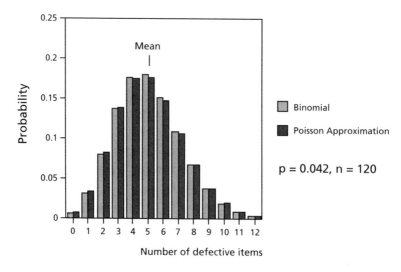

Figure 5.13: Poisson Approximation to Binomial Distribution

Probability Distributions of Discrete Variables

Figure 5.13 shows a comparison of the binomial distribution and the corresponding Poisson distribution, both for the same value of $\mu = np$. This might be for a case of sampling items coming off a production line when the value of p, the probability that any one item will be defective, is 0.042, and the value of n is the sample size, 120 items. As we can see, the agreement is good. This case meets the rule of thumb quite easily, so we would expect good agreement.

The Poisson distribution has only one parameter, μ, whereas the binomial distribution has two parameters, n and p. Probabilities according to the Poisson distribution are easier to calculate with a pocket calculator than for the binomial distribution, especially for very large values of n and very small values of p. However, this advantage is less important now that computer spreadsheets are readily available. We saw in section 5.3(f) of this chapter that the binomial distribution can be calculated easily using MS Excel.

Example 5.15

5% of the tools produced by a certain process are defective. Find the probability that in a sample of 40 tools chosen at random, exactly three will be defective. Calculate a) using the binomial distribution, and b) using the Poisson distribution as an approximation.

Answer: a) For the binomial distribution with $n = 40$, $p = 0.05$,

$$\Pr[R = 3] = {}_{40}C_3 (0.05)^3(0.95)^{37}$$
$$= \frac{(40)(39)(38)}{(3)(2)(1)} (0.05)^3(0.95)^{37}$$
$$= 0.185$$

b) For the Poisson distribution, $\mu = (n)(p) = (40)(0.05) = 2.00$.

$$\Pr[R = 3] = \frac{(2.00)^3 e^{-2.00}}{(3)(2)(1)} = 0.180$$

(d) Use of Computers

Values of Poisson probabilities can be found with the Excel function POISSON with parameters r, μ or λt, and an indication of whether or not a cumulative value is required. If the third parameter is TRUE, the function returns the cumulative probability that the number of random events will be less than or equal to r when either μ or its equivalent λt has the specified value. If the third parameter is FALSE, the function returns the probability that the number of events will be exactly r when $\mu = \lambda t$ has the value stated in the second parameter, For example, the cumulative probability of 12 or fewer random occurrences when $\mu = \lambda t = 10.5$ is given by POISSON(12,10.5,TRUE) as 0.742 (to three decimal points); the probability of exactly 12 random occurrences is given by POISSON(12,10.5,FALSE) as 0.103

Chapter 5

(again to three decimal points). As for the binomial distribution, use of the computer with Excel is especially labor-saving when cumulative probabilities are required.

Problems

1. The number of cars entering a small parking lot is a random variable having a Poisson distribution with a mean of 1.5 per hour. The lot holds only 12 cars.
   a) Find the probability that the lot fills up in the first hour (assuming that all cars stay in the lot longer than one hour).
   b) Find the probability that more than 3 cars arrive between 9 am and 11 am.

2. Customers arrive at a checkout counter at an average rate of 1.5 per minute. What distribution will apply if reasonable assumptions are made? List those assumptions. Find the probabilities that
   a) exactly two will arrive in any given minute;
   b) at least three will arrive during an interval of two minutes;
   c) at most 13 will arrive during an interval of six minutes.

3. Cumulative probability tables for the Poisson Distribution indicate that for $\mu = 2.5$, Pr $[R \leq 6] = 0.986$ and Pr $[R \leq 4] = 0.891$. Use these figures to calculate Pr $[R = 5$ or $6]$. Check using basic relations.

4. Cumulative probability tables indicate that for a Poisson distribution with $\mu = 5.5$, Pr $[R \leq 6] = 0.686$ and Pr $[R \leq 7] = 0.810$. Use these figures to calculate Pr $[R = 7]$. Check using a basic relation.

5. Records of an electrical distribution system in a particular area indicate that over the past twenty years there have been just six years in which lightning has not hit a transformer. Assume that the factors affecting lightning hits on transformers have not changed over that time, and that hits occur at random and independently.
   a) Then what would be the best estimate of the average number of hits on transformers per year?
   b) In how many of the next ten years would we expect to have more than two hits on transformers in a year?

6. A library employee shelves a large number of books every day. The average number of books misshelved per day is estimated over a long period to be 2.5.
   a) Calculate the probability that exactly three books are misshelved in a particular day.
   b) Calculate the probability that fewer than two books on one day and more than two books on the next day are misshelved.
   c) What assumptions have been made in these calculations?

Probability Distributions of Discrete Variables

7. The numbers of lightning strikes on power poles in a particular district have been recorded. Records show that in the past twenty-five years there have been seven years in which no lightning strikes on poles have occurred. Assume that strikes occur randomly and independently, and that the mean number of strikes per unit time does not change.
   a) What distribution applies?
   b) What is the probability that more than one strike will occur next year?
   c) What is the probability that exactly one strike will occur in the next two years?
   d) What is the best estimate of the standard deviation of number of strikes in one year?

8. The mean number of letters received each year by the university requesting information about the programs offered by a particular department is 98.8. Assume that letters are received randomly throughout a year which consists of 52 weeks.
   a) What is the probability of receiving no letters in a particular week?
   b) What is the probability of receiving two or more letters in a particular week?
   c) What is the probability of receiving no letters in any four-week period?
   d) What is the probability of having two weeks in a specified four-week period with no letters?

9. The number of grain elevator explosions due to spontaneous combustion has been 10 in the past 25 years for Great West Grain, a company with over a thousand grain elevators. Explosions occur randomly and independently.
   a) From these data make an estimate of the mean rate of occurrence of explosions in a year.
   b) On the basis of this estimate, what is the probability that there will be no explosions in the next five years?
   c) If there is at least one explosion a year for three years in a row, the insurance rates paid by the elevator company will double. What is the probability that this will happen over the next three years? Use the estimate from part (a).

10. The average number of traffic accidents in a certain city in a seven-day period is 28. All traffic accidents are investigated on the day of their occurrence by a police squad car. A maximum of three traffic accidents can be investigated by one squad car in a day. Assume that accidents occur randomly and independently.
    a) What is the probability that no accidents will have to be investigated on a given day?
    b) What is the probability that, on exactly two out of three successive days, more than two squad cars will have to be assigned to investigate traffic accidents?

11. Records for 13 summer weeks for each of the past 80 years in a particular district show that 32 weeks in total were very wet. Assume that wet weeks occur at random and independently and that the pattern does not change with time.

Chapter 5

a) What is the probability that no very wet weeks will occur in the next two years?
b) What is the probability that at least two very wet weeks will occur in the next two years?
c) What is the probability that exactly two very wet weeks will occur in the next two years?

12. In 104 days, 170 oil tankers arrive at a port for unloading. The tankers arrive randomly and independently. Probabilities are the same for every day of the week. A maximum of two oil tankers can be unloaded each day.
    a) What is the probability that no oil tankers will arrive on Tuesday?
    b) What is the probability that more than two will arrive on Friday? This will mean that not all can be unloaded on Friday, even if no oil tankers were left over from Thursday.
    c) Assuming that no oil tankers are left over from Tuesday, what is the probability that exactly one oil tanker will be left over from Wednesday and none will be left over from Thursday?
    d) What is the probability that more than three oil tankers will arrive in an interval of two days?

13. The probability of no floods during a year along the South Saskatchewan River has been estimated from considerable data to be 0.1353. Assume that floods occur randomly and independently.
    a) What is the expected number of floods during a year?
    b) What is the probability of two or more floods during exactly two of the next three years?
    c) What are the mean and standard deviation of the number of floods expected in a five-year period?

14. The number of new categories added each year to a major engineering handbook has been found to be a random variable, unaffected by the size of the handbook and its recent history. The probability that no new categories will be added in the annual update is 0.1353. This year's edition of the handbook contains 97 categories.
    a) How many categories is the next edition expected to contain?
    b) What is the probability that the edition two years from now will contain fewer than 100 categories?

15. In a plant manufacturing light bulbs, 1% of the production is known to be defective under normal conditions. A sample of 30 bulbs is drawn at random. Assume defective bulbs occur randomly and independently. What is the probability that:
    a) the sample contains no defective bulbs;
    b) more than 3 defective bulbs are in the sample.

    Do this problem both (1) using the binomial distribution, and (2) using the Poisson distribution. Compare the conditions of this problem to the rule of thumb stated in section 5.4(c).

Probability Distributions of Discrete Variables

16. Fifteen percent of piglets raised in total confinement under certain conditions will live less than three weeks after birth. Assume that deaths occur randomly and independently. Consider a group of eight newborn piglets.
    a) What probability distribution applies without any approximation to the number of piglets which will live less than three weeks?
    b) What is the expected mean number of deaths?
    c) What is the probability that exactly three piglets will die within three weeks of birth? Use the binomial distribution.
    d) Calculate the probability that exactly three piglets will die within three weeks of birth, but now use the Poisson distribution.
    e) Compare the conditions of this problem to the rule of thumb stated in section 5.4(c). Then would we expect the Poisson distribution to be a good approximation in this case?
    f) Use the binomial distribution to calculate the probability that fewer than three piglets will die within three weeks of birth.
    g) Use the Poisson distribution to calculate the probabilities that exactly 0, 1, and 2 piglets will die within three weeks of birth, and then that fewer than 3 piglets will die within three weeks of birth.

17. Tests on the brakes and steering gear of 200 cars indicate that the probability of defective brakes is 0.17 and the probability of defective steering is 0.14.
    a) If defective brakes and defective steering are independent of one another, what is the probability of finding both on the same car?
    b) Consider probability distributions which might apply to the occurrence of both defective brakes and defective steering among the 200 cars. Assume occurrences of both are random and independent of other occurrences. What probability distribution would be expected fundamentally if the probability of "success" is constant from trial to trial? What probability distribution would be applicable as a more convenient approximation, and why? Give the parameters of both distributions.
    c) Apply the approximate distribution to find the probability that at least eleven cars of 200 would have both defective brakes and defective steering if they are independent of one another.
    d) If in fact 11 of the 200 cars have both defective brakes and defective steering, is it reasonable to conclude that defective brakes and defective steering are independent of one another?

Computer Problems

C18. The number of cars entering a parking lot is a random variable having a Poisson Distribution with a mean of four per hour. The lot holds only 12 cars.
    a) Find the probability that the lot fills up in the first hour (assuming that all cars stay in the lot longer than one hour).
    b) Find the probability that fewer than 12 cars arrive during an eight-hour day.

Chapter 5

C19. Customers arrive at a checkout counter at an average rate of 1.5 per minute. What distribution will apply if reasonable assumptions are made? List those assumptions. Find the probability that at most 13 customers will arrive during an interval of six minutes.

C20. A library employee shelves a large number of books every day. The average number of books misshelved per day is estimated over a long period to be 2.5. Calculate the probability that between five and fifteen books (including both limits) are misshelved in a four-day period.

C21. The average number of vehicles arriving at an intersection under certain conditions is constant, but vehicles arrive independently and the actual number arriving in any interval of time is determined by chance. The average rate at which vehicles arrive at the intersection is 360 vehicles per hour. Traffic lights at this intersection go through a complete cycle in 40 seconds. During the green light only seven vehicles can pass through the intersection.

   a) What is the probability that exactly seven vehicles arrive during one cycle?
   b) What is the probability that fewer than seven vehicles arrive during one cycle?
   c) What is the probability that exactly eight vehicles arrive during one cycle, so that one vehicle is held for the next cycle (assuming there were no hold-overs from the previous cycle)?
   d) What is the probability that one vehicle is held over from cycle 1 as in part (c) and all the vehicles pass through on the following cycle?

C22. Grain loading facilities at a port have capacity to load five ships per day. Past experience of many years indicates that on the average 28 ships come in to pick up grain in a seven-day period. Ships arrive randomly and independently.

   a) What is the probability that on a given day the capacity of the dock will be exceeded by at least one ship, given that no ship was waiting at the beginning of the day?
   b) What is the probability that exactly four ships will show up at the port in a two-day period?
   c) By how much should the capacity of the loading docks be expanded so that the probability that a ship will not be able to dock on a given day will be less than 1%?

C23. The ABC Auto Supply Depot orders stock at the middle of the month and receives the goods at the first of the next month. The average number of requests for fuel pump XY33 is four per month. If on April 15, two of these fuel pumps are in stock and an additional five are ordered to be received by May 1, what is the probability that the ABC Depot will not be able to supply all the requests for XY33 in the month of May? Requests for pumps are random and independent of one another. Requests are not carried over from one month to the next.

Probability Distributions of Discrete Variables

C24. A manufacturer offers to sell a device for counting lightning flashes during thunder storms. The device can record up to five distinct flashes per minute.
   a) If the average flash intensity experienced during a thunder storm at a recording location is nine flashes in six minutes, what is the probability that at least one flash will not be recorded in a one-minute period? What assumptions are being made?
   b) Given this intensity, what is the probability of experiencing six lightning flashes in a two- minute period?
   c) What is the highest average intensity in flashes per hour for which the recorder can be used, if the probability of not recording all flashes in a minute must be less than 10%?

C25. The probability of no floods during a year along the South Saskatchewan River has been estimated from considerable data to be 0.1353. Assume that floods occur randomly and independently. What is the probability of seven or fewer floods during a five-year period?

C26. The cars passing a certain point as a function of time were counted during a traffic study of a city road. It was found that there was 10% probability of observing more than ten cars in an eight-minute interval.
   a) Find the probability that exactly five cars will pass in a four-minute interval What assumptions are being made?
   b) Find the probability that fewer than two cars will pass in each of three consecutive intervals.
   c) Find the probability that fewer than two cars will pass in exactly two of three consecutive intervals.
   d) How long an interval should be used so that the probability of observing more than nine cars becomes 40%?

C27. Rainstorms around Saskatoon occur at the mean rate of six in four weeks during the spring season. If one storm occurs in the week after spring snowmelt is over, the probability of flooding is 0.30; if two storms occur that week, the probability goes to 0.60. If more than two occur, the probability becomes 0.75. If no storms occur, the probability is 0. Overall, if no flooding has occurred by the end of the first week, the probability of flooding becomes 0.10 if one rainstorm occurs in the next two weeks, and 0.15 if two or more rainstorms occur in the next two weeks. Assume that rainstorms occur independently and randomly.
   (a) What is the probability of at least four rainstorms in the first three weeks?
   (b) What is the probability of flooding in those three weeks?

5.5 Extension: Other Discrete Distributions

Although the binomial distribution and the Poisson distribution are probably the most common and useful discrete distributions, a number of others are found useful in some engineering applications. Among them are the negative binomial distribution

Chapter 5

and the geometric distribution. Both these distributions are for the same conditions as for the binomial distribution *except* that trials are repeated until a fixed number of "successes" have occurred. The negative binomial distribution gives the probability that the *k*th success occurs on the *n*th trial, where both k and n are fixed quantities. The geometric distribution is a special case of the negative binomial distribution; it gives the probability that the *first* "success" occurs on the *n*th trial. We have already mentioned the multinomial distribution in part (i) of section 5.3. As discussed there, it can be considered a generalization of the binomial distribution when there are more than two possible outcomes for each trial. The negative binomial distribution, the geometric distribution, and the multinomial distribution are described more fully in the book by Walpole and Myers (see the List of Selected References in section 15.2 of this book).

The Bernoulli distribution is a special case of the binomial distribution when the number of trials is one. Thus, the only possible outcomes for the Bernoulli distribution are zero and one. $\Pr[R = 0] = (1 - p)$, and $\Pr[R = 1] = p$.

The hypergeometric probability distribution applies to a situation where there are only two possible outcomes to each trial, but the probability of "success" varies from one trial to another in accordance with sampling from a finite population *without replacement*. The total number of trials and the size of the population are then both parameters. This distribution is described in various references including the book by Mendenhall, Wackerly and Scheaffer (again see section 15.2). The book by Barnes (see that same section of this book) gives a guideline for approximating the hypergeometric distribution by the binomial distribution: the sample size should be less than one tenth of the size of the finite set of items being sampled.

Use of Computers: When a person has become familiar with the fundamental ideas of discrete random variables, it is often convenient to use a number of Excel's statistical functions, including the following:

HYPGEOMDIST( ) returns probabilities according to the hypergeometric distribution.

NEGBINOMDIST( ) returns probabilities according to the negative binomial distribution.

CRITBINOM( ) returns the limiting value of a parameter of the binomial distribution to meet a requirement. This is useful in quality assurance.

In most cases the most convenient way to use functions on Excel, including selection of arguments for the parameters, is probably to paste the required function into the appropriate cell on a worksheet. The detailed procedure varies from one version of Excel to another. On Excel 2000, for example, we click the cell where we want to enter the function, then from the Insert menu we choose the function category (for example, Statistical), then click the function (for example, HYPOGEOMDIST). Further details are given in part (b) of Appendix B.

Probability Distributions of Discrete Variables

These functions should not be used until the reader is familiar with the main ideas of this chapter.

5.6 Relation Between Probability Distributions and Frequency Distributions

This chapter has been concerned with *probability* distributions for discrete random variables. Chapter 3 included descriptions and examples of *frequency* distributions for discrete random variables. Probability distributions and frequency distributions are similar, but of course there are important differences between them. The probability distributions we have been considering are theoretical and depend on assumptions, whereas frequency distributions are usually empirical, the result of experiments. Probability distributions show *predictable* variations with the values of the variable. Frequency distributions show additional *random* variations, that is, variations which depend on chance.

In this section we will first look at comparisons of some probability distributions with simulated frequency distributions for the same parameters. Then we will discuss fitting binomial distributions and Poisson distributions to experimental frequency distributions.

Random numbers can be used to simulate frequency distributions corresponding to various discrete random variables. That is, random numbers can be combined with the parameters of a probability distribution to produce a simulated frequency distribution. The simulated frequency distributions discussed in this section were prepared using Excel, but the detailed procedures are not relevant to the present discussion.

(a) Comparison of a Probability Distribution with Corresponding Simulated Frequency Distributions

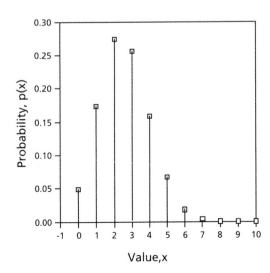

Figure 5.14: Probability Distribution: Binomial with $n = 10$ and $p = 0.26$

Chapter 5

Figure 5.14 shows a *probability* distribution for a binomial distribution with $n = 10$ and $p = 0.26$. Corresponding to this is Figure 5.15, which is for the same values of n and p but shows two simulated relative frequency distributions. These are for samples of size eight—that is, samples containing eight items each. As we have seen before, relative frequencies are often used as estimates of probabilities. However, with this small sample size the relative frequencies do not agree at all well with the corresponding probabilities, and they do not agree with one another.

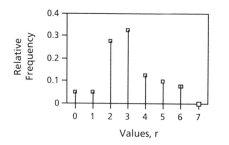 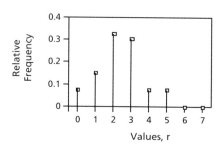

Figure 5.15: Simulated Frequency Distributions for Eight Repetitions

If the sample size is increased, agreement becomes better. Figure 5.16 shows two simulated relative frequency distributions for samples of size forty, still for a binomial distribution with $n = 10$ and $p = 0.26$. The graphs of Figure 5.16 still differ from one another because of random fluctuations, but they are much more similar to one another in shape than the graphs of Figure 5.15. Comparison to Figure 5.14 shows that the general shape of the probability distribution is beginning to come through.

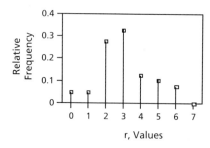 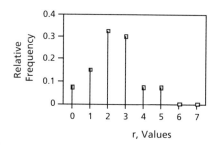

Figure 5.16: Simulated Relative Frequency Distributions for Forty Repetitions

Thus, we can see that the relative frequency distributions are both more consistent with one another and more similar to the corresponding probability distributions when they represent forty repetitions rather than eight repetitions. This seems reasonable. Huff points out that inadequate sample size often leads to incorrect or misleading conclusions. He gives some dramatic examples of this in his book *How to Lie with Statistics* (see section 15.2 for reference).

Probability Distributions of Discrete Variables

(b) Fitting a Binomial Distribution

We often want to compare a set of data from observations with a theoretical probability distribution. Can the data be represented satisfactorily by a theoretical distribution? If so, the data can be represented very succinctly by the parameters of the theoretical distribution. Specifically, let us consider whether a set of data can be represented by a binomial distribution.

The binomial distribution has two parameters, n and p. In any practical case we will already know n, the number of trials. How can we estimate p, the probability of "success" in a single trial? An intuitive answer is that we can estimate p by the fraction of all the trials which were "successes," that is, the proportion or relative frequency of "success." It is possible to show mathematically that this intuitive answer is correct, an unbiased estimate of the parameter p.

Example 5.16

In Example 3.2 we considered the number of defective items in groups of six items coming off a production line in a factory. We found there were 14 defectives in sixty groups giving a total sample of 360 items, so the proportion defective was 14/360 = 0.0389.

Let us try to fit the observed frequency distribution of Table 3.2 by a binomial distribution. We have $n = 6$ and p is estimated (probably not very accurately) to be 0.0389. Then the probability of exactly r defective items in a sample of six items according to the binomial distribution is given by equation 5.9 as

$$\Pr[R = r] = {}_6C_r (0.0389)^r (0.9611)^{(6-r)}$$

This prediction of probability by the binomial distribution should be compared with the observed relative frequencies for various numbers of defectives. These can be obtained simply by dividing the frequencies of Table 3.2 by the total frequency of 60. Since 60 groups is not a very large number we should not expect the agreement to be very close.

The results are shown in Table 5.2 and Figure 5.17: a theoretical binomial probability of 0.788 can be compared with an observed relative frequency of 0.600, and so on.

Table 5.2: Comparison of Binomial Probability with Observed Relative Frequency

Number of Defectives, r	Binomial Probability, $\Pr[R = r]$	Observed Frequency, f	Observed Relative Frequency, $f / \Sigma f$
0	0.788	48	0.600
1	0.191	10	0.167
2	0.019	2	0.033
3	0.001	0	0
4	3×10^{-5}	0	0
5	5×10^{-7}	0	0
6	3×10^{-9}	0	0

Chapter 5

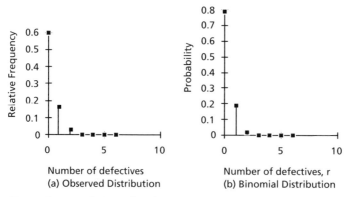

Figure 5.17: Comparison of Relative Frequencies with Binomial Probabilities

We can see that the comparison is reasonably good. In section 13.3 we will see a more quantitative comparison.

(c) Fitting a Poisson Distribution

We may have a set of data which we suspect can be represented by a Poisson distribution. If it is, we can describe it very compactly by the parameters of that distribution. In addition, there may be some implication (for example, regarding randomness) if the data can be represented by a Poisson distribution. Thus, we need to know how to find a Poisson distribution that will fit a set of data.

The Poisson distribution has only one parameter, μ or λt. As we have seen in Chapter 3, the sample mean, $\bar{x}$, is an unbiased estimate of the population mean, μ. Therefore, the first step in fitting a Poisson distribution to a set of data is to calculate the mean of the data. Then the relation for the Poisson distribution is used to calculate the probabilities of various numbers of occurrences if that distribution holds. These probabilities can be compared to the relative frequencies found by dividing the actual frequencies by the total frequency.

Example 5.17

The number of cars crossing a local bridge was counted for forty successive 6-minute intervals from 1:00 to 5:00 A.M. The numbers can be summarized as follows:

x_i, number of cars in 6-minute Interval	f_i, frequency
0	2
1	7
2	10
3	8
4	6
5	3
6	3
7	1
>8	0

Fit a Poisson Distribution to these data.

Probability Distributions of Discrete Variables

Answer: First, let us calculate the sample mean as an estimate of the population mean, μ.

x_i	f_i	xf_i
0	2	0
1	7	7
2	10	20
3	8	24
4	6	24
5	3	15
6	3	18
7	1	7
>8	0	0
Total	40	115

Then $\bar{x} = \dfrac{\sum f_i x_i}{\sum f_i} = \dfrac{115}{40} = 2.875$. Then take $\mu = \lambda t = 2.875$ in 6 minutes.

Then $\lambda = \dfrac{\lambda t}{t} = \dfrac{2.875}{6} = 0.479$ cars / minute.

According to the Poisson Distribution, then, Pr $[R=r]$ = $(2.875)^r e^{-2.875}$/ r!. It was mentioned previously that once one of the Poisson probabilities is calculated, others can be calculated conveniently using the recurrence relation of equation 5.14,

$$\text{Pr}[R=r+1] = \left(\dfrac{\lambda t}{r+1}\right) \text{Pr}[R=r].$$

Calculation of Poisson probabilities and relative frequencies gives the following results:

r	f_i	Pr $[R=r]$	Relative Frequency
0	2	0.0564	0.0500
1	7	0.1622	0.1750
2	10	0.2332	0.2500
3	8	0.2234	0.2000
4	6	0.1606	0.1500
5	3	0.0923	0.0750
6	3	0.0442	0.0750
7	1	0.0182	0.0250
>8	0	0.0095	0
Total	40		

The frequencies from the problem statement are compared with the calculated expected frequencies in Figure 5.18. It can be seen that the agreement between

Chapter 5

recorded and fitted frequencies appears to be very good, in fact better than we might expect.

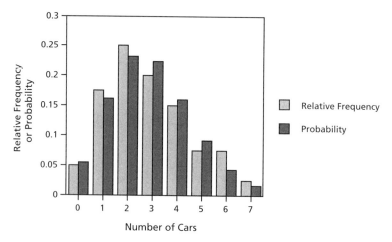

Figure 5.18: Comparison of Relative Frequencies with Probabilities for the Poisson Distribution

In section 13.3 we will see how to make a quantitative evaluation of the goodness of fit of two distributions. This example will be continued at that point.

Examples 5.16 and 5.17 have compared probabilities to relative frequencies. An alternative procedure is to calculate expected frequencies by multiplying each probability by the total frequency. Then the expected frequencies are compared with the observed frequencies. That procedure is logically equivalent to the comparison we have made here.

Problems

1. A sampling scheme for mechanical components from a production line calls for random samples, each consisting of eight components. Each component is classified as either good or defective. The results of 50 such samples are summarized in the table below.

Number of Defectives	Observed Frequency
0	30
1	17
2	3
>2	0

   From these data estimate the probability that a single component will be defective. Calculate the probabilities of various numbers of defectives in a sample of eight components, and prepare a table to compare predicted probabilities according to the binomial distribution with observed relative frequencies for various numbers of defectives in a sample.

Probability Distributions of Discrete Variables

2. Electrical components are produced on a production line, then inspected. Each component is classified as good or defective. 360 successive components were grouped into samples, each containing six components. The results are summarized in the table below.

Number of Defectives	Observed Frequency
0	34
1	24
2	2
>2	0

   From these data estimate the probability that a single component will be defective. Calculate the probabilities of various numbers of defectives in a sample of six components, and prepare a table to compare predicted probabilities according to the binomial distribution with observed relative frequencies for various numbers of defectives in a sample.

3. A study of four blocks containing 52 one-hour parking spaces was carried out and the results are given in the following table.

Number of vacant one-hour parking spaces per observation period	0	1	2	3	4	5	≥6
Observed frequency	31	45	20	15	7	3	0

   Assuming that the data follow a Poisson distribution, determine:
   a) the mean number of vacant parking spaces,
   b) the standard deviation both (i) from the given data and (ii) from the theoretical distribution, and
   c) the probability of finding one or more vacant one-hour parking spaces, calculating from the theoretical distribution.

4. In analysis of the treated water from a sewage treatment process, liquid containing harmful cells was placed on a slide and examined systematically under a microscope. One hundred counts of the number of harmful cells in 1 mm by 1 mm squares were made, with the following frequencies being obtained.

Count	0	1	2	3	4	5	6	7	8	9	10	11	12
Frequency	1	3	8	14	17	19	14	12	6	2	2	2	0

   Fit a Poisson distribution to these data. Calculate expected Poisson frequencies to compare with the observed frequencies. Is the fit reasonably good?

5. An air filter has been designed to remove particulate matter. A test calls for 40 specimens of air to be tested. Of 40 specimens, it was found that there were no particles in 15 specimens, one particle in 10 specimens, two particles in 8 specimens, three particles in 5 specimens, and four particles in 2 specimens.
   a) What type of distribution should the data follow? What are the necessary assumptions?

Chapter 5

b) Estimate the mean and standard deviation of the frequency distribution from the given data.
c) What is the theoretical standard deviation for the probability distribution?
d) Using probabilities calculated from the theoretical distribution, what is the probability that among ten specimens there would be eight or more with no particles?

6. A section of an oil field has been divided into 48 equal sub-areas. Counting the oil wells in the 48 sub-areas gives the following frequency distribution:

Number of oil wells	0	1	2	3	4	5	6	7
Number of sub-areas	5	10	11	10	6	4	0	2

Is there any evidence from these data that the oil wells are not distributed randomly throughout the section of the oil field?

CHAPTER 6

Probability Distributions of Continuous Variables

For this chapter the reader needs a good knowledge of integral calculus and the material in sections 2.1, 2.2, 5.1, and 5.2.

If a variable is continuous, between any two possible values of the variable are an infinite number of other possible values, even though we cannot distinguish some of them from one another in practice. It is therefore not possible to count the number of possible values of a continuous variable. In this situation calculus provides the logical means of finding probabilities.

6.1 Probability from the Probability Density Function

(a) Basic Relationships

The probability that a continuous random variable will be between limits a and b is given by an integral, or the area under a curve.

$$\Pr[a < X < b] = \int_a^b f(x)\,dx \tag{6.1}$$

The function $f(x)$ in equation 5.1 is called a *probability density function*. The probability that the *continuous* random variable, X, is between a and b corresponds to the *area* under the curve representing the probability density function between the limits a and b. This is the cross-hatched area in Figure 6.1. Compare this relation with the relation for the probability that a *discrete* random variable is between limits a and b, which is the sum of the probability functions for all values of the variable X between a and b, $\sum_{a \leq x_i \leq b} p(x_i)$.

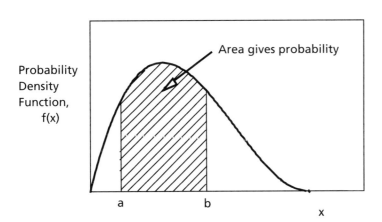

Figure 6.1: Probability for a Continuous Random Variable

Chapter 6

The cumulative distribution function for a continuous random variable is given by the integral of the probability density function between $x = -\infty$ and $x = x_1$, where x_1 is a limiting value. This corresponds to the area under the curve from $-\infty$ to x_1. The cumulative distribution function is often represented by $F(x_1)$ or $F(x)$.

$$\Pr[X \leq x_1] = F(x_1) = \int_{-\infty}^{x_1} f(x)dx \tag{6.2}$$

This expression should be compared with the expression for the cumulative distribution function for a *discrete* random variable, which is given by equation 5.1 to be $\sum_{x \leq x_1} p(x_i)$. Thus, a summation of individual probabilities (for a discrete case) corresponds to an integral of the probability density function with respect to the variable (for a continuous case).

$$\text{i.e.,} \quad \sum p(x_i) \quad \sim \quad \int f(x)dx \tag{6.3}$$
$$\text{(Discrete)} \qquad \text{(Continuous)}$$

To include all conceivable values of the variable X, the limits in equation 6.2 become from $x = -\infty$ to $x = +\infty$. The probability of a value is that interval must be 1. Then we have

$$F(\infty) = \int_{-\infty}^{+\infty} f(x)dx = 1 \tag{6.4}$$

In many cases only values of the variable in a certain interval are possible. Then outside that interval, the probability density function is zero. Intervals in which the probability density function is identically zero can be omitted in the integration.

Since any probability must be between 0 and 1, as we have seen previously, the probability density function must always be positive or zero, but not negative.

$$f(x) \geq 0 \tag{6.5}$$

Example 6.1

A probability density function is given by:

$f(x) = 0$ for $x < 0$
$f(x) = \dfrac{3}{8}x^2$ for $0 < x < 2$
$f(x) = 0$ for $x > 2$

A graph of this density function is shown in Figure 6.2.

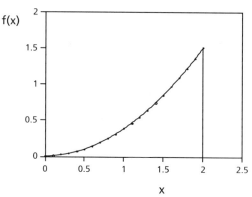

Figure 6.2: A Simple Probability Density Function

Probability Distributions of Continuous Variables

It is not hard to show that $f(x)$ meets the requirements for a probability density function. First, since x^2 is always positive for any real value of x, $f(x)$ is always greater than or equal to zero. Second, the integral of the probability density function from $-\infty$ to $+\infty$ is equal to 1, as we can show by integration:

$$F(\infty) = \int_{-\infty}^{+\infty} f(x)\,dx = \int_{-\infty}^{0}(0)\,dx + \int_{0}^{2}\frac{3}{8}x^2\,dx + \int_{2}^{\infty}(0)\,dx$$

$$= 0 + \left[\left(\frac{3}{8}\right)\left(\frac{1}{3}x^3\right)\right]_{0}^{2} + 0$$

$$= 0 + \left(\frac{3}{8}\right)\left(\frac{1}{3}\right)(2^3) + 0$$

$$= 1$$

(b) A Simple Illustration: Waiting Time

A student arrives at a bus stop and waits for the bus. He knows that the bus comes every 15 minutes (which we will assume is exact), but he doesn't know when the next bus will come. Let's assume the bus is as likely to come in any one instant as in any other within the next 15 minutes. Let the time the student has to wait for the bus be x minutes. Let us first explore the probabilities intuitively, and then apply equations 6.1 and 6.2.

i) What is the probability that the waiting time will be less than or equal to 15 minutes?

Since we know that the bus comes every 15 minutes, this probability must be 1.

ii) What is the probability that the waiting time will be less than 5 minutes?

Since the bus is as likely to come in any one instant as in any other to a maximum of 15 minutes, the probability that the waiting time is less than 5 minutes must be $\frac{5}{15} = \frac{1}{3}$.

Similarly, the probability that the waiting time is less than 10 minutes must be $\frac{10}{15} = \frac{2}{3}$.

iii) Then we can generalize the expression for probability. The probability that the waiting time will be less than x minutes, where $0 \leq x \leq 15$, must be $\frac{x}{15}$.

iv) What is the probability that the waiting time will be between 5 minutes and 10 minutes? This must be:

$$\Pr[5 < x < 10] = \Pr[x < 10] - \Pr[x < 5]$$

$$= \frac{10}{15} - \frac{5}{15} = \frac{5}{15} \text{ or } \frac{1}{3}.$$

Chapter 6

Comparison to equation 6.1 with $a = 5$ and $b = 10$ indicates that:

$$\Pr[5 < X < 10] = \int_{5}^{10} f(x)\,dx = \frac{10}{15} - \frac{5}{15}$$

What simple expression for $f(x)$ will integrate with respect to x to give $\frac{x}{15}$?
It must be $\frac{1}{15}$.

Then the probability density function must be given by:

$f(x) = 0 \quad$ for $x < 0 \quad$ (since waiting time can't be negative).
$f(x) = \frac{1}{15} \quad$ for $0 < x < 15$
$f(x) = 0 \quad$ for $x > 15 \quad$ (since waiting time can't be more than 15 minutes)

Let's check the integral of $f(x)$ for x between 0 and 15, the only interval for which $f(x)$ is not equal to zero. We have $\int_{0}^{15} \frac{1}{15}\,dx = \frac{15-0}{15} = 1$ (as required), so the constant value, $\frac{1}{15}$, is correct.

v) By comparison to equation 6.2 the probability that the waiting time will be less than 5 minutes must be:

$$F(5) = \int_{-\infty}^{5} f(x)\,dx$$

$$= \int_{-\infty}^{0} 0\,dx + \int_{0}^{5} \frac{1}{15}\,dx$$

$$= 0 + \left(\frac{1}{15}\right)(5)$$

$$= \frac{5}{15} \text{ or } \frac{1}{3}$$

This agrees with part ii.

vi) Using the expressions for the probability density function from part iv, the general expression for the cumulative distribution function for this illustration must be:

$$F(x_1) = \int_{-\infty}^{x_1} 0\,dx = 0 \qquad \text{for } x_1 < 0$$

$$F(x_1) = 0 + \int_{0}^{x_1} \frac{1}{15}\,dx = \frac{x_1}{15} \qquad \text{for } 0 < x_1 < 15$$

$$F(x_1) = 0 + \int_{0}^{15} \frac{1}{15}\,dx + \int_{15}^{x_1} 0\,dx$$

Probability Distributions of Continuous Variables

$$= 0 + \frac{15}{15} + 0$$

$F(x_1) = 1$ for $x_1 > 15$

The probability density function and the cumulative distribution function are shown graphically in Figure 6.3.

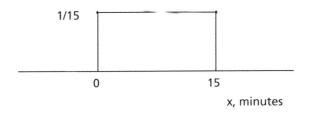

Figure 6.3 (a): Probability Density Function for Waiting Time for a Bus

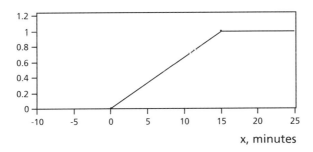

Figure 6.3 (b): Cumulative Distribution Function for Waiting Time for a Bus

(c) *Example 6.2*

A probability density function is given by:
$f(x) = 0$ for $x < 1$
$f(x) = b / x^2$ for $1 < x < 5$
$f(x) = 0$ for $x > 5$

a) What is the value of b?
b) From this obtain the probability that X is between 2 and 4.
c) What is the probability that X is exactly 2?
d) Find the cumulative distribution function of X.

Answer:

a) To satisfy equation 6.4:

$$\int_{-\infty}^{1} 0\, dx + \int_{1}^{5} \frac{b}{x^2}\, dx + \int_{5}^{\infty} 0\, dx = 1$$

Chapter 6

Therefore $\int_1^5 b x^{-2} \, dx = 1$

$$\left[-b x^{-1}\right]_1^5 = 1$$

$$-b\left(\frac{1}{5} - 1\right) = 1$$

$$\frac{4}{5}b = 1$$

$$b = 1.25$$

(In Example 6.1 the constant $\frac{3}{8}$ was obtained in the same way).

Then a graph of the density function for this example is shown below:

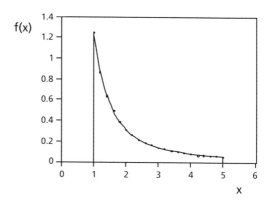

Figure 6.4:
Graph of Function for Example 6.2

b) $\Pr[2 < X < 4] = \int_2^4 1.25 \, x^{-2} dx$

$$= \left[-1.25 \, x^{-1}\right]_2^4$$

$$= (-1.25)\left(\frac{1}{4} - \frac{1}{2}\right)$$

$$= 0.3125$$

c) $\Pr[X = 2 \text{ exactly}] = \int_2^2 1.25 \, x^{-2} \, dx$

$$= \left[-1.25 \, x^{-1}\right]_2^2$$

$$= 0$$

Probability Distributions of Continuous Variables

Note: The result obtained here is important and applies to all continuous random variables. The probability that any continuous random variable is *exactly* equal to a single quantity is zero. We will see this again in Example 7.2.

d) For $x_1 < 1$: $\quad F(x_1) = \int_{-\infty}^{x_1} 0\, dx = 0$

For $1 < x_1 < 5$: $\quad F(x_1) = 0 + \int_{1}^{x_1} 1.25\, x^{-2}\, dx$

$$= (-1.25)\left[x^{-1}\right]_{1}^{x_1}$$

$$= (-1.25)\left(\frac{1}{x_1} - 1\right)$$

$$= 1.25\left(1 - \frac{1}{x_1}\right)$$

For $5 < x_1 < \infty$: $\quad F(x_1) = \int_{-\infty}^{x_1} f(x)\, dx$

$$= \int_{0}^{1} 0\, dx + \int_{1}^{5} 1.25 x^{-2}\, dx + \int_{3}^{x_1} 0\, dx$$

$$= 0 + (-1.25)\left[x^{-1}\right]_{1}^{5} + 0$$

$$= (-1.25)\left(\frac{1}{5} - 1\right)$$

$$= 1$$

Then to summarize, the cumulative distribution function of X is:

$$\begin{array}{ll} 0 & \text{for } x_1 < 1 \\ 1.25\left(1 - \dfrac{1}{x_1}\right) & \text{for } 0 < x_1 < 5 \\ \text{and} \quad 1 & \text{for } x_1 > 5 \end{array}$$

Problems

1. A probability density function for x in radians is given by:

   $f(x) = 0 \qquad$ for $\quad x < -\pi/2$

   $f(x) = \dfrac{1}{2} \cos x \qquad$ for $\quad -\pi/2 < x < \pi/2$

   $f(x) = 0 \qquad$ for $\quad x > \pi/2$

Chapter 6

a) Find the probability that X is between 0 and $\pi/4$.
b) Find an expression for the corresponding cumulative distribution function, $F(x)$, for $-\pi/2 \leq x \leq \pi/2$.
c) If $x = \pi/2$, what is the value of $f(x)$? Explain why this is or is not a reasonable result.
d) What is the probability that X is exactly $\pi/4$? Explain why this is or is not a reasonable result.
e) Repeat part (a) using $F(x)$.

2. A probability density function is given by:

$f(x) = 0$ for $x < -2$
$f(x) = 1/3$ for $-2 < x < 0$
$f(x) = \dfrac{1}{3}\left(1 - \dfrac{x}{2}\right)$ for $0 < x < 2$
$f(x) = 0$ for $x > 2$

a) What is the probability that X is between 0 and +1?
b) Find the cumulative distribution function of X for each interval. Is the cumulative distribution function for $x > 2$ reasonable? Why?
c) Sketch the cumulative distribution function, showing scales.
d) Use the results of part b to find the probability that X is between 0 and 1.
f) Find the median of this probability distribution.

3. A radar telemetry tracking station requires a vast quantity of high-quality magnetic tape. It has been established that the distance X (in meters) between tape-surface flaws has the following probability density functions:

$f(x) = 0.005\, e^{-0.005\, x}$ $x \geq 0$

$f(x) = 0$ otherwise

a) Plot a graph of $f(x)$ versus x for $0 \leq x \leq 800$.
b) Find the cumulative probability distribution function,

$$F(x_i) = \int_{-\infty}^{x_1} f(x)\, dx \text{ for } x_1 > 0.$$

c) Suppose one flaw in the tape-surface has been identified. Calculate:
  (i) the probability that an additional flaw will be found within the next 100 m of tape.
  (ii) the probability that an additional flaw will not be found for at least 200 m.
  (iii) the probability that an additional flaw will be found between 100 and 200 m from the flaw already identified.

4. A continuous random variable X has the following probability density function:

$f(x) = k\, x^{1/3}$ for $0 < x < 1$
$f(x) = 0$ for $x < 0$ and $x > 1$

a) Find k.

b) Find the cumulative distribution function.
c) Find the probability that $0.3 < X < 0.6$.

6.2 Expected Value and Variance

We saw in Chapter 5 that the mathematical expectation or expected value of a discrete random variable is a mean result for an infinitely large number of trials, so it is a mean value that would be approximated by a large but finite number of trials. This holds also for a continuous random variable. For a discrete random variable the expected value is found by adding up the product of each possible outcome with its probability, giving

$$\mu = E(X) = \sum_{\text{all } x_i} (x_i) \Pr[x_i].$$

For a continuous random variable this becomes (using equation 6.3) the corresponding integral involving the probability density function:

$$\mu = E(X) = \int_{-\infty}^{+\infty} x\, f(x)\, dx \tag{6.6}$$

We saw in Chapter 5 also that the variance of a discrete random variable is the expectation of $(x_i - \mu)^2$. This carries over to a continuous random variable and becomes:

$$\sigma_x^2 = E\big[(x-\mu)^2\big] = \int_{-\infty}^{+\infty} (x-\mu)^2 f(x)\, dx \tag{6.7}$$

The alternative form given by equation 5.7

$$\sigma_x^2 = E(X^2) - \mu_x^2 \tag{6.8}$$

still holds and is generally faster for calculations. For continuous random variables

$$E(X^2) = \int_{-\infty}^{+\infty} x^2 f(x)\, dx \tag{6.9}$$

Example 6.3

The random variable of Example 6.1 has the probability density function given by:

$f(x) = 0$ for $x < 0$
$f(x) = \dfrac{3}{8} x^2$ for $0 < x < 2$
$f(x) = 0$ for $x > 2$

a) Find the probability that X is between 1 and 2.
b) Find the cumulative distribution function of X.
c) Find the expected value of X.
d) Find the variance and standard deviation of X.

Chapter 6

Answer:

a) $\Pr[1 < X < 2] = \int_1^2 f(x)\,dx$

$$= \int_1^2 \frac{3}{8}x^2\,dx$$

$$= \left[\left(\frac{3}{8}\right)\left(\frac{1}{3}x^3\right)\right]_1^2$$

$$= \left(\frac{1}{8}\right)(2^3 - 1^3)$$

$$= \frac{7}{8}$$

b) $\Pr[x \leq x_1] = F(x_1) = \int_{-\infty}^{x_1} f(x)\,dx$

If $x_1 < 0$, $F(x_1) = \int_{-\infty}^{x_1} (0)\,dx = 0$

If $0 < x_1 < 2$, $F(x_1) = \int_{-\infty}^{0} (0)\,dx + \int_0^{x_1} \frac{3}{8}x^2\,dx$

$$= 0 + \left[\left(\frac{3}{8}\right)\left(\frac{1}{3}x^3\right)\right]_0^{x_1}$$

$$= \frac{1}{8}x_1^3$$

If $x_1 > 2$, $F(x_1) = \int_{-\infty}^{0} (0)\,dx + \int_0^{2} \frac{3}{8}x^2\,dx + \int_2^{x_1} (0)\,dx$

$$= 0 + \left[\left(\frac{3}{8}\right)\left(\frac{1}{3}x^3\right)\right]_0^{2} + 0$$

$$= 1$$

Then the cumulative distribution function is:

$F(x_1) = 0$ for $x_1 < 0$

$F(x_1) = \dfrac{1}{8}x_1^3$ for $0 < x_1 < 2$

$F(x_1) = 1$ for $x_1 > 2$

Probability Distributions of Continuous Variables

c) $\mu_x = E(X) = \int_{-\infty}^{+\infty} x\, f(x)\, dx$

$= \int_{-\infty}^{0} (x)(0)\, dx + \int_{0}^{2} (x)\left(\frac{3}{8}x^2\right) dx + \int_{2}^{\infty} (x)(0)\, dx$

$= \int_{0}^{2} \frac{3}{8} x^3\, dx$

$= \left[\left(\frac{3}{8}\right)\left(\frac{1}{4} x^4\right)\right]_{0}^{2}$

$= \left(\frac{3}{32}\right)(16 - 0)$

$= 1.5$

d) $E(X^2) = \int_{-\infty}^{+\infty} x^2\, f(x)\, dx$

$= \int_{-\infty}^{0} (x^2)(0)\, dx + \int_{0}^{2} (x^2)\left(\frac{3}{8}x^2\right) dx + \int_{2}^{8} (x^2)(0)\, dx$

$= \int_{0}^{2} \frac{3}{8} x^4\, dx$

$= \left[\left(\frac{3}{8}\right)\left(\frac{1}{5} x^5\right)\right]_{0}^{2}$

$= \left(\frac{3}{(8)(5)}\right)(32 - 0)$

$= \frac{96}{40} = 2.4$

Then $\sigma_x^2 = E(X^2) - \mu_x^2$

$= 2.4 - (1.5)^2$

$= 0.150$

and $\sigma_x = \sqrt{0.150} = 0.387$

Chapter 6

Example 6.4

In the illustration of section 6.1(b) the probability density function for the waiting time was given by

$$f(x) = 0 \quad \text{for } x < 0$$
$$f(x) = \frac{1}{15} \quad \text{for } 0 < x < 15$$
$$f(x) = 0 \quad \text{for } x > 15$$

a) Find the expected value of the waiting time, X minutes.
b) Find the variance and standard deviation of the waiting time.
c) What is the probability that the waiting time is within two standard deviations of its expected mean value?

Answer:

a) $E(X) = \int_{-\infty}^{+\infty} x f(x) dx$

$$= \int_0^{15} (x)\left(\frac{1}{15}\right) dx$$

$$= \left[\left(\frac{1}{15}\right)\left(\frac{x^2}{2}\right)\right]_0^{15}$$

$$= \frac{1}{(15)(2)}(225 - 0)$$

$$= \frac{15}{2}$$

$$= 7.5$$

Then the expected value of the waiting time, or the mean, μ_x, of the probability distribution, is 7.5 minutes. This seems reasonable, as it is halfway between the minimum waiting time, 0 minutes, and the maximum waiting time, 15 minutes.

b) $E(X^2) = \int_{-\infty}^{+\infty} x^2 f(x) dx$

$$= \int_0^{15} (x^2)\left(\frac{1}{15}\right) dx$$

$$= \left[\left(\frac{1}{15}\right)\left(\frac{x^3}{3}\right)\right]_0^{15}$$

$$= \frac{1}{(15)(3)}(15^3 - 0)$$

$$= 75$$

$$\sigma_x^2 = E(X^2) - \mu_x^2$$
$$= 75 - (7.5)^2$$
$$= 18.75$$

Then the variance of the waiting time is 18.75 minute², and the standard deviation is $\sqrt{18.75} = 4.33$ minutes.

c) The interval which is within two standard deviations of the expected value is $(\mu_x - 2\sigma_x)$ to $(\mu_x + 2\sigma_x)$, or from $7.5 - (2)(4.33) = -1.16$ to $7.5 + (2)(4.33) = 16.16$ minutes.

Then we have:

$$\Pr[(\mu_x - 2\sigma_x) < X < (\mu_x + 2\sigma_x)] = \Pr[-1.16 < X < 16.16]$$
$$= \int_{-1.16}^{0} 0\, dx + \int_{0}^{15} \frac{1}{15}\, dx + \int_{15}^{16.16} 0\, dx$$
$$= 0 + 1 + 0$$
$$= 1$$

The probability that the waiting time for this particular probability distribution is within two standard deviations of its expected mean value is 1 or 100%. We will find that other distributions often give different results. For example, a different result is obtained for the normal distribution, as we will see in the next chapter.

Problems

1. Given $f(x) = b/x^2$ for $1 < x < 3$
   $f(x) = 0$ for $x < 1$ and $x > 3$
   a) Determine the value of b that will make $f(x)$ a probability density function.
   b) Find the cumulative probability distribution function and use it to determine the probability that X is greater than 2 but less than 3.
   c) Find the probability that X is exactly equal to 2.
   d) Find the mean of this probability distribution.
   e) Find the standard deviation of this probability distribution.

2. An electrical voltage is determined by the probability density function
   $f(x) = \dfrac{1}{2\pi}$ for $0 \leq x \leq 2\pi$
   $f(x) = 0$ for all other values of x

   (This is a uniform distribution.)

   a) Find its cumulative distribution function for all values of x.
   b) Find the mean of this probability distribution.
   c) Find its standard deviation.

Chapter 6

    d) What is the probability that the voltage is within two standard deviations of its mean?

3. An electrical voltage is determined by the probability density function

    $f(x) = 1$ for $1 \leq x \leq 2$
    $f(x) = 0$ for all other values of x

(This is a uniform distribution.)

    a) Find its cumulative distribution function for all values of x.
    b) Find the mean of this probability distribution.
    c) Find its standard deviation.
    d) What is the probability that the voltage is within one standard deviation of its mean?

4. The time between arrivals of trucks at a warehouse is a continuous random variable. The probability of time between arrivals is given by the probability density function for which

    $f(t) = 4 e^{-4t}$ for $t \geq 0$
    $f(t) = 0$ for $t < 0$

where t is time in hours. (This is an exponential distribution. See section 6.3)

    a) What is the probability that the time between arrivals of the first and second trucks is less than 5 minutes?
    b) Find the mean time between arrivals of trucks, μ hours.
    c) Find the standard deviation of time between arrivals of trucks, σ hours.
    d) What is the probability that the waiting time between arrivals of trucks will be between $(\mu - \sigma)$ hours and $(\mu + \sigma)$ hours?
    e) What is the probability that the time between arrivals of trucks at the warehouse will be between $(\mu - 2\sigma)$ hours and $(\mu + 2\sigma)$ hours?

5. The probability of failure of a mechanical device as a function of time is given by the following probability density function:

    $f(t) = 3 e^{-3t}$ for $t \geq 0$
    $f(t) = 0$ for $t < 0$

where t is time in months. (This is an exponential distribution. See section 6.3)

    a) Find the mean of the probability distribution. This is the mean lifetime of the device.
    b) Find the standard deviation of the probability distribution.
    c) What is the probability that the device will fail within one standard deviation of its mean lifetime?
    d) What is the probability that the device will fail within two standard deviations of its mean lifetime?

Probability Distributions of Continuous Variables

6.3 Extension: Useful Continuous Distributions

The normal distribution is the continuous distribution which is by far the most used by engineers; it will be considered in Chapter 7. However, a number of others are also used very widely. Some are based on the normal distribution, and the corresponding tests assume that the underlying population is at least approximately normally distributed. We will encounter some of these continuous distributions in Chapters 9, 10 and 13 because they correspond to statistical tests used very frequently. These are the *t*-distribution, the *F*-distribution, and the chi-squared distribution.

The other continuous distributions which should be mentioned here are the uniform distribution, the exponential distribution, the Weibull distribution, the beta distribution, and the gamma distribution. Others are important in various specialized applications.

The *uniform distribution* is very simple. Its probability density function is a constant in a particular interval (say for $a < X < b$) and zero outside that interval. We have already seen an example of it in the waiting time for a bus, used as a simple illustration of a continuous distribution in section 6.1, and it has appeared in some of the problems. It is sometimes used to model errors in electrical communication with pulse code modulation. Electrical noise on the other hand, is often modeled by a normal distribution.

The *exponential distribution* has the following probability density function:

$$f(x) = \lambda e^{-\lambda x} \quad \text{for } x \geq 0$$
$$f(x) = 0 \quad \text{for } x < 0 \quad (6.10)$$

where λ is a constant closely related to the mean and standard deviation.

For $x > 0$ the cumulative distribution function for the exponential distribution is found easily by integration:

$$F(x_1) = \Pr[0 < X < x_1]$$
$$= \int_0^{x_1} \lambda e^{-\lambda x} \, dx \quad (6.11)$$
$$= 1 - e^{-\lambda x_1}$$

The exponential distribution is related to the Poisson distribution, although the exponential distribution is continuous whereas the Poisson distribution is discrete. The Poisson distribution gives the probabilities of various numbers of random events in a given interval of time or space when the possible number of discrete events is much larger than the average number of events in the given interval. If the variable is time, the exponential distribution gives the probability distribution of the *time between successive random events* for the same conditions as apply to the Poisson distribution.

Chapter 6

The following expression can be found in tables of integrals:

$$\int_0^\infty x^n e^{-ax} dx = n! a^{-(n+1)} \qquad (6.12)$$

Use of it greatly reduces the labor of finding expected values and variances for the exponential distribution.

The exponential distribution is used for studies of reliability, which will be discussed very briefly in section 6.4, and of queuing theory. Queuing theory gives probability as a function of waiting time in a queue for service. An example might be: what is the probability that the time between arrival of one customer and of the next at a service counter will be more than a stated time, such as three minutes?

The *Weibull distribution*, the *beta distribution*, and the *gamma distribution* are more complicated, mainly because each has two independent parameters. Both the Weibull distribution and the gamma distribution give the exponential distribution with particular choices of one of their two parameters. These distributions are discussed more fully in the books by Miller, Freund, and Johnson and by Ross (see List of Selected References, section 15.2), and all but the gamma distribution are discussed in the book by Vardeman.

6.4 Extension: Reliability

What is the probability that an engineering device will function as specified for a particular length of time under specified conditions? How will this probability be modified if we put further components in series or in parallel with one another? These are the sorts of questions which are addressed in the study of *engineering reliability*.

Reliability is applied in many areas of engineering, including design of mechanical devices, electronic equipment, and power transmission systems. Although failures of supply of electricity to factories, offices, and residences were once frequent, they have become much less frequent as engineers have devoted more attention to reliability. The concepts of reliability have been exceedingly important to manned flights in space.

The study of reliability makes use of the exponential distribution, the gamma distribution, and the Weibull distribution. Theory has been developed for many applications.

A general reference book on the use of reliability in engineering is by Billinton and Allan (see List of Selected References in section 15.2).

CHAPTER 7

The Normal Distribution

This chapter requires a good knowledge of the material covered in sections 2.1, 2.2, 3.1, 3.2, and 4.4. Chapter 6 is also helpful as background.

The *normal distribution* is the most important of all probability distributions. It is applied directly to many practical problems, and several very useful distributions are based on it. We will encounter these other distributions later in this book.

7.1 Characteristics

Many empirical frequency distributions have the following characteristics:

1. They are approximately symmetrical, and the mode is close to the centre of the distribution.
2. The mean, median, and mode are close together.
3. The shape of the distribution can be approximated by a bell: nearly flat on top, then decreasing more quickly, then decreasing more slowly toward the tails of the distribution. This implies that values close to the mean are relatively frequent, and values farther from the mean tend to occur less frequently. Remember that we are dealing with a random variable, so a frequency distribution will not fit this pattern exactly. There will be random variations from this general pattern.

Remember also that many frequency distributions do not conform to this pattern. We have already seen a variety of frequency distributions in Chapter 4, and many other types of distribution occur in practice.

Example 4.2 showed data on the thickness of a particular metal part of an optical instrument as items came off a production line. A histogram for 121 items is shown in Figure 4.4, reproduced here.

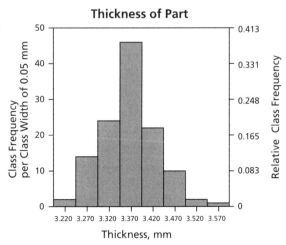

Figure 4.4: Histogram of Thickness of Metal Part

Chapter 7

We can see that the characteristics stated above are present, at least approximately, in Figure 4.4. Random variation (and the arbitrary division into classes for the histogram) could reasonably be responsible for deviation from a smooth bell shape.

A theoretical distribution that has the stated characteristics and can be used to approximate many empirical distributions was devised more than two hundred years ago. It is called the "normal probability distribution," or the *normal distribution*. It is sometimes called the Gaussian distribution, but other mathematicians developed it earlier than Gauss did. It was soon found to approximate the distribution of many errors of measurement.

7.2 Probability from the Probability Density Function

The probability density function for the normal distribution is given by:

$$f(x) = \frac{1}{\sigma\sqrt{2\pi}} e^{-\frac{(x-\mu)^2}{2\sigma^2}} \qquad (7.1)$$

where μ is the mean of the theoretical distribution, σ is the standard deviation, and $\pi = 3.14159\ldots$ This density function extends from $-\infty$ to $+\infty$. Its shape is shown in Figure 7.1 below. The first scale on Figure 7.1 gives values of $\frac{x-\mu}{\sigma}$, and the scale below it gives corresponding values of x. Thus, $\frac{x-\mu}{\sigma} = 0$ corresponds to $x = \mu$, and $\frac{x-\mu}{\sigma} = -3$ corresponds to $x = \mu - 3\sigma$.

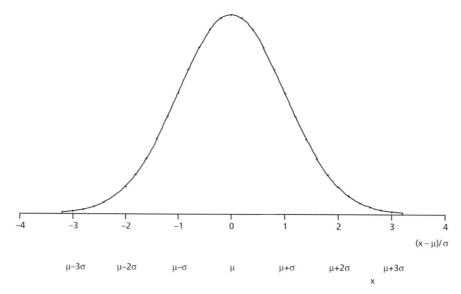

Figure 7.1: Shape of the Normal Distribution

The Normal Distribution

Because the normal probability density function is symmetrical, the mean, median and mode coincide at $x = \mu$. Thus, the value of μ determines the location of the center of the distribution, and the value of σ determines its spread.

We have seen that probabilities for a continuous random variable are given by integration of the probability density function. Then normal probabilities are given by integration of the function shown in equation 7.1, or the areas under the corresponding curve.

The probability that a variable, X, is between x_1 and x_2 according to the normal distribution is given by:

$$\Pr[x_1 < X < x_2] = \int_{x_1}^{x_2} \frac{1}{\sigma\sqrt{2\pi}} e^{-\frac{(x-\mu)^2}{2\sigma^2}} dx \qquad (7.2)$$

as shown in Figure 7.2.

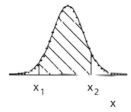

Figure 7.2: Probability of X Between x_1 and x_2

A corresponding cumulative probability is given by:

$$\Pr[-\infty < X < x] = F(x) = \int_{-\infty}^{x} \frac{1}{\sigma\sqrt{2\pi}} e^{-\frac{(x-\mu)^2}{2\sigma^2}} dx \qquad (7.3)$$

However, the integral of equations 7.2 and 7.3 cannot be evaluated analytically in closed form. It is evaluated to any required precision numerically and shown in tables or given by computer software. The constant, $\frac{1}{\sigma\sqrt{2\pi}}$, in equations 7.2 and 7.3 is determined by the requirement that $F(\infty) = 1$ (see equation 6.4).

Equations 7.1, 7.2 and 7.3 represent an infinite number of normal distributions with various values of the parameters μ and σ. A simpler form in a single curve is obtained by a change of variable.

$$\text{Let} \quad z = \frac{x - \mu}{\sigma} \qquad (7.4)$$

Then z is a ratio between $(x - \mu)$ and σ. It represents the number of standard deviations between any point and the mean. Since x, μ, and σ all have the same units in any particular case, z is dimensionless.

Chapter 7

Since μ and σ are constants for any particular distribution, differentiation of equation 7.4 gives:

$$dz = \frac{1}{\sigma} dx$$
$$dx = \sigma \, dz \tag{7.5}$$

Substitution of equations 7.4 and 7.5 in equation 7.2 gives:

$$\Pr[x_1 < X < x_2] = \int_{z_1}^{z_2} \frac{1}{\sigma\sqrt{2\pi}} e^{-\frac{z^2}{2}} \sigma \, dz$$

$$= \int_{z_1}^{z_2} \frac{1}{\sqrt{2\pi}} e^{-\frac{z^2}{2}} dz \tag{7.6}$$

where, according to equation 7.4, $z_1 = \dfrac{x_1 - \mu}{\sigma}$ and $z_2 = \dfrac{x_2 - \mu}{\sigma}$.

Figure 7.3 shows the normal distribution in terms of z, the number of standard deviations from the mean. It can be seen that almost all the area under the curve is between $z = -3$ and $z = +3$. Therefore, the practical width of the normal distribution is about six standard deviations.

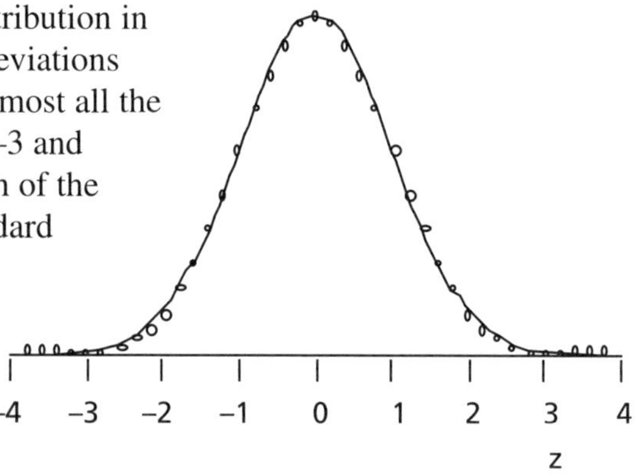

Figure 7.3: Normal Distribution as a Function of z

The standard normal cumulative distribution function, $\Phi(z)$, as a function of z, is defined as follows:

$$\Phi(z_1) = \Pr[-\infty < Z < z_1] = \Pr[Z < z_1]$$

$$= \int_{-\infty}^{z_1} \frac{1}{\sqrt{2\pi}} e^{-\frac{z^2}{2}} dz \tag{7.7}$$

It corresponds to the area under the curve in Figure 7.4.

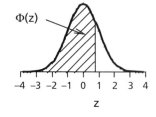

Figure 7.4: Standard Cumulative Distribution Function for the Normal Probability Distribution

The Normal Distribution

If the change of variable shown in equation 7.4 is applied and the curve shown in Figure 7.1 is integrated according to equation 7.3 to obtain a cumulative normal distribution, the result is an *s*-shaped curve, as shown in Figure 7.5.

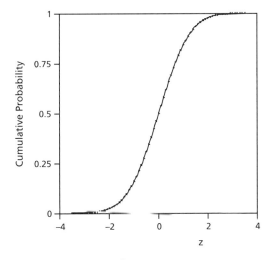

Figure 7.5: Cumulative Normal Probability

7.3 Using Tables for the Normal Distribution

Table A1 in Appendix A gives values of the cumulative normal probability as a function of z, the number of standard deviations from the mean. Part of Table A1 is shown below.

**Part of Table A1
Cumulative Normal Probability**

$$\Phi(z) = \Pr[Z < z]$$

$\Delta z=$	−0.09	.	−0.07	−0.06	−0.05	.	−0.01	−0.00	
- -									- -
z_0									z_0
−3.7	0.0001	.	0.0001	0.0001	0.0001	.	0.0001	0.0001	−3.7
...	...	.	...	...	...	.	...	...	...
−0.8	0.1867	.	0.1922	0.1949	0.1977	.	0.2090	0.2119	−0.8
−0.7	0.2148	.	0.2206	**0.2236**	0.2266	.	0.2389	0.2420	−0.7
−0.6	0.2451	.	0.2514	0.2546	0.2578	.	0.2709	0.2743	−0.6
...	...	.	...	...	...	.	...	...	...
−0.0	0.4641	.	0.4721	0.4761	0.4801	.	0.4960	0.5000	−0.0

Chapter 7

Table A1 gives values of z_0 (–3.7, –3.6, ... –0.1, –0.0 ; 0.0, 0.1, ... 3.7, 3.8) along the lefthand side and righthand side of the table over two pages. The numbers along the top of the table give smaller *increments*, $\Delta z = -0.09, -0.08, ..., -0.01, 0.00$ on the first page, and on the second page 0.00, 0.01, ..., 0.08, 0.09. The value of z for a particular row and column is the *sum* of the value of z_0 for that row (along the sides) plus the increment, Δz, for that column (along the top of the table).

$$z = z_0 + \Delta z \qquad (7.8)$$

To illustrate, see the part of Table A1 shown above. Say we want $\Phi(-0.76)$: we look for the row labeled $z_0 = -0.7$ along the sides and the column labeled $\Delta z = -0.06$ along the top (since $-0.76 = (-0.7) + (-0.06)$) and read $\Phi(-0.76) = 0.2236$.

The diagram at the top of the table towards the right indicates that $\Phi(z)$ corresponds to the area under the curve to the left of a particular value of z (here $z = -0.76$).

Suppose that instead we want $\Phi(+0.76)$. This is given on the second page of Table A1 in Appendix A. As before, we look for the applicable row, labeled $z_0 = 0.7$ along the sides, and the column labeled $\Delta z = 0.06$ (since $0.76 = 0.7 + 0.06$). For this value of z we read from the table that $\Phi(0.76) = 0.7764$.

Because the distribution is symmetrical, there must be a simple relation between $\Phi(-0.76)$ and $\Phi(+0.76)$, or in general between $\Phi(-z)$ and $\Phi(+z)$. That relation is:

$$\Phi(-z_1) = 1 - \Phi(+z_1) \qquad (7.9)$$

or in this case $\Phi(-0.76) = 1 - \Phi(+0.76) = 1 - 0.7764 = 0.2236$. Of course that means that $\Phi(-0.00) = \Phi(+0.00) = 0.5000$, so half of the total area under the curve is to the left of $z = 0$, the mean and median and mode of the distribution. If you think about it, that makes sense.

Example 7.1

a) What is the probability that Z for a normal probability distribution is between –0.76 and +0.76?

b) What is the probability that Z for a normal probability distribution is smaller than –0.76 or larger than +0.76?

Answer:

A sketch such as that shown in Figure 7.6 is very helpful in visualizing the required integral and finding appropriate values from the table.

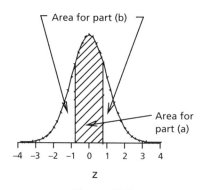

Figure 7.6: Probabilities for Example 7.1

a) Pr $[-0.76 < Z < +0.76]$ corresponds to the middle area cross-hatched in Figure 7.6. The calculation of probabilities is as follows:

$$\begin{aligned} \text{Pr}\,[-0.76 < Z < +0.76] &= \text{Pr}[Z > 0.76] - \text{Pr}\,[Z > -0.76] \\ &= \Phi(0.76) - \Phi(-0.76) \\ &= 0.7764 - 0.2236 \quad \text{(from before)} \\ &= 0.5528 \end{aligned}$$

b) Pr $[(Z < -0.76) \cup (Z > +0.76)]$ corresponds to the outer areas in the sketch above.

$$\begin{aligned} \text{Pr}\,[(Z < -0.76) \cup (Z > +0.76)] &= [\Phi(-0.76)] + [1 - \Phi(+0.76)] \\ &= 0.2236 + [1 - 0.7664] \\ &= 0.4472 \end{aligned}$$

Check: Between them, parts (a) and (b) cover all possible results:

Then Pr$\{[-0.76 < Z < +0.76] + [(Z < -0.76) \cap (Z > +0.76)]\} = 0.5528 + 0.4472$
$$= 1.0000 \quad \text{(check)}$$

Because the normal distribution is used so frequently, it is important to become familiar with Table A1.

The reader should note that other forms of tables for the normal distribution are also in common use. One form gives the probability of a result in one tail of the distribution, that is Pr $[Z > z_1]$ for $z_1 \geq 0$, or Pr $[Z < z_1]$ for $z_1 \leq 0$. A variation gives the probability corresponding to both tails together. Another type gives the probability of a result between the mean and z_2 standard deviations from the mean, that is Pr $[Z < z_2]$ for $z_2 \geq 0$, or Pr $[Z > z_2]$ for $z_2 \leq 0$. These different forms of tables must not be confused. Confusion is reduced because a small graph at the top of a table almost always indicates which area corresponds to the values given.

Study the following examples carefully.

Example 7.2

A city installs 2000 electric lamps for street lighting. These lamps have a mean burning life of 1000 hours with a standard deviation of 200 hours. The normal distribution is a close approximation to this case.

a) What is the probability that a lamp will fail in the first 700 burning hours?

$$z_1 = \frac{x_1 - \mu}{\sigma} = \frac{700 - 1000}{200} = -1.50$$

From Table A1 for $z_1 = -1.50 = (-1.5) + (-0.00)$,

$$\begin{aligned} \text{Pr}\,[X < 700] &= \text{Pr}\,[Z < -1.50] \\ &= \Phi(-1.50) \\ &= 0.0668 \end{aligned}$$

Chapter 7

Then Pr [burning life < 700 hours] = 0.0668

or 0.067.

b) What is the probability that a lamp will fail between 900 and 1300 burning hours?

$$z_1 = \frac{x_1 - \mu}{\sigma} = \frac{900 - 1000}{200} =$$
$$= -0.50 = (-0.5) + (-0.00)$$
$$z_2 = \frac{x_2 - \mu}{\sigma} = \frac{1300 - 1000}{200} =$$
$$= +1.50 = (+1.5) + (0.00)$$

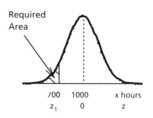

Figure 7.7: Probabilities for Example 7.2(a)

From Table A1, $\Phi(z_1) = \Phi(-0.50) = 0.3085$

and $\Phi(z_2) = \Phi(1.50) = 0.9332$

Then Pr [900 hours < burning life < 1300 hours]

$$= \Phi(z_2) - \Phi(z_1)$$
$$= 0.9332 - 0.3085$$
$$= 0.6247 \text{ or } 0.625.$$

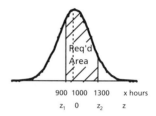

Figure 7.8: Probabilities for Example 7.2(b)

c) How many lamps are expected to fail between 900 and 1300 burning hours?

This is a continuation of part (b). The expected number of failures is given by the total number of lamps multiplied by the probability of failure in that interval. Then the expected number of failures = (2000) (0.6247) = 1249.4 or 1250 lamps. Because the burning life of each lamp is a random variable, the actual number of failures between 900 and 1300 burning hours would be only approximately 1250.

d) What is the probability that a lamp will burn for *exactly* 900 hours?

Since the burning life is a continuous random variable, the probability of a life of *exactly* 900 burning hours (not 900.1 hours or 900.01 hours or 900.001 hours, etc.) is zero. Another way of looking at it is that there are an infinite number of possible lifetimes between 899 and 901 hours, so the probability of any *one* of them is one divided by infinity, so zero. We saw this before in Example 6.2.

e) What is the probability that a lamp will burn between 899 hours and 901 hours before it fails?

Since this is an interval rather than a single exact value, the probability of failure in this interval is not infinitesimal (although in this instance the probability is small).

$$z_1 = \frac{x_1 - \mu}{\sigma} = \frac{899 - 1000}{200} = -0.505$$

$$z_2 = \frac{901 - 1000}{200} = -0.495$$

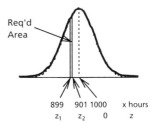

Figure 7.9: Probabilities for Example 7.2(e)

We could apply linear interpolation between the values given in Table A1. However, considering that in practice the parameters are not known exactly and the real distribution may not be exactly a normal distribution, the extra precision is not worthwhile.

Pr [899 hours < burning life < 901 hours]

$$\approx \Phi(-0.49) - \Phi(-0.50)$$
$$= \Phi(-0.4 - 0.09) - \Phi(-0.5 - 0.00)$$
$$= 0.3121 - 0.3085$$
$$= 0.0036 \text{ or } 0.4\%$$

(0.3% would also be a reasonable approximation).

f) After how many burning hours would we expect 10% of the lamps to be left?

This corresponds to the time at which

Pr [burning life > x_1 hours] = 0.10,

so Pr [burning life < x_1 hours] = 1 − 0.10 = 0.90.

Thus, Pr [$Z < z_1$] = 0.90

or $\Phi(z_1) = 0.90$

From Table A1,

$\Phi(1.2 + 0.08) = 0.8997$

and $\Phi(1.2 + 0.09) = 0.9015$

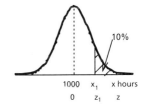

Figure 7.10: Probabilities for Example 7.2(f)

Once again, we could apply linear interpolation but the accuracy of the calculation probably does not justify it.

Since (0.90 − 0.8997) << (0.9015 − 0.90), let us take $z_1 = 1.28$. Then we have

$$z_1 = \frac{x_1 - \mu}{\sigma} = 1.28$$

$$\frac{x_1 - 1000}{200} = 1.28$$

$$x_1 = (200)(1.28) + 1000 = 1256$$

Chapter 7

Then after 1256 hours of burning, we would expect 10% of the lamps to be left. And again, because the burning time is a random variable, performing the experiment would give a result which would be close to 1256 hours but probably not exactly that, even if the normal distribution with the given values of the mean and standard deviation applied exactly.

g) After how many burning hours would we expect 90% of the lamps to be left?

We won't draw another diagram, but imagine looking at Figure 7.10 from the back.

Pr $[Z < z_2] = 0.10$ or $\phi(z_2) = 0.10$. From Table A1 we find

$$\phi(-1.2 - 0.08) = 0.1003$$
$$\phi(-1.2 - 0.09) = 0.0985$$

so $z_2 \approx -1.28$. (Do you see any resemblance to the answer to part (f)? Look again at equation 7.9.)

$$z_2 = \frac{x_2 - \mu}{\sigma} = \frac{x_2 - 1000}{200} = -1.28$$
$$x_2 - 1000 = -256$$
$$x_2 = 744$$

After 744 hours we would expect 90% of the lamps to be left.

Example 7.3

In another city 2500 electric lamps are installed for street lighting. The lamps come from a different manufacturer and have a mean burning life of 1050 hours. We know from past experience that the distribution of burning lives approximates a normal distribution. The 250th lamp fails after 819 hours. Approximately what is the standard deviation of burning lives for this set of lamps?

Answer:

$$\Phi(z_1) = \frac{250}{2500} = 0.100$$

From Table A1, $\Phi(-1.2 - 0.09) = 0.0985$

and $\Phi(-1.2 - 0.08) = 0.1003$

Then $z_1 = \dfrac{x_1 - \mu}{\sigma} \cong -1.28$

$$\frac{819 - 1050}{\sigma} = -1.28$$

$$\sigma = \frac{-231}{-1.28} = 180$$

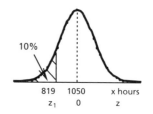

Figure 7.11: Probabilities for Example 7.3

Then the standard deviation of burning hours is approximately 180 hours. (As well as random variation, the term "approximately" covers a "correction for continuity" which we will encounter a little later.)

Example 7.4

The strengths of individual bars made by a certain manufacturing process are approximately normally distributed with mean 28.4 and standard deviation 2.95 (in appropriate units). To ensure safety, a customer requires at least 95% of the bars to be stronger than 24.0.

a) Do the bars meet the specification?
b) By improved manufacturing techniques the manufacturer can make the bars more uniform (that is, decrease the standard deviation). What value of the standard deviation will just meet the specification if the mean stays the same?

Answer:

a) $z_1 = \dfrac{x_1 - \mu}{\sigma}$

$= \dfrac{24.0 - 28.4}{2.95} = -1.49$

$\Phi(-1.49) = \Phi(-1.4 - 0.09) = 0.0681$
(from Table A1)

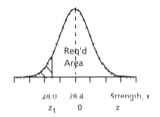

Figure 7.12: Probabilities for Example 7.4(a)

The probability that the bars will be stronger than 24.0 is $1 - 0.0681 = 0.9319$ or 93.2%. Since this is less than 95%, the bars do not meet the specification.

b) For this part, σ is the unknown.

From Table A1 we look for a value of z for which $\Phi(z_2) = 0.05$. We find $\Phi(-1.65) = 0.0495$ and $\Phi(-1.64) = 0.0505$. Then z_2 must be between -1.65 and -1.64. Since in this case the desired value of $\Phi(z_2)$ is halfway between $\Phi(-1.65)$ and $\Phi(-1.64)$, interpolation is very easy, giving $z_2 = -1.645$.

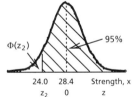

Then $z_2 = \dfrac{x_2 - \mu}{\sigma}$

$-1.645 = \dfrac{24.0 - 28.4}{\sigma}$

$\sigma = \dfrac{-4.4}{-1.645} = 2.67$

Figure 7.13: Probabilities for Example 7.4(b)

If the standard deviation can be reduced to 2.67 while keeping the mean constant, the specification will just be met.

Chapter 7

Example 7.5

An engineer decides to buy four new snow tires for his car. He finds that Retailer A is offering a special cash rebate, which depends on how much snow falls during the first winter. If this snowfall is less than 50% of the mean annual snowfall for his city, his rebate will be 50% of the list price. If the snowfall that winter is more than 50% but less than 75% of the mean annual snowfall, his rebate will be 25% of the list price. If the snowfall is more than 75% of the mean annual snowfall, he will receive no rebate. The engineer finds from a reference book that the annual snowfall for his city has a mean of 80 cm and standard deviation of 20 cm and approximates a normal distribution. The list price for the brand and size of tires he wants is $80.00 per tire.

The engineer checks other retailers and finds that Retailer B sells the same brand and size of tires with the same warranty for the same list price but offers a discount of 5% of the list price regardless of snowfall that year.

a) Compare the expected costs of the two deals. Which expected cost is less?
b) How much is the difference for four new snow tires? Neglect the relative advantages of a cash rebate as compared to a discount.

Answer: a) For Retailer A: $\mu = 80$ cm, $\sigma = 20$ cm.

50% of μ is 40 cm, and 75% of μ is 60 cm

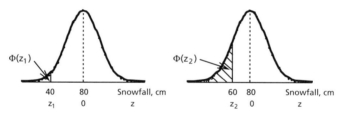

Figure 7.14: Probabilities for Example 7.5(a)

$$z_1 = \frac{x_1 - \mu}{\sigma} = \frac{40 - 80}{20} = -2.00$$

Pr [snowfall < 50% of μ] = Pr [$Z < -2.00$]

$$= \Phi(-2.00)$$
$$= 0.0228 \quad \text{(from Table A1)}$$

$$z_2 = \frac{x_2 - \mu}{\sigma} = \frac{60 - 80}{20} = -1.00$$

Pr [snowfall < 75% of μ] = Pr [$Z < -1.00$]

$$= \Phi(-1.00)$$
$$= 0.1587 \quad \text{(from Table A1)}$$

The Normal Distribution

Then $\Pr[50\% \text{ of } \mu < \text{snowfall} < 75\% \text{ of } \mu] = \Phi(-1.00) - \Phi(-2.00)$
$$= 0.1587 - 0.0228$$
$$= 0.1359$$

Then expected rebate from Retailer A is:

$(50\%) (\Pr[\text{snowfall} < 50\% \text{ of } \mu]) + (25\%) (\Pr[50\% \text{ of } \mu < \text{snowfall} < 75\% \text{ of } \mu])$
$$= (50\%)(0.0228) + (25\%)(0.1359)$$
$$= (1.14 + 3.40)\%$$
$$= 4.54\% \text{ of list price}$$

Discount from Retailer B is 5% of list price, so the discount from Retailer B is larger than the expected rebate from Retailer A. Therefore, the expected cost of buying from Retailer B is a little less than the expected cost of buying from Retailer A.

b) Cost of four new snow tires is as follows.

List price: (4) ($80.00) = $320.00

After rebate from Retailer A, expected cost = (1 − 0.0454) ($320.00) = $305.48

After discount from Retailer B, cost = (1 − 0.05) ($320.00) = $304.00

Then the difference in expected cost for four new snow tires is $1.48.

Some Quantitative Relationships

We can also use Table A1 to make more quantitative comments concerning *probabilities of results inside or outside chosen intervals* on Figure 7.4.

Since $\Pr[-2 < Z < +2] = \Phi(+2.0 + 0.00) - \Phi(-2.0 - 0.00)$
$$= 0.9772 - 0.0228$$
$$= 0.9544$$

[Check: $\Phi(-z_1) = 1 - \Phi(+z_1)$ (from eq. 7.9)
0.0228 = 1 − 0.9772 √]

Thus, 95.4% of all values are expected to be within two standard deviations from the mean of a normal distribution. By subtraction from 100%, 4.6% of all values are expected to be outside that interval.

Similarly, $\Pr[-3 < Z < +3] = \Phi(+3.0 + 0.00) - \Phi(-3.0 - 0.00)$
$$= 0.9987 - 0.0013)$$
$$= 0.9974$$

So 99.7% of all values are expected to be within three standard deviations from the mean. Only 0.3% of all values are expected to be farther from the mean than three standard deviations. Then, although the normal distribution extends in principle from −∞ to +∞, the practical width is about six standard deviations. If there is some

Chapter 7

practical limit on a variable (most commonly, that the variable never becomes negative), it will have little effect if the limiting value is at least three standard deviations from the mean.

Problems

(The following problems can be solved either with a pocket calculator and tables, or using a computer, as will be discussed in section 7.4.)

1. Diameters of bolts produced by a particular machine are normally distributed with mean 0.760 cm and standard deviation 0.012 cm. Specifications call for diameters from 0.720 cm to 0.780 cm.
   a) What percentage of bolts will meet these specifications?
   b) What percentage of bolts will be smaller than 0.730 cm?

2. The annual snowfall in Saskatoon is a normally distributed variable with a mean of 80 cm and a standard deviation of 20 cm.
   a) What is the probability that the snowfall in any year will exceed 30 cm?
   b) What is the probability that the snowfall in any year will be between 55 and 90 cm?

3. The diameters of screws in a batch are normally distributed with mean equal to 2.10 cm and standard deviation equal to 0.15 cm.
   a) What proportion of screws are expected to have diameters greater than 2.50 cm?
   b) A specification calls for screw diameters between 1.75 cm and 2.50 cm. What proportion of screws will meet the specification?

4. Diameters of ball bearings produced by a company follow a normal distribution. If the mean diameter is 0.400 cm and the standard deviation is 0.001 cm, what percentage of the bearings can be used on a machine specifying a size of 0.399 ±0.0015 cm? What is the upper bound of the size range that has a lower bound of 0.398 cm and includes 80% of the bearings?

5. An engineer working for a manufacturer of electronic components takes a large number of measurements of a particular dimension of components from the production line. She finds that the distribution of dimensions is normal, with a mean of 2.340 cm and a coefficient of variation of 2.4%.
   a) What percentage of measurements will be less than 2.45 cm?
   b) What percentage of dimensions will be between 2.25 cm and 2.45 cm?
   d) What value of the dimension will be exceeded by 98% of the components?

6. The probability that a river flow exceeds 2,000 cubic meters per second is 15%. The coefficient of variation of these flows is 20%. Assuming a normal distribution, calculate
   a) the mean of the flow.
   b) the standard deviation of the flow.
   c) the probability that the flow will be between 1300 and 1900 m^3/s.

The Normal Distribution

7. Bags of fertilizer are weighed as they come off a production line. The weights are normally distributed, and the coefficient of variation is 0.085%. It is found that 2% of the bags are under 50.00 kg.
   a) What is the mean weight of a bag of fertilizer?
   b) What percentage of the bags weigh more than 50.020 kg?
   c) What is the upper quartile of the weights?

8. The variation of copper content in a particular ore body follows a normal distribution. The coefficient of variation is 18%. The probability that the copper content exceeds 18.2 is 0.240.
   a) What is the mean copper content?
   b) What is the standard deviation of the copper content?
   c) What is the probability that the copper content will be less than 11.2?

9. 30% of the soil samples obtained from a proposed construction site gave test results for compressive strength of more than 3.5 tons per square foot. The coefficient of variation of the strengths is known to be 20%. Calculate:
   a) the mean soil strength,
   b) the standard deviation of soil strengths,
   c) the probability of soil strengths falling between 2.7 and 4.0 tons per square foot. State any assumptions made.

10. For a certain type of fluorescent light in a large building, the cost per bulb of replacing bulbs all at once is much less than if they are replaced individually as they burn out. It is known that the lifetime of these bulbs is normally distributed, and that 60% last longer than 2500 hours, while 30% last longer than 3000 hours.
    a) What are the approximate mean and standard deviation of the lifetimes of the bulbs?
    b) If the light bulbs are completely replaced when more than 20% have burned out, what is the time between complete replacements?

11. It is known that 10% of concrete samples have compressive strength less than 30.0 MN/m^2 and 20% have compressive strength greater than 36.0 MN/m^2. If the minimum acceptable strength is specified to be 28.0 MN/m^2, what is the probability that a sample will have a strength less than the specified minimum? What assumption is being made?

12. Of the Type A electrical resistors produced by a factory, 85.0% have resistance greater than 41 ohms, and 3.7% of them have resistance greater than 45 ohms. The resistances follow a normal distribution. What percentage of these resistors have resistance greater than 44 ohms?

13. A manufactured product has a length that is normally distributed with a mean of 12 cm. The product will be unusable if the length is 11½ cm or less.
    a) If the probability of this has to be less than 0.01, what is the maximum allowable standard deviation?

Chapter 7

b) Assuming this standard deviation, what is the probability that the product's length will be between 11.75 and 12.35 cm?

14. The probability of a river flow exceeding 2,000 cubic meters per second is 15% and the coefficient of variation of these flows is 20%. Assuming a normal distribution calculate
    (a) the mean of the flow,
    (b) the standard deviation of the flow,
    (c) the probability that the flow will be between 1300 and 1900 meters^3/sec.

15. A water quality parameter monitored in a lake is normally distributed with a mean of 24.3. It is also known that there is 70% probability that the parameter will exceed 17.6.
    a) Find the standard deviation of the parameter.
    b) If the parameter exceeds the 95th percentile, an investigation of a local industry begins. What is this critical value?

16. The time of snowpack formation is the time of the first snowfall which stays for the winter. In one Canadian city the mean time of snowpack formation is midnight of November 24, the 329th day of the year, and this time is approximately normally distributed. The standard deviation of the time of snowpack formation is 16.0 days. What is the probability that snowpack formation will occur before midnight October 20, the 294th day of the year, for two years in a row?

17. In a university scholarship program, anyone with a grade point average over 7.5 receives a $1,000 scholarship, anyone with an average between 7.0 and 7.5 receives $500, anyone with an average between 6.5 and 7.0 receives $100, and all others receive nothing. A particular class of 500 students has an overall average of 4.8 with a standard deviation of 1.2. Calculate the cost to the university of supplying scholarships for this class. State any assumption.

18. Steel used for water pipelines is sometimes coated on the inside with cement mortar to prevent corrosion. In a study of the mortar coatings of a pipeline used in a water transmission project, the mortar thicknesses were measured for a very large number of specimens. The mean and the standard deviation were found to be 0.62 inch and 0.13 inch, respectively, and the thickness was found to be normally distributed.
    a) In what percentage of the pipelines is the thickness of mortar less than 0.5 inch?
    b) If four pipes are selected at random, what is the probability that two or more have mortar thickness less than 0.5 inch?
    c) 100 pipes are taken and their mortar thicknesses are measured individually. If the mortar thickness of a pipe is found to be less than 0.5 inch, 10% less is paid to the manufacturer for that pipe. If the normal price of a pipe is $125.00, what is the expected cost of 100 pipes?

The Normal Distribution

19. On a particular farm, profit depends on rainfall. The rainfall is normally distributed with a mean of 31 cm and a standard deviation of 9 cm. Farm profits are:
    a) $100,000 if rainfall is over 44 cm,
    b) $150,000 if rainfall is between 29 and 44 cm,
    c) $130,000 if rainfall is between 22 and 29 cm,
    d) $ 65,000 if rainfall is between 15 and 22 cm, and
    e) –$ 80,000 if rainfall is less than 15 cm

    Find the expected farm profit.

20. The time a student takes to arrive at a solution for a statistics problem depends upon whether he or she recognizes certain simplifying comments in the problem statement. The probability of this recognition is 0.7. If the student recognizes the comments, the solution time is normally distributed with a mean time of 20 minutes and standard deviation of 4.3 minutes. If the student does not recognize the simplifying comments, the solution time is normally distributed with a mean time of 43 minutes with a standard deviation of 10.2 minutes.
    a) What is the expected solution time in a large class of students?
    b) What is the probability that a student chosen at random will require more than 28.2 minutes?
    c) What is the probability that he or she will require more than 43 minutes?

21. An irrigation pump is located on a reservoir whose mean water level is 550 m with a standard deviation of 10 m. The water level affects the output of the pump. If the level is below 538 m, then the expected pump output is 250 L / min with a standard deviation of 45 L / min; if the level is between 538 and 555 m, then the expected pump output is 325 L / min with a standard deviation of 52 L / min; and if the level is greater than 555 m, then the expected pump output is 375 L / min with a standard deviation of 48 L / min. The variation in the output at any given water level is due to variations in the electrical power supply and wave action on the reservoir. All variables are normally distributed.
    a) What are the probabilities of the levels being
        i. less than 538 m?
        ii. between 538 m and 555 m?
        iii. greater than 555 m?
    b) What is the expected pumping rate?
    c) If the cost of pumping is $25 / hr when the flow rate is less than 350 L / min, and $35 / hr when the flow rate exceeds 350 L / min, calculate the average cost of pumping.

7.4 Using the Computer

Instead of using tables such as Table A1, cumulative normal probabilities can be obtained from computer software such as Excel. Standard cumulative normal probabilities, $\Phi(z)$, can be obtained by the Excel function =NORMSDIST(z), where

Chapter 7

$z = \dfrac{x - \mu}{\sigma}$ is the standard normal variable. The inverse function is also available on Excel. If we know a value of the cumulative normal probability, $\Phi(z)$, and want to find the value of z to which it applies, we can use the function =NORMSINV(cumulative probability). In both function names the letter "s" stands for the standard form—that is, a relation between Φ and z rather than between Φ and x. Both function names can be pasted into the required cell choosing the statistical category and then the required function, as discussed in section 5.5. Alternatively, they can be typed.

These Excel functions can be used to solve Examples 7.1 to 7.5 and the Problems following section 7.3. To illustrate, here is an alternative solution of Example 7.4. Sketches of the probability relations shown in Figures 7.11 and 7.12 are still needed to check that the calculated probabilities are reasonable.

Example 7.4 (Solution Using Excel)

The strengths of individual bars made by a certain manufacturing process are approximately normally distributed with mean 28.4 and standard deviation 2.95 (in appropriate units). To ensure safety, a customer requires at least 95% of the bars to be stronger than 24.0.

a) Do the bars meet the specification?
b) By improved manufacturing techniques, the manufacturer can make the bars more uniform (i.e., decrease the standard deviation). What value of the standard deviation will just meet the specification if the mean stays the same?

Answer: a) $z_1 = \dfrac{x_1 - \mu}{\sigma}$ with $\mu = 28.4$, $\sigma = 2.95$, and $x_1 = 24$. Then the function =(24–28.4)/2.95 was entered in cell C2 with the label z_1 in cell A2. Explanations are in column B. Since $\Phi(z_1)$ is given by NORMSDIST(z_1), the function =NORMSDIST(C2) was entered in cell C3, and the label Phi(z1) was entered in cell A3. The percentage probability that the bars will be stronger than 24.0 is given by the function =(1–C2)*100%, which was entered in cell C4, and the corresponding label Pr%(stronger) was entered in cell A4. The result of the calculation was 93.2 (formatted to 1 decimal place using the Format menu). The answer to part (a) of the problem was placed in row 5.

(b) Now we require $\Phi(z_2) = 1 - 0.95$. Therefore the label Phi(z2) was entered in cell A7, and the function =1 – 0.95 was entered in cell C7. The label z2 was entered in cell A8, and the function, =NORMSINV(C7), was entered in cell C8. The result was –1.645. Since $z_2 = \dfrac{x_2 - \mu}{\sigma}$, the function =(24.0–28.4)/C8 was entered in cell C9, and the label Reqd SD was entered in cell A9. The result was 2.675 (formatted to 3 decimal places using the Format menu). The answer to part (b) was placed in rows 10 and 11.

The Normal Distribution

The Excel work sheet is shown below in Table 7.1. Answers to the specific questions are in rows 5, 10 and 11.

Table 7.1: Work Sheet for Example 7.4

	A	B	C
1	Ex 7.4 (a)		
2	z1	(24−28.4)/2.95=	−1.4915254
3	Phi(z1)	NORMSDIST(C1)=	0.06791183
4	Pr%(stronger)	(1−C2)*100%=	93.2
5	> Since 93.2% < 95%, the bars do not meet the specification.		
6	(b)		
7	Phi(z2)	1−0.95=	0.05
8	z2	NORMSINV(C7)=	−1.644853
9	Reqd SD	(24−28.4)/C8=	2.675
10	> If std dev can be reduced to 2.675 and the mean		
11	stays the same, the specification will just be met.		

7.5 Fitting the Normal Distribution to Frequency Data

We will find great advantages in fitting a normal distribution to a set of frequency data if the two distributions agree reasonably well. We can summarize the data very compactly in that case by giving the mean and standard deviation. Powerful statistical tests that assume that the underlying distribution is normal become available for our use.

In this section we will examine fitting a normal distribution to grouped frequency data and to discrete frequency data. This approach will be extended in section 7.6 to approximating another distribution (specifically a binomial distribution for certain circumstances) by a normal distribution. Then in section 7.7 we will look at fitting a normal distribution to cumulative frequency data.

Since a normal distribution is described completely by two parameters, its mean and standard deviation, usually the first step in fitting the normal distribution is to calculate the mean and standard deviation for the other distribution. Then we use these parameters to obtain a normal distribution comparable to the other distribution.

(a) Fitting to a Continuous Frequency Distribution

First, then, we need to estimate the parameters of the normal distribution that will fit the frequency distribution in which we are interested. We have seen in Chapter 3 how to estimate the mean and standard deviation of the population from which a sample came. Then we can compare the normal distribution having those parameters to the corresponding grouped frequency data.

Chapter 7

Example 7.6

Example 4.2 gave measurements of the thickness of a particular metal part of an optical instrument on 121 successive items from a production line. Taking these data as a sample, calculations shown in Example 4.2 gave the estimate of the mean of the population to be $\bar{x} = 3.369$ mm, and the estimate of the standard deviation of the population to be $s = 0.0629$ mm.

We saw in section 7.1 that the shape of the histogram for these data seems to be at least approximately consistent with a normal distribution. Therefore we will compare the class frequencies found in Example 4.2 with the expected frequencies for a normal distribution with mean and standard deviation as stated above. The first step in this comparison is to calculate cumulative normal probabilities, $\phi(z)$, at the class boundaries using Table A1 or the equivalent Excel function.

Class Boundary, x mm	$z = \dfrac{x-\mu}{\sigma}$	$\Phi(z)$
3.195	−2.77	0.0028
3.245	−1.97	0.0244
3.295	−1.18	0.1190
3.345	−0.38	0.3520
3.395	+0.41	0.6591
3.445	+1.21	0.8869
3.495	+2.00	0.9772
3.545	+2.80	0.9974
3.595	+3.59	0.9998

According to the normal distribution:

Pr $[X < 3.195]$			=	0.0028
Pr $[3.195 < X < 3.245]$	=	0.0244 − 0.0028	=	0.0216
Pr $[3.245 < X < 3.295]$	=	0.1190 − 0.0244	=	0.0946
Pr $[3.295 < X < 3.345]$	=	0.3520 − 0.1190	=	0.2330
Pr $[3.345 < X < 3.395]$	=	0.6591 − 0.3520	=	0.3071
Pr $[3.395 < X < 3.445]$	=	0.8869 − 0.6591	=	0.2278
Pr $[3.445 < X < 3.495]$	=	0.9772 − 0.8869	=	0.0903
Pr $[3.495 < X < 3.545]$	=	0.9974 − 0.9772	=	0.0202
Pr $[3.545 < X < 3.595]$	=	0.9998 − 0.9974	=	0.0024
Pr $[X > 3.595]$	=	1 − 0.9998	=	0.0002

The Normal Distribution

The expected frequency for each interval is obtained by multiplying the corresponding probability by the total frequency, 121. The results are:

Lower Boundary	Upper Class Boundary	Probability	Expected Frequency	Observed Frequency
—	3.195	0.0028	0.3	0
3.195	3.245	0.0216	2.6	2
3.245	3.295	0.0946	11.4	14
3.295	3.345	0.2330	28.2	24
3.345	3.395	0.3071	37.2	46
3.395	3.445	0.2278	27.6	22
3.445	3.495	0.0903	10.9	10
3.495	3.545	0.0202	2.4	2
3.545	3.595	0.0024	0.3	1
3.595	—	0.0002	0.0	0

Expected and observed frequencies are compared in Figure 7.15.

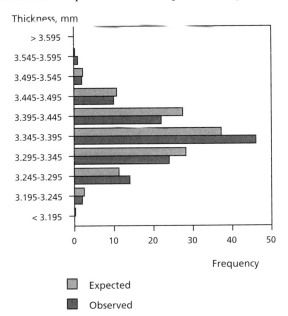

Figure 7.15: Comparison of Observed Frequencies with Expected Frequencies according to Fitted Normal Distribution

We can see in Figure 7.5 that actual frequencies are sometimes above and sometimes below the theoretical expected frequencies according to the normal distribution. The differences might well be explained by random variations, so we can conclude that the frequency distribution seems to be consistent with a normal distribution.

Chapter 7

(b) Fitting to a Discrete Frequency Distribution

If the distribution to which we compare a normal distribution is discrete, because the normal distribution is *continuous* we need a correction for continuity. The correction for continuity will be examined in the next section, in which the discrete binomial distribution is approximated by a normal distribution.

7.6 Normal Approximation to a Binomial Distribution

It is often desirable to use the normal distribution in place of another probability distribution. In particular, it is convenient to replace the binomial distribution with the normal when certain conditions are met. Remember, though, that the binomial distribution is discrete, whereas the normal distribution is continuous.

The shape of the binomial distribution varies considerably according to its parameters, n and p. If the parameter p, the probability of "success" (or a defective item or a failure, etc.) in a single trial, is sufficiently small (or if $q = 1 - p$ is sufficiently small), the distribution is usually unsymmetrical. If p or q is sufficiently small and if the number of trials, n, is large enough, a binomial distribution can be approximated by a Poisson distribution. This was discussed in section 5.4 (c).

On the other hand, if p is sufficiently close to 0.5 and n is sufficiently large, the binomial distribution can be approximated by a normal distribution. Under these conditions the binomial distribution is approximately symmetrical and tends toward a bell shape. A larger value of n allows greater departure of p from 0.5; a binomial distribution with very small p (or p very close to 1) can be approximated by a normal distribution if n is very large. If n is large enough, sometimes both the Poisson approximation and the normal approximation are applicable. In that case, use of the normal approximation is usually preferable because the normal distribution allows easy calculation of cumulative probabilities using tables or computer software.

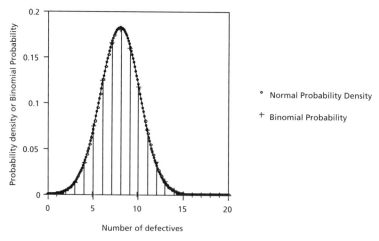

Figure 7.16: Comparison of a Binomial Distribution with a Normal Distribution Fitted to It

The Normal Distribution

Figure 7.16 compares a binomial distribution with a normal distribution. The parameters of the binomial distribution are $p = 0.4$ and $n = 20$ (for instance, we might take samples of 20 items from a production line when the probability that any one item will require further processing is 0.4). To fit a normal distribution we need to know the mean and the standard deviation. Remember that the mean of a binomial distribution is $\mu = np$, and that the standard deviation for that distribution is $\sigma = \sqrt{np(1-p)}$. To fit a normal distribution to this binomial distribution, we must have $\mu = np = (20)(0.4) = 8$, and $\sigma = \sqrt{np(1-p)} = \sqrt{(20)(0.4)(0.6)} = 2.191$. In Figure 7.6 the continuous curve passing through small circles represents the density function for the fitted normal distribution, while the vertical lines topped by small crosses represent binomial probabilities. The agreement appears to be very good.

But we have a difficulty to deal with. That is, the normal distribution is *continuous*, whereas the binomial distribution is *discrete*. Probabilities according to the binomial distribution are different from zero *only* when the number of defectives is a whole number, not when the number is between the whole numbers. On the other hand, if we integrate the normal distribution only for limits infinitesimally apart around the whole numbers, the area under the curve will be infinitesimally small. Then the corresponding probability will be zero.

The common-sense solution is to integrate for wider steps, which together cover the whole range. We set limits for integration of the normal distribution halfway between possible values of the discrete variable. This modification is called the *correction for continuity*. In Figure 7.6 the limits for integration of the normal distribution would be from 5.5 to 6.5 to compare with a binomial probability at 6 defects. For comparison with the binomial value at 7, the limits would be from 6.5 to 7.5, and so on.

The numerical comparison of probabilities using the correction for continuity is shown in Example 7.7. Approximating binomial probabilities in this way is called the *normal approximation to a binomial distribution*.

Example 7.7

Corresponding to the case shown in Figure 7.6, let's calculate probabilities according to the binomial distribution and for the normal distribution which fits it approximately. In a sample of 20 items when the probability that any one item requires further processing is 0.4, the binomial distribution gives probabilities that various numbers of items will require more processing. This is then a binomial distribution with $n = 20$ and $p = 0.4$.

Answer: Sample calculations will be shown for the probability of *six* items requiring further processing in a sample of 20, and then all the results will be compared.

Chapter 7

By the binomial distribution,

$$\Pr[R = 6] = {}_{20}C_6 (0.4)^6 (0.6)^{14} = \frac{(20)(19)(18)(17)(16)(15)}{(6)(5)(4)(3)(2)} (0.4)^6 (0.6)^{14}$$
$$= 0.124$$

By the normal approximation,

$$\Pr[R = 6] \approx \Pr[5.5 < X < 6.5] = \Phi\left(\frac{6.5 - 8}{2.191}\right) - \Phi\left(\frac{5.5 - 8}{2.191}\right)$$
$$= \Phi(-0.68) - \Phi(-1.14)$$
$$= 0.121$$

The values for the normal approximation shown above were read from tables with z evaluated to two decimal places. Evaluating z to three decimal places and using linear interpolation, or using computer software such as the function NORMSDIST from Excel, would give $0.2468 - 0.1269 = 0.120$ for the probability of six defectives. In Table 7.2 the normal approximations have been calculated with z evaluated to three decimal places and with linear interpolation to give a more accurate error of approximation, but interpolation is not ordinarily required.

Table 7.2: Comparison of Binomial Distribution and Normal Approximation

Number for Further Processing	Binomial Probability	Normal Approximation	Error of Approximation
0	0.00004	0.00026	−0.0002
1	0.0005	0.0012	−0.0007
2	0.0031	0.0045	−0.0014
3	0.012	0.014	−0.0016
4	0.035	0.035	−0.0001
5	0.075	0.072	+0.003
6	0.124	0.120	+0.005
7	0.166	0.163	+0.003
8	0.180	0.180	−0.001
9	0.160	0.163	−0.003
10	0.117	0.120	−0.003
11	0.071	0.072	−0.0009
12	0.035	0.035	+0.0004
13	0.015	0.014	+0.0006
14	0.0049	0.0045	+0.0003
15	0.0013	0.0012	+0.0001

16	0.0003	0.0003	+0.0000
17	0.00004	0.00005	−0.0000
18	5×10^{-6}	0.00001	−0.0000
19	3×10^{-7}	$<10^{-6}$	
20	1×10^{-8}	$<10^{-6}$	

The largest error in Table 7.2 is 0.005, 0.124 vs. 0.120 for six defectives.

As a rough rule, the normal approximation to the binomial distribution is usually reasonably good *if both np and (n)(1–p) are greater than 5*. In Example 7.7, np is equal to $(20)(0.4) = 8$ and $(n)(1 - p)$ is equal to $(20)(0.6) = 12$, so the rough rule is satisfied with some to spare. The rough rule should be used in solving problems in this book.

The rule is only a *rough* guide because the two parameters, n and p, affect the agreement separately. For the same value of the product np, the normal approximation to the binomial distribution is better when p is closer to 0.5. We can illustrate that by comparing the binomial distribution with the corresponding normal approximation just at $np = 5$, the limit given by the rough rule, at three combinations of n and p. Figure 7.17 shows these comparisons.

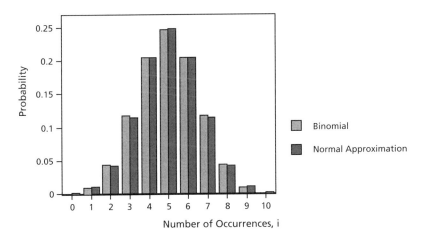

Figure 7.17(a): Comparison at $n = 10$ and $p = 0.5$

Chapter 7

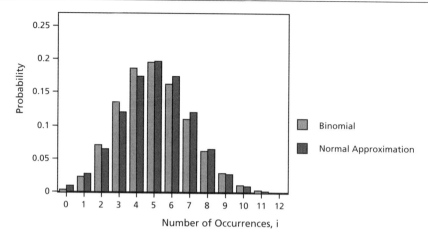

Figure 7.17(b): Comparison at $n = 25$ and $p = 0.2$

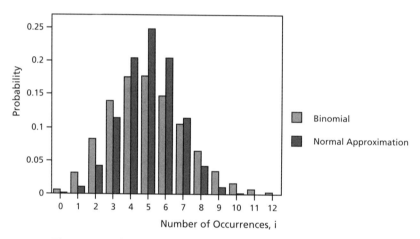

Figure 7.17(c): Comparison at $n = 250$ and $p = 0.02$

We can see from Figure 7.17 that the discrepancies are smallest at $n = 10$ and $p = 0.5$, intermediate at $n = 25$ and $p = 0.2$, and largest at $n = 250$ and $p = 0.02$, even though all are at $np = 5$ and $n(1 - p) > 5$. At $n = 10$ and $p = 0.5$ the largest absolute discrepancy is 0.002; at $n = 25$ and $p = 0.2$ the largest absolute discrepancy is 0.011; and at $n = 250$ and $p = 0.02$ the largest absolute discrepancy is 0.071.

Example 7.8

A coin is biased. We are told that the probability of heads on any one toss is 40% and the corresponding probability of tails is 60%. The coin is tossed 120 times, giving 56 heads and 64 tails. From what we were told about the bias, we expect $(120)(0.40) = 48$ heads. If the given information is correct, what is the probability of getting either

The Normal Distribution

56 or more heads, or 40 or fewer heads (i.e., a result as far from the expected result as 56 heads or farther in either direction)? Is the result so unlikely that we should doubt that the probability of heads on a single toss is only 40%?

Answer: This problem could be solved using the binomial distribution directly:

Pr $[R = 56] = {}_{120}C_{56} (0.4)^{56} (0.6)^{64}$, and similarly for $R = 57, 58, \ldots 120$ and $R = 0, 1, 2, \ldots, 39, 40$, then adding up probabilities. However, these calculations are very laborious. It would be less work to calculate the sum of Pr $[R = 41]$, Pr $[R = 42]$, ... Pr$[R = 54]$, Pr $[R = 55]$ and subtract that sum from 1, but that would still be a lot of labor. It is much easier to apply the normal approximation, and results should be very little different. In this case $np = (120)(0.4) = 48$ and $(n)(1 - p) = (120)(0.6) = 72$, so the rough rule is very easily satisfied. For the normal approximation $\mu = np = (120)(0.4) = 48$ and $\sigma = \sqrt{(n)(p)(1-p)} = \sqrt{(120)(0.4)(0.6)} = 5.367$.

Using the correction for continuity, Pr $[R = 56]$ corresponds to the area under the normal probability curve between 55.5 and 56.5. So, Pr $[R > 55]$ corresponds to the area under the curve beyond 55.5. Similarly, Pr $[R < 41]$ corresponds to the area under the curve for $X < 40.5$. If $x_1 = 55.5$, $z_1 = \dfrac{x_1 - \mu}{\sigma} = \dfrac{55.5 - 48}{5.367} = 1.397$

Similarly, if $x_2 = 40.5$, $z_2 = \dfrac{40.5 - 48}{5.367} = -1.397$

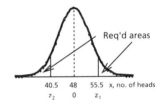

Then Pr $[R > 55, \text{Binomial}] \approx$ Pr $[Z > 1.397]$

$= 1 - \Phi(1.397)$

$\approx 1 - \Phi(1.40)$

$= 1 - 0.9192 = 0.081$.

Figure 7.18: Probabilities for Example 7.8

Then Pr [more than 55 heads] $\approx 8.1\%$.

Similarly, Pr [fewer than 41 heads] $\approx 8.1\%$. The probability of a result as far from the mean as 56 heads or farther in either direction, given that $p = 0.400$, is $(2)(8.1\%) = 16.2\%$. This would happen by chance about one time in six, so it is not very unlikely. Then the result of tossing the coin gives us no evidence that p is not equal to 0.400.

Approximations such as the normal approximation to the binomial distribution are not as important as they used to be because nearly exact values can be obtained using computer software. As we saw in section 5.5(b), both single and cumulative values for the binomial distribution can be obtained from Microsoft Excel. However, even when these nearly exact values are available, it may be desirable to use a convenient approximation.

Chapter 7

7.7 Fitting the Normal Distribution to Cumulative Frequency Data

(a) Cumulative Normal Probability and Normal Probability Paper

Instead of comparing a frequency distribution or probability distribution to a normal probability distribution using a histogram or the equivalent, often a better alternative is to compare graphically using *cumulative* probabilities. This has the advantage of giving an overall picture, showing the sum of deviations to any particular point. However, Figure 7.3 shows that the cumulative normal probability plotted against z gives an S-shaped curve. That would also be true plotted against x. It is not convenient to make graphical comparisons using an S-shaped curve.

However, the scale can be modified (or distorted) to give a more convenient comparison. The scale is modified in such a way that cumulative probability plotted against x or z will give a straight line for a normal distribution. A frequency distribution will still show random variations, but real departure from a normal distribution is much easier to spot. Thus, cumulative relative frequencies (on the modified scale) are plotted versus the variable, x, on a linear scale. If the data came from a normal distribution, this plot will give approximately a straight line. If the underlying distribution is appreciably different from a normal distribution, larger deviations and systematic variations will be present.

Graph paper using such a modified or distorted scale for cumulative relative frequency, and a uniform scale for the measured variable, is called *normal probability paper.* This special type of commercial graph paper, like the special types for logarithmic and log-log scales, is available from many suppliers. Commercial normal probability paper comes with a distorted scale for relative cumulative frequency along one axis and corresponding unequally spaced grid lines. The other scale (with corresponding grid lines) is uniform. Points are plotted by hand on this paper with co-ordinates corresponding to relative cumulative frequency (on the distorted scale) versus the value of the variable (on the linear scale). In most cases we will use data from a grouped frequency distribution. Since normal probability paper uses cumulative frequency or probability, data from a grouped frequency distribution should be plotted versus class boundaries, not class midpoints.

The points so plotted can be compared with the straight line representing a normal distribution fitted to the data and so having the same mean and standard deviation. Since the median of a normal distribution is equal to its mean, one point on this line should be at 50% relative cumulative frequency and $\bar{x}$, the estimated mean. Another point should be at 97.7% relative cumulative frequency and $(\bar{x} + 2s)$; a third should be at 2.3% relative cumulative frequency and $(\bar{x} - 2s)$.

The Normal Distribution

Example 7.9

Compare the data of Example 4.2 and Table 4.6 with a normal distribution using normal probability paper. The data are for measurements of the thickness of a metal part of an optical instrument. The histogram shown in Figure 4.4 seems qualitatively consistent with a normal distribution.

Answer: Table 7.3 was obtained using the data of Table 4.6:

Table 7.3: Data for Plot on Normal Probability Paper

Thickness, mm (class boundary)	Cumulative Frequency	Relative Cumulative Frequency, %
3.245	2	1.7
3.295	16	13.2
3.345	40	33.1
3.395	86	77.1
3.445	108	89.3
3.495	118	97.5
3.545	120	99.2

From Table 7.3 thickness was plotted (on a linear scale) against relative cumulative frequency (on a distorted scale) on normal probability paper as shown in Figure 7.19. From Example 4.2 the estimate of the mean is $\bar{x} = 3.3685$ mm, and the estimate of the standard deviation is $s = 0.0629$ mm. Then the straight line was drawn on Figure 7.9 to pass through the following points: 3.369 and 50.0% relative cumulative frequency; $3.3685 - (2)(0.0629) = 3.243$ mm and 2.3% relative cumulative frequency; $3.3685 + (2)(0.0629) = 3.494$ mm and 97.7% relative cumulative frequency.

The points seem to agree very well with the line, so it is reasonable to represent the data by a normal distribution. A more quantitative comparison will be given in Chapter 13, but the comparison using normal probability paper has the advantage of pointing out any part of the distribution where local departure from the line occurs.

(b) Computer Plot Equivalent to Normal Probability Paper

Instead of obtaining commercial probability paper and plotting points manually, it may be more convenient to make essentially the same visual comparison using a computer. However, it is not convenient to plot data directly to a nonuniform scale using a computer unless specialized software is available (but if the specialized software is available, it can certainly be used). The alternative is to plot $-z_{equivalent}$ (or $z_{equivalent}$) on a uniform scale against the experimental variable, also on a uniform scale. Remember that the relative cumulative frequency gives an approximation to

Chapter 7

cumulative normal probability if the data came from a population governed by the normal distribution. Remember also that a plot equivalent to use of normal probability paper would give approximately a straight line if the points follow a normal distribution; if that condition is met, z is approximately a linear function of the experimental variable, x. Then $z_{equivalent}$ is calculated from the inverse normal probability function of the relative cumulative frequency. For Excel, $z_{equivalent}$ is found from NORMSINV(relative cumulative frequency).

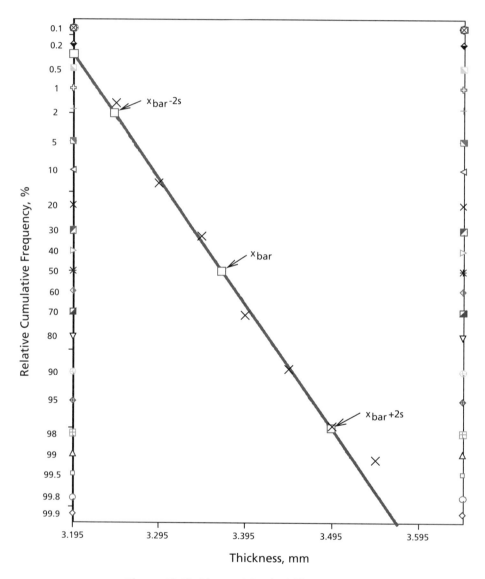

Figure 7.19: Normal Probability Paper

The Normal Distribution

Since $z = \frac{x - \bar{x}}{s}$, the straight line corresponding to the normal distribution is given by $x = \bar{x} + z\,s$, where x is the experimental variable and $\bar{x}$ and s are the sample mean and the estimate from the sample of the standard deviation. This straight line is plotted for comparison with the data. If they agree, then the data correspond approximately to a normal distribution.

Example 7.10

Data for measurements of the thickness of a metal part of an optical instrument from Example 4.2 have already been compared with a normal distribution in Example 7.6 (where observed frequencies were compared to expected frequencies) and Example 7.9 (where normal probability paper was used). Now we will calculate cumulative relative frequency and $z_{equivalent}$ for plotting against thickness at the corresponding upper class boundary. The calculations are shown in Table 7.4, and Figure 7.10 shows the resulting graph.

Table 7.4: Points for Computer Equivalent of Normal Probability Paper

Thickness at Class Boundary mm	Relative Cumulative Frequency, %	$-z_{equivalent} =$ $-\text{NORMSINV}(rcf)$
3.245	1.65	2.131
3.295	13.22	1.116
3.345	33.06	.438
3.395	71.07	−.556
3.445	89.26	−1.24
3.495	97.52	−1.964
3.545	99.17	−2.397

The straight line for the normal distribution can be located by plotting any two points on the line $z = \frac{x - \bar{x}}{s}$. Since $\bar{x} = 3.3685$ and $s = 0.0629$, at $x = 3.195$ the line must pass through $-z = \frac{3.3685 - 3.195}{0.0629} = 2.758$, and at $x = 3.595$ the line must pass through $-z = \frac{3.3685 - 3.595}{0.0629} = -3.601$. This line is also shown on Figure 7.10.

An extra scale has been added to Figure 7.10 giving percentage relative cumulative frequencies corresponding to the tick marks on the uniform vertical scale. An alternative, which will be adopted in some later examples, is to move the x-scale to the righthand side and to mark the percentage relative cumulative frequencies for the lefthand, uniformly spaced, tick marks. The relative cumulative frequencies at these tick marks are given by cumulative normal probabilities of the corresponding values of z.

Chapter 7

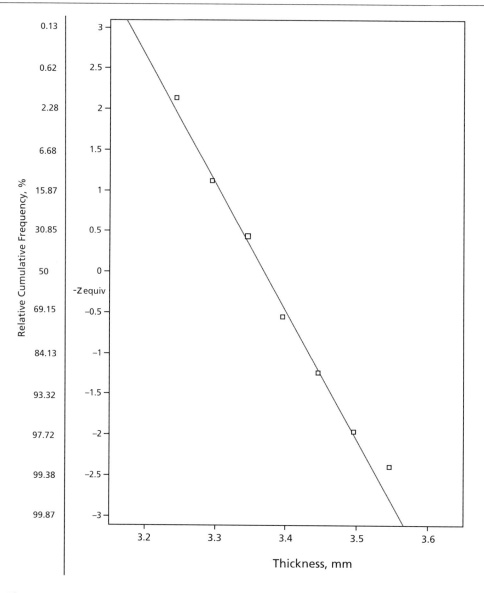

Figure 7.20: Computer Equivalent to Normal Probability Paper of Figure 7.19

(c) Plotting Individual Points Using a Computer

Rather than using the grouped frequency approach, we may want to plot all the individual points in a form suitable for visual comparison with a normal distribution. If the data set is small, we might do that by hand on normal probability paper, but most often we would use a computer. We saw in section 3.3 that each individual point can be considered a separate quantile. If the points are arranged in order of magnitude from the smallest to the largest, the i th item in order of magnitude among

The Normal Distribution

a total of n items represents a relative cumulative frequency of $(i - 0.5)/n$. If the data follow a normal distribution, $z_{equivalent}$ calculated from NORMSINV(relative cumulative frequency) will be approximately a straight-line function of the independent variable. The points can be compared to a straight line calculated from the sample mean and sample estimate of standard deviation according to the normal distribution. If the data of Example 4.2 which were used in Examples 7.6, 7.9, and 7.10 are plotted in this way, the result is shown in Figure 7.11. Some of the calculations are shown in Table 7.5.

Table 7.5: Calculations for Comparison Using Individual Points

Thickness at Class Boundary x mm	x^2	Order number, i	Relative Cumulative Frequency $=(i-0.5)/n$	$z_{equivalent}$ =NORMSINV (rel cum fr)
3.21	10.3041	1	0.0041	−2.6411
3.24	10.4976	2	0.0124	−2.2446
3.25	10.5625	3	0.0207	−2.0403
3.26	10.6276	4	0.0289	−1.8968
3.26	10.6276	5	0.0372	−1.7843
3.26	10.6276	6	0.0455	−1.6906
...	...	...	...	...
3.49	12.1801	118	0.9711	1.8968
3.51	12.3201	119	0.9793	2.0403
3.51	12.3201	120	0.9876	2.2446
3.57	12.7449	121	0.9959	2.6411

$$\bar{x} = \sum x_i / n = 407.59 / 121 = 3.3685$$

$$s^2 = \left[\sum x_i^2 - \left(\sum x_i\right)^2 / n\right] / (n-1)$$

$$= \left[1373.4471 - (407.59)^2 / 121\right] / 120$$

$$s = 0.0629$$

At $x = 3.15$, line passes through $z = (3.15 - 3.3685)/0.0629 = -3.47$

At $x = 3.6$, line passes through $z = (3.6 - 3.3685)/0.0629 = 3.68$

Chapter 7

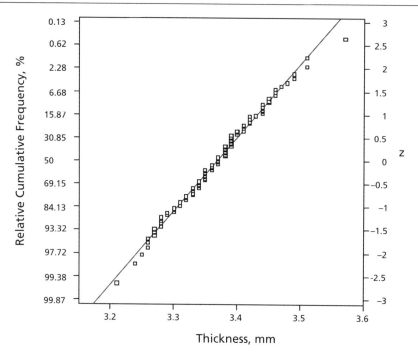

Figure 7.21: Computer Comparison Using Individual Points

The vertical groups in Figure 7.21 occur because of multiple measurements. For example, there are 11 points corresponding to a thickness of 3.35 mm (measured to two decimals).

(d) Extension: Probability Plotting in General

The discussion in this book of special plots for comparing probability distributions or frequency distributions is limited to comparisons for normal distributions, but there are other types for various situations. Other books give details of these methods.

A generalization of this method is called quantile-quantile plotting, or Q-Q plotting. It can be used to compare one relative frequency distribution with another, so giving an empirical comparison. It can also be used to compare a set of data with any of several theoretical probability distributions, including the exponential and Weibull distributions, which were discussed briefly in section 5.3.

A good discussion of probability plotting and its application in industry can be found in the book by Vardeman. See the List of Selected References in section 15.2.

7.8 Transformation of Variables to Give a Normal Distribution

Later in this book we will see statistical tests which assume that the underlying distribution is a normal distribution. Although there are other tests that do not require

The Normal Distribution

this or any similar assumption, in general these other tests are less sensitive than tests that do assume an underlying normal distribution. Furthermore, normal probabilities are convenient and familiar.

Therefore, if the original variable shows a distribution which is not a normal distribution, it is very useful to try to change the variable so that the new form will follow a normal distribution. This strategy is often successful if the original distribution showed a single mode somewhere between the smallest and largest values of the variable, but the original distribution was not symmetrical. If the original distribution was x, forms of the new variable to try include $\log x$, $1/x$, $\sqrt{x}$, and $\sqrt[3]{x}$: whatever will do the job.

The most common transformation for this purpose is replacing x by $\ln x$, $\log_{10} x$ or logarithm of x to any other base. If one of these works, the others will, too. This change of variable arises naturally in some cases, such as changing hydrogen-ion concentration to pH, or changing noise intensity or power to decibels. It is found useful for data of hydrology, fatigue failures, and particle size distribution.

Example 7.11

The size distribution of particles from a grinder was measured using a scanning electron microscope. The size distribution, as cumulative percentage of number of particles as a function of particle size in millimeters, is shown below.

Particle Size x mm	Relative Cumulative Frequency by Number, %
5.9	7.3
9.6	29.8
13.8	41.2
18.3	58.2
24.8	72.6
30.1	86.8
39.2	95.6
62.7	96.8
84.7	97.7
97.3	98.3
127.2	99
170	99.7

These data are also shown graphically on the equivalent of normal probability paper in Figure 7.22.

Chapter 7

Figure 7.22: Particle Size Data before Transformation

We can see that the pattern of the points shows a great deal of curvature, indicating that the distribution is far from a normal distribution. In fact, the distribution is not symmetrical, since the mean size is 19.9 μm and the median is appreciably different at approximately 16 μm.

Figure 7.23 shows the transformed data. The linear particle size, x mm, has been replaced by $y = \ln x$. Again the data are shown on the computer equivalent of normal probability paper. The straight line on Figure 7.23 has been fitted to the points. Thus, the transformed data can be approximated by a normal distribution, represented by the straight line.

Distributions of the random variable x for which $\log x$ is normally distributed occur often enough so that they are given a special name. They are called *lognormal* distributions.

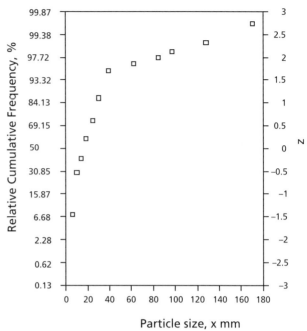

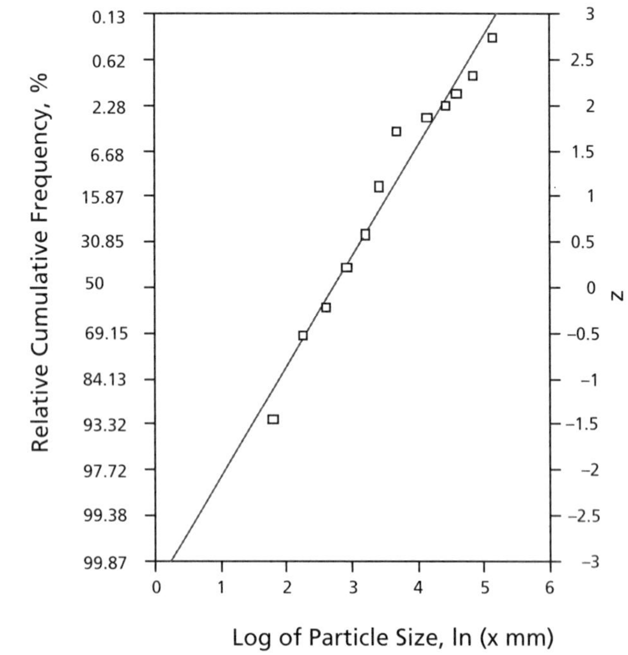

Figure 7.23: Particle Size Data after Transformation

The Normal Distribution

Problems

1. Four identical fair coins are tossed.
   a) Calculate and draw a graph of the probability distribution of the number of heads.
   b) What is the probability of obtaining three or more heads?
   c) Fit a normal distribution to the probability distribution of the number of heads. Sketch this distribution on top of the distribution drawn in part (a).
   d) What is the probability of obtaining three or more heads according to the normal approximation?
   e) (i) Would the normal approximation improve if coins which were not "fair" were used? Explain your answer.
   (ii) Would the normal approximation improve if a larger number of identical coins were used? Explain your answer.

2. A large shipment of books contains 2% which have imperfect bindings. Calculate the probability that out of 400 books,
   a) exactly 10 will have imperfect bindings (using two different approximations);
   b) more than 10 will have imperfect bindings (choosing one of the approximations for this calculation).

3. It is known that 3% of the plastic parts made by an injection molding machine are defective. If a sample of 30 parts is taken at random from this machine's production, calculate:
   a) the probability that exactly 3 parts will be defective.
   b) the probability that fewer than 4 parts will be defective.
   Do a) and b) using: (1) binomial distribution, (2) Poisson approximation, and (3) normal approximation.
   c) If the sample size is increased to 150 parts use the normal and Poisson approximations to calculate the probability of:
   1) more than 5 defectives
   2) between 6 and 8 defectives, inclusive.

4. The managers of an electronics firm estimate that 70% of the new products they market will be successful.
   a) If the company markets 20 products in the next two years, calculate using the binomial formulae and using the normal approximation:
   (i) the probability that exactly four new products will not be successful;
   (ii) the probability that no more than four new products will not be successful
   b) If the company markets 100 products over the next five years, what is the probability of:
   (i) more than 15 unsuccessful products?
   (ii) more than 70 but less than 85 successful new products?

193

Chapter 7

5. Under certain conditions twenty percent of piglets raised in total confinement will die during the first three weeks after birth. Consider a group of 20 newborn piglets.
   a) Calculate the probability that exactly 10 piglets will live to three weeks of age. Do by:
      (i) Binomial distribution
      (ii) Poisson approximation
      (iii) Normal approximation.
   b) Calculate the probability that no more than 15 piglets will live to three weeks of age. Do by:
      (i) Binomial distribution
      (ii) Poisson approximation
      (iii) Normal approximation.
   c) For both parts (a) and (b), discuss the validity of the approximations to the binomial distribution.

6. The proportion of males in a particular area is 0.52 . A sample of 50 people is taken at random.
   a) What probability distribution fits this case without any approximation? Why?
   b) Is a Poisson approximation suitable? Why?
   c) Is a normal approximation suitable? Why?
   d) Use whichever of (b) or (c) is more exact to find the probability that a sample of 50 people will contain at least 29 males but no more than 34 males.

7. A conservative candidate captured 48 percent of the popular vote in her riding in the last federal election. In a sample of 50 people from the candidate's riding, 35 claim to have voted for the conservative candidate. What is the probability in a sample of this size that 35 or more persons would have voted for this candidate? that 13 or fewer persons would have voted for this candidate? That either one or the other of these alternatives would have occurred? Then is there any reason to suspect that this sample may not be representative of the total population in the riding, or that some of the individuals in the sample are not being truthful about the way they voted?

8. Consider the following data on average daily yields of coke from coal in a coke oven plant:

Class Boundaries	Frequency
67.95 – 68.95	1
68.95 – 69.95	8
69.95 – 70.95	22
70.95 – 71.95	22
71.95 – 72.95	9
72.95 – 73.95	8
73.95 – 74.95	2

The mean and standard deviation for this population are estimated from the data to be 71.25 and 1.2775, respectively.
   a) Draw the frequency histogram for these grouped frequency data, and sketch a normal distribution, fitted to the data, superimposed on the histogram.
   b) Plot the grouped frequency data on normal probability paper or its computer equivalent. Draw the appropriate straight line to represent the normal distribution. Comment on the apparent fit or lack of fit between the data and the fitted normal distribution.
   c) Estimate the probability of average daily coke yields less than 70.95 using:
      (i) the grouped frequency data,
      (ii) the normal probability paper or its computer equivalent,
      (iii) tabulated values for the normal distribution.

9. It is known that the negative of the logarithm of the soil permeability, $y = -\log(k)$, in a particular soil type, Type A, follows a normal distribution. It is known that $\Pr[y > 7.2] = 30\%$, and $\Pr[y < 5.6] = 5\%$.
   a) Find the mean and the standard deviation of y.
   b) If 40% of the total plot of interest has soil Type A and 60% has Type B, for which y has a mean of 7.5 and a standard deviation of 0.45, for what percentage of the plot is y greater than 7.35?

10. Data were collected on the insulation thickness (mm) in transformer windings, and the data were grouped as follows:

Class Boundaries	Frequency
17.5 – 22.5	2
22.5 – 27.5	5
27.5 – 32.5	8
32.5 – 37.5	16
37.5 – 42.5	17
42.5 – 47.5	6
47.5 – 52.5	2
52.5 – 57.5	4

The mean and standard deviation estimated from these data are 37.25 mm and 8.084 mm, respectively.
   a) Plot the grouped frequency data and the fitted normal data (at $\bar{x}$, $\bar{x} - 2s$, and $\bar{x} + 2s$) on normal probability paper. Comment on the goodness of fit.
   b) Estimate the probability of insulation thickness being greater than 27.5 mm using:
      (i) the frequency grouped data;
      (ii) normal probability paper or its computer equivalent;
      (iii) calculated values for the normal distribution.

11. Data on heights of 60 adult males can be grouped as shown below. Heights are in cm.

Chapter 7

Class Bounds	Class Frequency
144.95 – 148.95	1
148.95 – 152.95	1
152 95 – 156.95	3
156.95 – 160.95	13
160.95 – 164.95	16
164.95 – 168.95	15
168.95 – 172.95	11

The mean and standard deviation of the population as estimated from this data are 163.68 cm and 5.39 cm respectively.

a) Plot the data on normal probability paper or its computer equivalent. Mark the points corresponding to $\bar{x}$, $\bar{x} - 2s$, and $\bar{x} + 2s$, with corresponding cumulative probabilities.
b) Comment qualitatively on how well the normal distribution fits the data.
c) Calculate the probability that an adult male from the population will be less than 160.95 cm high, (i) using the grouped frequency table; and (ii) using the fitted normal distribution.

12. Data on percentage relative humidity in a vegetable storage building were grouped as follows:

Class Lower Bound	Class Upper Bound	Frequency
59.5	64.5	4
64.5	69.5	3
69.5	74.5	8
74.5	79.5	16
79.5	84.5	17
84.5	89.5	5
89.5	94 5	3
94.5	99.5	4
	Total	60

The mean and standard deviation of the population as estimated from the data are 79.2 and 8.56, respectively.

a) Plot the data on normal probability paper and superimpose a line through points corresponding to $\bar{x}$, $\bar{x} - 2s$, and $\bar{x} + 2s$, with probabilities according to the normal distribution. An alternative is to plot cumulative relative frequency vs. cumulative Normal probability as discussed in section 7.7.
b) Estimate the probability of relative humidities being between 74.5 and 84.5 using
   i) tabulated data
   ii) tabulated values for the normal distribution
   iii) the straight line on normal probability paper or the alternative plot using a computer.

CHAPTER 8

Sampling and Combination of Variables

Some parts of this chapter require a good understanding of sections 3.1 and 3.2 and of Chapter 7.

Very frequently, engineers take samples from industrial systems. These samples are used to infer some of the characteristics of the populations from which they came. What factors must be kept in mind in taking the samples? How big does a sample need to be? These are some of the questions to be considered in this chapter. In answering some of them, we will need to consider the other major area of this chapter, the combination of variables.

We often need to combine two or more distributions, giving a new variable that may be a sum or difference or mean of the original variables. If we know the variance of the original distributions, can we calculate the variance of the new distribution? Can we predict the shape of the new distribution? Although in some cases the relationships are rather difficult to obtain, in some very important and useful cases simple relationships are available. We will be considering these simple relationships in this chapter.

8.1 Sampling

Remember that the terms "population" and "sample" were introduced in Chapter 1. A *population* might be thought of as the entire group of objects or possible measurements in which we are interested. A *sample* is a group of objects or readings taken from a population for counting or measurement. From the observations of the sample, we *infer* properties of the population. For example, we have already seen that the sample mean, $\bar{x}$, is an unbiased estimate of the population mean, μ, and that the sample variance, s^2, is an unbiased estimate of the corresponding population variance, σ^2. Further discussion of statistical inference will be found later in this chapter and later chapters.

However, if these inferences are to be useful, the sample must truly represent the population from which it came. The sample must be *random*, meaning that all possible samples of a particular size must have equal probabilities of being chosen from the population. This will prevent bias in the sampling process. If the effects of one or more factors are being investigated but other factors (not of direct interest) may interfere, sampling must be done carefully to avoid bias from the interfering

Chapter 8

factors. The effects of the interfering factors can be minimized (and usually made negligible) by randomizing with great care both choice of the parts of the sample that receive different treatments and the order with which items are taken for sampling and analysis.

Illustration Suppose there is an interfering factor that, unknown to the experimenters, tends to produce a percentage of rejects that increases as time goes on. Say the experimenters apply the previous method to the first thirty items and a modified method to the next thirty items. Then, clearly, comparing the number of rejects in the first group to the number of rejected items in the second group would not be fair to the modified method. However, if there is a *random* choice of either the standard method or the modified method for each item as it is produced, effects of the interfering factor will be greatly reduced and will probably be negligible. The same would be true if the interfering factor tended to produce a different pattern of rejected items.

The most common methods of randomization are some form of drawing lots, use of tables of random numbers, and use of random numbers produced by a computer. Discussion of these methods will be left to Chapter 11, Introduction to the Design of Experiments.

8.2 Linear Combination of Independent Variables

Say we have two independent variables, X and Y. Then a linear combination consists of the sum of a constant multiplied by one variable and another constant multiplied by the other variable. Algebraically, this becomes $W = aX + bY$, where W is the combined variable and a and b are constants.

The mean of a linear combination is exactly what we would expect: $\overline{W} = a\overline{X} + b\overline{Y}$. Nothing further needs to be said.

If we multiply a variable by a constant, the variance increases by a factor of the constant squared: variance(aX) = a^2 variance(X). This is consistent with the fact that variance has units of the square of the variable. Variances must increase when two variables are combined: there can be no cancellation because variabilities accumulate. Variance is always a positive quantity, so variance multiplied by the square of a constant would be positive. Thus, the following relation for combination of two independent variables is reasonable:

$$\sigma_w^2 = a^2\sigma_x^2 + b^2\sigma_y^2 \tag{8.1}$$

More than two independent variables can be combined in the same way.

If the independent variables X and Y are simply added together, the constants a and b are both equal to one, so the individual variances are added:

$$\sigma_{(X+Y)}^2 = \sigma_x^2 + \sigma_y^2 \tag{8.2}$$

Sampling and Combination of Variables

Thus, since the circumference of a board with rectangular cross-section is twice the sum of the width and thickness of the board, the variance of the sum of width and thickness is the sum of the variances of the width and thickness, and the variance of the circumference is $2^2(\sigma_{width}^2 + \sigma_{thickness}^2)$ if the width and thickness are independent of one another. Equation 8.2 can be extended easily to more than two independent variables.

If the variable W is the sum of n independent variables X, each of which has the same probability distribution and so the same variance σ_x^2, then

$$\sigma_w^2 = \left(\sigma_x^2\right)_1 + \left(\sigma_x^2\right)_2 + \cdots + \left(\sigma_x^2\right)_n = n\sigma_x^2 \tag{8.3}$$

Example 8.1

Cans of beef stew have a mean content of 300 g each with a standard deviation of 6 g. There are 24 cans in a case. The mean content of a case is therefore $(24)(300) = 7200$ g. What is the standard deviation of the contents of a case?

Answer. Variances are additive, standard deviations are not. The variance of the content of a can is $(6 \text{ g})^2 = 36 \text{ g}^2$. Then the variance of the contents of a case is $(24)(36 \text{ g}^2) = 864 \text{ g}^2$. The standard deviation of the contents of a case is $\sqrt{864}$ g = 29.4 g.

If the variable Y is subtracted from the variable X, where the two variables are independent of one another, their variances are still added:

$$\sigma_{(X-Y)}^2 = \sigma_X^2 + \sigma_Y^2 \tag{8.4}$$

This is consistent with equation 8.1 with $a = 1$ and $b = -1$. An example using this relationship will be seen later in this chapter.

If the variables being combined are not independent of one another, a correction term to account for the correlation between them must be included in the expression for their combined variance. This correction term involves the covariance between the variables, a quantity which we do not consider in this book. See the book by Walpole and Myers with reference given in section 15.2.

8.3 Variance of Sample Means

We have already seen the usefulness of the variance of a population. Now we need to investigate the variance of a *sample mean*. It is an indication of the reliability of the sample mean as an estimate of the population mean.

Say the sample consists of n independent observations. If each observation is multiplied by $\frac{1}{n}$, the sum of the products is the mean of the observations, $\overline{X}$. That is,

Chapter 8

$$\overline{X} = \frac{1}{n}X_1 + \frac{1}{n}X_2 + \cdots + \frac{1}{n}X_n$$

$$= \left(\frac{1}{n}\right)(X_1 + X_2 + \cdots + X_n)$$

Now consider the variances. Let the first observation come from a population of variance σ_1^2, the second from a population of variance σ_2^2, and so on to the *n*th observation. Then from equation 8.1 the variance of $\overline{X}$ is

$$\sigma_{\overline{X}}^2 = \left(\frac{1}{n}\right)^2 \sigma_1^2 + \left(\frac{1}{n}\right)^2 \sigma_2^2 + \cdots + \left(\frac{1}{n}\right)^2 \sigma_n^2$$

But the variables all came from the same distribution with variance σ^2, so

$$\sigma_{\overline{X}}^2 = \left(\frac{1}{n}\right)^2 \sigma^2 + \left(\frac{1}{n}\right)^2 \sigma^2 + \cdots + \left(\frac{1}{n}\right)^2 \sigma^2$$

$$= \left(\frac{1}{n}\right)^2 (n\sigma^2)$$

or

$$\sigma_{\overline{X}}^2 = \frac{\sigma^2}{n} \tag{8.5}$$

That is, the variance of the mean of *n independent* variables, taken from a probability distribution with variance σ^2, is $\frac{\sigma^2}{n}$. The quantity *n* is the number of items in the sample, or the *sample size*. The square root of the quantity $\frac{\sigma^2}{n}$, that is $\frac{\sigma}{\sqrt{n}}$, is called the *standard error of the mean* for this case. Notice that as the sample size increases, the standard error of the mean decreases. Then the sample mean, $\overline{X}$, has a smaller standard deviation and so becomes more reliable as the sample size increases. That seems reasonable.

But equation 8.5 applies only if the items are chosen independently as well as randomly. The items in the sample are statistically independent only if sampling is done *with replacement*, meaning that each item is returned to the system and the system is well mixed before the next item is chosen. If sampling occurs without replacement, we have seen that probabilities for the items chosen later depend on the identities of the items chosen earlier. Therefore, the relation for variance given by equation 8.5 applies directly only for sampling with replacement.

In practical cases, however, sampling with replacement is often not feasible. If an item is known to be unsatisfactory, surely we should remove it from the system, not stir it back in. Some methods of testing destroy the specimen so that it *can't* be returned to the system.

Sampling and Combination of Variables

If sampling occurs without replacement a correction factor can be derived for equation 8.5. The result is that the standard error of the mean for random sampling *without replacement*, still with all samples equally likely to be chosen, is given by

$$\sigma_{\bar{X}} = \frac{\sigma}{\sqrt{n}} \sqrt{\frac{N-n}{N-1}} \qquad (8.6)$$

where N is the size of the population, the number of items in it, and n is the sample size.

If the population size is large in comparison to the sample size, equation 8.6 reduces approximately to equation 8.5. Often engineering measurements can be repeated as many times as desired, so the effective population size is infinite. In that case, equation 8.6 can be replaced by equation 8.5.

Example 8.2

A population consists of one 2, one 5, and one 9. Samples of size 2 are chosen randomly from this population with replacement. Verify equation 8.5 for this case.

Answer: The original population has a mean of 5.3333 and a variance of $(2^2 + 5^2 + 9^2 - 16^2/3) / 3 = 8.2222$, so a standard deviation of 2.8674. Its probability distribution is shown below in Figure 8.1.

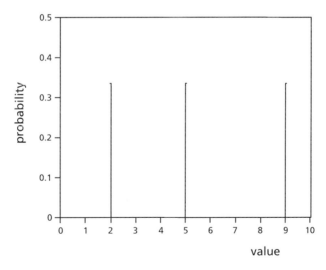

Figure 8.1: Probability Distribution of Population

Samples of size 2 with replacement can consist of two 2's, a 2 and a 5, a 2 and a 9, two 5's, a 5 and a 2, a 5 and a 9, two 9's, a 9 and a 2, and a 9 and a 5, a total of $3^2 = 9$ different results. These are all the possibilities, and they are all equally likely. Their respective sample means are 2, 3.5, 5.5, 5, 3.5, 7, 9, 5.5, and 7. The probability of each is $1/9 = 0.1111$. Since sample means of 3.5, 5.5, and 7 occur twice, the sampling distribution looks like the following:

Chapter 8

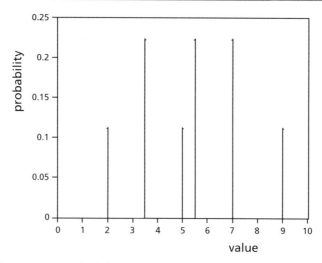

Figure 8.2: Probability Distribution of Samples of Size 2

The expected sample mean is $\mu_{\bar{x}} = 0.1111(2+5+9) + 0.2222(3.5+5.5+7) = 5.3333$, which agrees with the mean of the original distribution. The expected sample variance is $[(2^2+5^2+9^2) + 2(3.5^2+5.5^2+7^2) - 48^2/9] / 9 = 4.1111$, and the expected standard error of the mean is $\sqrt{4.1111} = 2.0276$. From equation 8.5 the predicted variance is $\frac{8.2222}{2} = 4.1111$. Thus, equation 8.5 is satisfied in this case.

The relationships for standard error of the mean are often used to determine how large the sample must be to make the result sufficiently reliable. The sample size is the number of times the complete process under study is repeated. For example, say the effect of an additive on the strength of concrete is being investigated. We decide using the relation for the standard error of the mean that a sample of size 8 is required. Then the whole process of preparing specimens with and without the additive (with other factors unchanged or changed in a chosen pattern) and measuring the strength of specimens must be repeated 8 times. If the specimens were prepared only once but *analysis* was repeated 8 times, the analysis would be sampled 8 times, but the effect of the additive would be examined only once.

Example 8.3

A population of size 20 is sampled without replacement. The standard deviation of the population is 0.35. We require the standard error of the mean to be no more than 0.15. What is the minimum sample size?

Answer: Equation 8.6 gives the relationship, $\sigma_{\bar{x}} = \frac{\sigma}{\sqrt{n}}\sqrt{\frac{N-n}{N-1}}$.

In this case σ is 0.35 and N is 20. What value of n is required if σ_x is at the limiting value of 0.15?

Sampling and Combination of Variables

Substituting, $0.15 = \dfrac{0.35}{\sqrt{n}} \sqrt{\dfrac{20-n}{20-1}}$.

$\sqrt{\dfrac{20-n}{n}} = \dfrac{0.15\sqrt{19}}{0.35} = 1.868$

$20 - n = 3.490\, n$

Then $n = \dfrac{20}{4.490} = 4.45$

But the sample size, the number of observations in the sample, must be an *integer*. It must be at least 4.45, so the minimum sample size is 5. A sample size of 4 would not satisfy the requirement.

Example 8.4

The standard deviation of measurements of a linear dimension of a mechanical part is 0.14 mm. What sample size is required if the standard error of the mean must be no more than (a) 0.04 mm, (b) 0.02 mm?

Answer: Since the dimension can be measured as many times as desired, the population size is effectively infinite. Then

$$\sigma_{\bar{x}} = \dfrac{\sigma}{\sqrt{n}}$$

(a) For $\sigma_{\bar{x}} = 0.04$ mm and $\sigma = 0.14$ mm,

$\sqrt{n} = \dfrac{0.14}{0.04} = 3.50$

$n = 12.25$

Then for $\sigma_{\bar{x}} \leq 0.04$ mm, the minimum sample size is 13.

(b) For $\sigma_{\bar{x}} = 0.02$ mm and $\sigma = 0.14$ mm,

$\sqrt{n} = \dfrac{0.14}{0.02} = 7.00$

$n = 49$

Then for $\sigma_{\bar{x}} \leq 0.02$ mm, the minimum sample size is 49.

Because of the inverse square relationship between sample size and the standard error of the mean, the required sample size often increases rapidly as the required standard error of the mean becomes smaller. At some point, further decreasing of the standard error of the mean by this method becomes uneconomic.

Chapter 8

Example 8.5

A plant manufactures electric light bulbs with a burning life that is approximately normally distributed with a mean of 1200 hours and a standard deviation of 36 hours. Find the probability that a random sample of 16 bulbs will have a sample mean less than 1180 burning hours.

Answer: If the bulb lives are normally distributed, the means of samples of size 16 will also be normally distributed. The sampling distribution will have mean $\mu_{\bar{x}} = 1200$ hours and standard deviation $\sigma_{\bar{x}} = \frac{36}{\sqrt{16}} = 9$ hours.

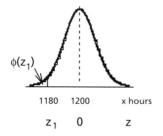

Figure 8.3: Distribution of Burning Lives

At 1180 hours we have $z_1 = \frac{1180 - 1200}{9} = -2.222$, and the cumulative normal probability, $\Phi(z_1) = 0.0132$ (from Table A1 with z_1 taken to two decimals) or 0.0131 (from the Excel function NORMSDIST). Then the probability that a random sample of 16 bulbs will have a sample mean less than 1180 hours is 0.013 or 1.3%.

A final example uses the difference of two normal distributions.

Example 8.6

An assembly plant has a bin full of steel rods, for which the diameters follow a normal distribution with a mean of 7.00 mm and a variance of 0.100 mm², and a bin full of sleeve bearings, for which the diameters follow a normal distribution with a mean of 7.50 mm and a variance of 0.100 mm². What percentage of randomly selected rods and bearings will not fit together?

Answer:

Figure 8.4 shows the overlap between the diameters of steel rods and sleeve bearings. However, it is not clear from this graph how to calculate an answer to the question.

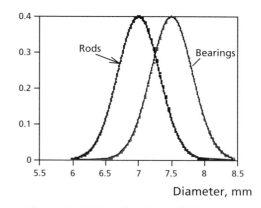

Figure 8.4: Distribution of Diameters of Rods and Bearings

Sampling and Combination of Variables

If for any selection of one rod and one bearing, the difference between the bearing diameter and the rod diameter is positive, the pair will fit. If not, they won't. (That may be a little oversimplified, but let us neglect consideration of clearance.)

Let d be the difference between the bearing diameter and the rod diameter. Because both the diameters of bearings and the diameters of rods follow normal distributions, the difference will also follow a normal distribution. The mean difference will be 7.50 mm – 7.00 mm = 0.50 mm. The variance of the differences will be the sum of the variances of bearings and rods: $\sigma_d^2 = 0.100 + 0.100 = 0.200$ mm². Then $\sigma_d = \sqrt{0.200} = 0.447$ mm. See Figure 8.5.

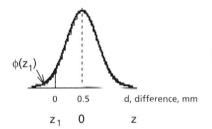

Figure 8.5: Distribution of Differences

$$z_1 = \frac{0 - 0.5}{0.447} = -1.118$$

$\Phi(z_1) = \Phi(-1.118) = 0.1314$ (from normal distribution table with z taken to two decimals) or 0.1318 (from the Excel function NORMSDIST).

Therefore, 13.2% or 13.1% of randomly selected sleeves and rods will not fit together.

8.4 Shape of Distribution of Sample Means: Central Limit Theorem

Let us look again at the distributions of Example 8.2. We started with a distribution consisting of three unsymmetrical spikes, shown in Figure 8.1. The probability distribution of sample means of size 2 in Figure 8.2 shows that values closer to the mean have become more likely.

Now let us look at a sampling distribution (probability distribution of sample means) for a sample of size 5 from the same original distribution. The number of equally likely samples of size 5 from a population of 3 items is $3^5 = 243$, so the complete set of results is much larger. Therefore, only some of the possible samples will be shown here.

Among the 243 equally likely resulting samples are the following:

Chapter 8

					Sample Mean
2	2	2	2	2	2
5	2	2	2	2	2.6
2	5	2	2	2	2.6
2	2	5	2	2	2.6
2	2	2	5	2	2.6
2	2	2	2	5	2.6
5	5	2	2	2	3.2
5	2	5	2	2	3.2
...	...	...	...	...	...
5	5	5	2	2	3.8
...	...	...	...	...	...
9	9	9	9	9	9

Because there are so many different sample means with varying frequencies, the sampling distribution is best shown as a histogram or a cumulative distribution. Figure 8.6 is the histogram for samples of size 5.

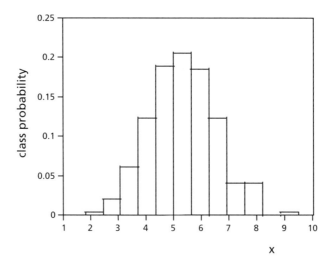

Figure 8.6: Sampling Distribution for Samples of Size 5

We can see that this histogram shows the largest class frequencies are near the mean, and they become generally smaller to the left and to the right. In fact, the distribution seems to be approximately a normal distribution. This is confirmed by the plot of normal cumulative probability against cumulative probability for the samples, which is shown in Figure 8.7. The distribution is not quite normal, but it is fairly close. It would come closer to normal if the sample size were increased.

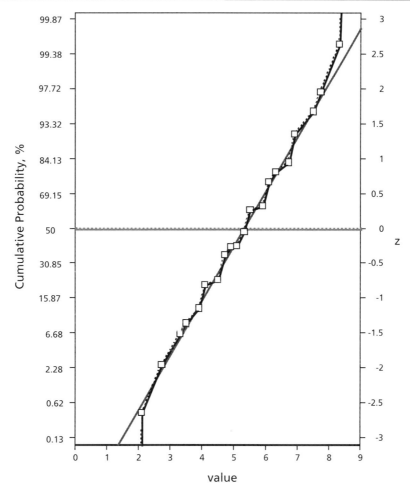

Figure 8.7: Comparison with Normal Distribution on Equivalent of Normal Probability Paper

In fact, this behavior as sample size increases is general. This is the *Central Limit Theorem*: if random and independent samples are taken from any practical population of mean μ and variance σ^2, as the sample size n increases the distribution of sample means approaches a normal distribution. As we have seen, the sampling distribution will have mean μ and variance $\dfrac{\sigma^2}{n}$. How large does the sample size have to be before the distribution of sample means becomes approximately the normal distribution? That depends on the shape of the original distribution. If the original population was normally distributed, means of samples of any size at all will be normally distributed (and sums and differences of normally distributed variables will also be normally distributed). If the original distribution was not normal, the means of samples of size two or larger will come closer to a normal distribution. Sample

Chapter 8

means of samples taken from almost all distributions encountered in practice will be normally distributed with negligible error if the sample size is *at least 30*. Almost the only exceptions will be samples taken from populations containing distant outliers.

The Central Limit Theorem is very important. It greatly increases the usefulness of the normal distribution. Many of the sets of data encountered by engineers are means, so the normal distribution applies to them if the sample size is large enough.

The Central Limit Theorem also gives us some indication of which sets of measurements are likely to be closely approximated by normal distributions. If variation is caused by several small, independent, random sources of variation of similar size, the measurements are likely close to a normal distribution. Of course, if one variable affects the probability distribution of the result in a form of conditional probability, so that the probability distribution changes as the variable changes, we cannot expect the result to be distributed normally. If the most important single factor is far from being normally distributed, the resulting distribution may not be close to normal. If there are only a few sources of variation, the resulting measurements are not likely to follow a distribution close to normal.

Problems

1. The mean content of a box of cat food is 2.50 kg, and the standard deviation of the content of a box is 0.030 kg. There are 24 boxes in a case, and there are 400 cases in a car load as it leaves the factory. What is the standard deviation of the amount of cat food contained in (i) a case, and (ii) a car load?

2. The design load on a hoist is 50 tonnes. The hoist is used to lift packages each having a mean weight of 1.2 tonnes. The weights of individual packages are known to be normally distributed with a standard deviation of 0.3 tonnes. If 40 packages are lifted at one time, what is the probability that the design load on the hoist will be exceeded?

3. Bags of sugar from a production line have a mean weight of 5.020 kg with a standard deviation of 0.078 kg. The bags of sugar are packed in cartons of 20 bags each, and the cartons are piled in lots of 12 onto pallets for shipping.
    a) What percentage of cartons would be expected to contain less than 100 kg of sugar?
    b) Find the upper quartile of sugar content of a carton.
    c) What mean weight of an individual bag of sugar will result in 95% of the pallets weighing more than 1200 kg.?

4. Fertilizer is sold in bags. The standard deviation of the content of a bag is 0.43 kg. Weights of fertilizer in bags are normally distributed. 40 bags are piled on a pallet and weighted.
    a) If the net weight of fertilizer in the 40 bags is 826 kg, (i) what percentage of the bags are expected to each contain less than 20.00 kg? (ii) Find the 10th percentile (or smallest decile) of weight of fertilizer in a bag.

Sampling and Combination of Variables

b) How many pallets, carrying 40 bags each, will have to be weighed so that there is at least 96% probability that the mean weight of fertilizer in a bag is known within 0.05 kg?

5. A trucking company delivering bags of cement to suppliers has a fleet of trucks whose mean unloaded weight is 6700 kg with a standard deviation of 100 kg. They are each loaded with 800 bags of cement which have a mean weight of 44 kg and a standard deviation of 3 kg. The trucks travel in a convoy of four and pass over a weigh scale en route.
   a) The government limit on loaded truck weight is 42,000 kg. Exceeding this limit by (i) less than 125 kg results in a fine of $200, (ii) between 125 and 200 kg results in a fine of $400 and (iii) over 200 kg yields a fine of $600. What is the expected fine per truck?
   b) In addition to the above, the government charges a special road tax if the mean loaded weight of the trucks in a convoy of four trucks is greater than 42,000 kg. What is the probability that any particular convoy will be charged this tax?

6. A population consists of one each of the four values 1, 3, 4, 6.
   a) Calculate the standard deviation of this population.
   b) A sample of size 2 is taken from this population without replacement. What is the standard error of the mean?
   c) A sample of size 2 is taken from this population with replacement. What is the standard error of the mean?
   d) If a population now consists of 1000 each of the same four values, what now will be the standard error of the mean (i) without replacement and (ii) with replacement?

7. A population consists of one each of the four numbers 2, 3, 4, 7.
   a) Calculate the mean and standard deviation of this population.
   b) (i) List all the possible and equally likely samples of size 2 drawn from this population without replacement. Calculate the sample means. (ii) Find the mean and standard deviation of the sample means from (i). (iii) Use mean and standard deviation for the original population to calculate the mean of the sample means and the standard error of the mean. Compare with the results of (ii).
   c) Repeat part (b) (i) to (iii), but now for samples drawn with replacement.

8. a) A random sample of size 2 is drawn without replacement from the population that consists of one each of the four numbers 5, 6, 7, and 8. (i) Calculate the mean and standard deviation for the population. (ii) List all the possible and equally likely random samples of size 2 and calculate their means. (iii) Calculate the mean and standard deviation of these samples and compare to the expected values.
   b) Consider now the same population but sample with replacement. (i) List all the possible random samples of size 2 and calculate their means. (ii) Calcu-

Chapter 8

late the mean and standard deviation of these sample means and compare to the results obtained by use of the theoretical equations.

9. The resistances of four electrical specimens were found to be 12, 15, 17 and 20 ohms. List all possible samples of size two drawn from this population (a) with replacement and (b) without replacement. In each case find: (i) the population mean, (ii) the population standard deviation, (iii) the mean of the sample means (i.e., of the sampling distribution of the mean), (iv) the standard deviation of the sample means (i.e., the standard error of the mean) Show how to obtain (iii) and (iv) from (i) and (ii) using the appropriate relationships.

10. A population consists of one of each of the four numbers 3, 7, 11, 15.
    a) Find the mean and standard deviation of the population.
    b) Consider all possible samples of size 2 which can be drawn without replacement from this population. Find the mean of each possible sample. Find the mean and the standard deviation of the sampling distribution of means. Compare the standard error of the means with an appropriate equation from this book.
    c) Repeat (b), except that. the samples are chosen with replacement.

11. Steel plates are to rest in corresponding grooves. The mean thickness of the plates is 2.100 mm, and the mean width of the grooves is 2.200 mm. The standard deviation of plate thicknesses is 0.024 mm, and the standard deviation of groove widths is 0.028 mm. We find that unless the clearance (difference between groove width and plate thickness) for a particular pair is at least 0.040 mm, there is risk of binding. Assume that both plate thicknesses and groove widths are normally distributed. If plates and grooves are matched randomly, what percentage of pairs will have clearances less than 0.04 mm?

12. Rods are taken from a bin in which the mean diameter is 8.30 mm and the standard deviation is 0.40 mm. Bearings are taken from another bin in which the mean diameter is 9.70 mm and the standard deviation is 0.35 mm. A rod and a bearing are both chosen at random. If diameters of both are normally distributed, what is the probability that the rod will fit inside the bearing with at least 0.10 mm clearance?

13. A coffee dispensing machine is supposed to dispense a mean of 7.00 fluid ounces of coffee per cup with standard deviation 0.25 fluid ounces. The distribution approximates a normal distribution. What is the probability that, when 12 cups are dispensed, their mean volume is more than 7.15 fluid ounces? Explain why the normal distribution can be assumed in this calculation.

14. According to a manufacturer, a five-liter can of his paint will cover 60 square meters on average (when his instructions are followed), and the standard deviation for coverage by one can will be 3.10 m². A painting contractor buys 40 cans and finds that the average coverage for these 40 cans is only 58.8 m².
    a) No information is available on the distribution of coverage by a can of paint.

Sampling and Combination of Variables

What rule or theorem indicates that the normal distribution can be used for the mean coverage by 40 cans? What numerical criterion is satisfied?

b) What is the probability that the sample mean will be this small or smaller when the true population mean is 60.0 m^2—that is, if the manufacturer's claim is true?

15. The amount of copper in an ore is estimated by analyzing a sample made up of n specimens. Previous experience indicates that this gives no systematic error and the standard deviation of analysis on individual specimens is 2.00 grams. How many measurements must be made to reduce the standard error of the sample mean to no more than 0.6 grams?

16. The resistance of a group of specimens was found to be 12, 15, 17 and 20 ohms. Consider all possible samples of size two drawn from this group:
    a) with replacement
    b) without replacement.

    In each case find:
    (i) the population mean
    (ii) the population standard deviation
    (iii) the mean of the sample means (i.e., of the sampling distribution of the mean)
    (iv) the standard deviation of the sample means (i.e., the standard error of the mean)

    Show how to obtain (iii) and (iv) from (i) and (ii) using the appropriate formulas.

CHAPTER 9

Statistical Inferences for the Mean

*This chapter requires a good knowledge of Chapter 7.
Some parts require a knowledge of sections 8.3 and 8.4.*

We have already seen that samples can be used to infer some information about the population from which the sample came, specifically the mean and variance of the population. Now we intend to infer further information, which should be quantitative. This chapter will be concerned with statistical inferences for the mean, and later chapters will be concerned with statistical inferences for other statistical quantities. The ideas, approaches and nomenclature concerning statistical inference developed in this chapter will mostly be applicable to the other quantities.

There are two main questions for which statistical inference may provide answers in this and later chapters. Suppose we have collected a representative sample that gives some information concerning a mean or other statistical quantity. One question on the basis of this information would be: are the sample quantity and a corresponding population quantity close enough together so that it is reasonable to say that the sample might have come from the population? Or are they far enough apart so that they likely represent different populations? We should make the answer to those questions as quantitative as we can. Another question might be: for what interval of values can we have a specific level of confidence that the interval contains the true value of the parameter of interest? We will find that the answers to these two questions are related to one another.

Furthermore, we can divide statistical inferences for the mean into two main categories. In one category, we already know the variance or standard deviation of the population, usually from previous measurements, and the normal distribution can be used for calculations. In the other category, we find an estimate of the variance or standard deviation of the population from the sample itself. The normal distribution is assumed to apply to the underlying population, but another distribution related to it will usually be required for calculations. We will start with the first category because it is simpler and allows us to develop the main ideas needed. The second category is found more often in practice.

Statistical Inferences for the Mean

9.1 Inferences for the Mean when Variance Is Known

We may have some previous data giving the variance or standard deviation of the population, and it may be reasonable to assume that the previous value of the variance still applies. In that case, how can we make quantitative inferences for the mean?

9.1.1 Test of Hypothesis

Here we are testing the *hypothesis* that a sample is similar enough to a particular population so that it might have come from that population. In that case, if the hypothesis is true, all disagreement between sample and population is due to random variation, and we say that the sample is consistent with the population. But is that hypothesis reasonable or plausible? Specifically, we make the *null hypothesis* that the sample came from a population having the stated value of the population characteristic, which is the mean in this case. Then we do calculations to see how reasonable such a hypothesis is. We have to keep in mind the alternative if the null hypothesis is not true, as the alternative will affect the calculations.

Illustration: Say the percentage metal in the tailings stream from a flotation mill in the metallurgical industry has been found to follow a normal distribution. When the mill is operating normally, the mean percentage metal in the stream is 0.370 and the standard deviation is 0.015. These are assumed to be population values, μ and σ. Now a plant operator takes a single specimen as a sample and finds a percentage metal of 0.410. Does this indicate that something in the process has changed, or is it still reasonable to say that the mill is operating normally? To put that question a little differently, is it plausible to say that this sample value or a more extreme one might occur *by chance* while the population mean percentage metal is still 0.370?

Our null hypothesis is that nothing has changed, so the population mean is still 0.370. What is the alternative hypothesis? Are we concerned with possible changes in both directions, positive and negative, or are changes in only one direction important? In the situation of measurements of percentage metal in a waste stream, we would likely be concerned with deviations in both directions. Then the null hypothesis would be questioned in relation to a value as far away from 0.370 as 0.410 or farther in either direction. This is often called a *two-sided or two-tailed test*. In that case the specific null hypothesis is that $\mu = 0.370$, and the specific *alternative hypothesis* is that $\mu \neq 0.370$. In other cases we may be interested in changes in only one direction, so a different alternative hypothesis would apply. We use the symbols H_0 for the null hypothesis and H_a for the alternative hypothesis. Notice that the null hypothesis and alternative hypothesis must always be stated in terms of population values, such as the population mean, μ.

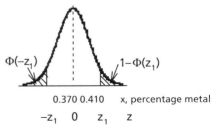

Figure 9.1: Test of Hypothesis

Chapter 9

Now we calculate the probability of getting a sample value this far from the population mean or farther using the normal distribution and assuming that the null hypothesis is true. For a two-sided test, deviations in both directions have to be considered. The situation now is shown in Figure 9.1. (It is always a good idea to sketch a simple diagram like Figure 9.1 in this type of problem. We should mark on it as much information as we have available at this point in the solution.) Because the test is two-sided, we need to calculate the probability corresponding to both tails.

The test statistic is $z = \dfrac{x - 0.370}{0.015}$, the test distribution is normal, and large values of $|z|$ will give evidence against the null hypothesis. z_1 will be the value of z corresponding to the sample observation, $x = 0.410$.

Now we are ready for calculations using the sample observation.

We have $z_1 = \dfrac{0.410 - 0.370}{0.015} = 2.67$.

From Table A1 or the function NORMSDIST on Excel,

$\Phi(2.67) = 0.9962$, so

$1 - \Phi(2.67) = 1 - 0.9962 = 0.0038$,

and $\Phi(-2.67) = 0.0038$.

Then assuming that the null hypothesis is true, the probability of a sample value this far away from the population mean or farther by chance in either direction is 0.0038 + 0.0038 = 0.0076 or 0.8%.

Is it reasonable to think that this is the one time in about 130 that a result this far away from the mean or farther would occur by chance? It might be, but more likely the population mean has changed, contrary to the null hypothesis. Then the conclusion is to reject the null hypothesis.

The observed level of significance or *p-value* is the probability of obtaining a result as far away from the expected value as the observation is, or farther, purely by chance, when the null hypothesis is *true*. That would be 0.8% in the numerical illustration. Notice that a smaller observed level of significance indicates that the null hypothesis is less likely. If this observed level of significance is small enough, we conclude that the null hypothesis is not plausible.

The procedure for tests of significance can be summarized as follows:
1. State the null hypothesis in terms of a population parameter, such as μ.
2. State the alternative hypothesis in terms of the same population parameter.
3. State the test statistic, substituting quantities given by the null hypothesis but not the observed values. What values of the test statistic will indicate that the difference may be significant? State what statistical distribution is being used.

Statistical Inferences for the Mean

4. Show calculations assuming that the null hypothesis is true.
5. Report the observed level of significance, or else compare the value of the test statistic with a critical value as discussed below.
6. State a conclusion. That might be either to accept the null hypothesis, or else to reject the null hypothesis in favor of the alternative hypothesis. If the evidence is not strong enough to reject the null hypothesis, it is tentatively accepted, but that might be changed by further evidence. By statistical analysis we cannot *prove* that the null hypothesis is correct. Instead of saying that the null hypothesis is accepted, it is often better to say just that the null hypothesis is not rejected.

In many instances we choose a critical level of significance before observations are made. The most common choices for the critical level of significance are 10%, 5%, and 1%. If the observed level of significance is smaller than a particular critical level of significance, we say that the result is statistically significant at that level of significance. If the observed level of significance is not smaller than the critical level of significance, we say that the result is not statistically significant at that level of significance.

Example 9.1

It is very important that a certain solution in a chemical process have a pH of 8.30. The method used gives measurements which are approximately normally distributed about the actual pH of the solution with a known standard deviation of 0.020. We decide to use 5% as the critical level of significance.

a) Suppose a single determination shows pH of 8.32. The null hypothesis is that the true pH is 8.30 (H_0: pH = 8.30). The alternative hypothesis is H_a: pH ≠ 8.30 (this is a two-sided test because there is no indication that changes in only one direction are important).

The test statistic is $z = \dfrac{\text{pH} - 8.30}{0.020}$, and large values of $|z|$ will make the null hypothesis implausible. The normal distribution applies. See Figure 9.2.

$$z_1 = \frac{8.32 - 8.30}{0.020} = 1.00$$

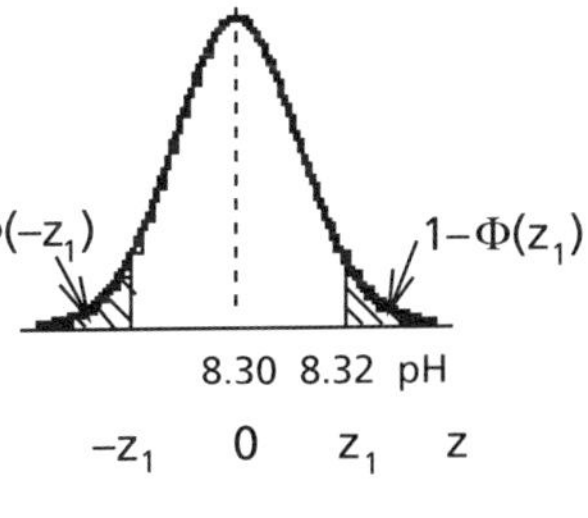

$\Phi(z_1) = 0.8413$ (from Table A1 or the function NORMSDIST on Excel), so $1 - \Phi(z_1) = 0.1587$.

Figure 9.2: Test of Hypothesis

$\Phi(-z_1) = 0.1587$ (from same sources).

Then the observed level of significance is $(2)(0.1587) = 0.317$ or 31.7%.

Since this is larger than 5%, we do not reject the null hypothesis. We do not have enough evidence from this calculation to say that the pH is not equal to

8.32. We could say that the difference from a pH of 8.30 is not statistically significant at the 5% level of significance.

b) Suppose that now our sample consists of 4 determinations giving values of 8.31, 8.34, 8.32, 8.31. The sample mean is $\bar{x}_1 = 8.32$.

The null hypothesis is H_0: $\mu = 8.30$.

The alternative hypothesis is H_a: $\mu \neq 8.30$ (still a two-sided test).

The test statistic is $z = \dfrac{\bar{x} - 8.30}{\left(0.020/\sqrt{4}\right)}$.

The normal distribution applies.

The diagram is the same as before (Figure 9.2) except that pH is replaced by $\bar{x}$.

Now $z_1 = \dfrac{8.32 - 8.30}{0.010} = 2.00$.

$\Phi(z_1) = 0.9772$ (from Table A1 or the function NORMSDIST on Excel),

$1 - \Phi(z_1) = 0.0228$. $\Phi(-z_1) = 0.0228$ (from the same sources).

Then the observed level of significance is $(2)(0.0228) = 0.046$ or 4.6%.

Since this is (just) less than 5%, we reject the null hypothesis and accept the alternative hypothesis that $\mu \neq 8.30$. At the 5% level of significance we conclude that the true mean pH is no longer 8.30.

We might want to examine the rejection region for this problem. See Figure 9.3.

For samples of size 4 the rejection region at 5% critical level of significance is the union of $\left[\bar{x} > 8.30 + z_2\left(\dfrac{0.020}{\sqrt{4}}\right)\right]$ and $\left[\bar{x} < 8.30 - z_2\left(\dfrac{0.020}{\sqrt{4}}\right)\right]$,

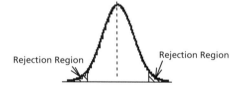

Figure 9.3: Rejection Region

where $1 - F(z_2) = 0.025$, so $F(z_2) = 0.975$, and $F(-z_2) = 0.025$. The value of z_2 can be found from Table A1 or from the Excel function NORMSINV to be 1.96. Since $z_1 = 2.00$, just larger than 1.96, again the conclusion would be to reject the null hypothesis.

A result this close to the boundary between the rejection region and the acceptance region would likely result in further sampling in practice.

A comparison of parts (a) and (b) of Example 9.1 shows the effect of sample size.

Example 9.2

The strength of steel wire made by an existing process is normally distributed with a mean of 1250 and a standard deviation of 150. A batch of wire is made by a new process, and a random sample consisting of 25 measurements gives an average

Statistical Inferences for the Mean

strength of 1312. Assume that the standard deviation does not change. Is there evidence at the 1% level of significance that the new process gives a larger mean strength than the old?

Answer:

$H_0: \mu = 1250$

$H_a: \mu > 1250$ (a *one-tailed* test because the question asks about a larger mean strength)

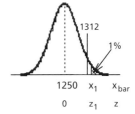

Figure 9.4: One-tailed Test

The test statistic is $z = \dfrac{\bar{x} - 1250}{(150/\sqrt{25})}$, and large values of z will indicate that H_0 is not plausible. The critical value of z for a one-tailed test for 1% level of significance corresponds to $1 - F(z_1) = 0.01$, so $F(z_1) = 1 - 0.01 = 0.99$.

From Table A1 or NORMSINV this corresponds to $z_1 = 2.33$.

Then the critical value of x is

$$x_1 = 1250 + (2.33)\left(\frac{150}{\sqrt{25}}\right) = 1320.$$

The sample mean is 1312. These quantities are shown in Figure 9.4.

Since 1312 < 1320, the result is not in the rejection region; we have insufficient evidence to reject the null hypothesis. There is not enough evidence at the 1% level of significance to say that the new process gives a larger mean strength than the old. We may want to obtain more evidence by a larger sample. But for now, we can say just that the increase in mean strength is not statistically significant at the 1% level of significance.

We may decide to take some action on the basis of the test of significance, such as adjusting the process if a result is statistically significant. But we can never be completely certain we are taking the right action. There are two types of possible error which we must consider.

A *Type I error* is to reject the null hypothesis when it is true. In the case of a mean, this occurs when the null hypothesis is correct, but an observation or sample mean is so far from the expected mean by chance that the null hypothesis is rejected. The probability of a Type I error is equal to the level of significance.

A *Type II error* is to accept the null hypothesis when it is false. If we are applying a test of significance to a mean, the null hypothesis would usually be that the population mean has not changed. If in fact the population mean has changed, the null hypothesis is false. But the sample mean might still *by chance* come close enough to the original sample mean so that we would accept the null hypothesis, giving a Type II error. This is more likely to occur if the population mean has changed only a little. Thus, the probability of a Type II error depends on how much

Chapter 9

the population mean has changed in comparison to the standard error of the mean. How much change do we want to be fairly certain of detecting? We should take this into account when we choose the critical level of significance.

If the critical level of significance is made smaller, the probability of a Type I error becomes smaller, but the probability of a Type II error becomes larger. To make the probability of a Type II error smaller, we should choose a larger value for the critical level of significance. Rational choice of a critical level of significance then depends on balancing the two types of error. Notice that we may be able to reduce both errors either by decreasing the variance of the underlying system (i.e., making the measurements more reproducible) or by increasing the sample size. Further discussion of choosing a critical level of significance can be found in various reference books, such as the book by Vardeman or the one by Walpole and Myers. See section 15.2 for references.

We must distinguish clearly between statistical significance, as shown by a test of hypothesis, and practical significance, which is determined by an economic analysis. An alternative may give a result which is significantly better than the previous choice *statistically*, but the difference may be too small to be worthwhile economically. For example, say a mechanical device gives a small improvement in an automobile's gasoline mileage. That improvement may be statistically significant, but it may not be enough to justify its cost economically.

Problems

1. When a manufacturing process is operating properly, the mean length of a certain part is known to be 6.175 inches, and lengths are normally distributed. The standard deviation of this length is 0.0080 inches. If a sample consisting of 6 items taken from current production has a mean length of 6.168 inches, is there evidence at the 5% level of significance that some adjustment of the process is required?

2. A taxi company has been using Brand A tires, and the distribution of kilometers to wear-out has been found to be approximately normal with $\mu = 114{,}000$ and $\sigma = 11{,}600$. Now it tries 12 tires of Brand B and finds a sample mean of $\bar{x} = 117{,}200$. Test at the 5% level of significance to see whether there is a significant difference (positive or negative) in kilometers to wear-out between Brand A and Brand B. Assume the standard deviation is unchanged. Show all steps of the procedure described before Example 9.1.

3. The average daily amount of scrap from a particular manufacturing process is 25.5 kg with a standard deviation of 1.6 kg. A modification of the process is tried in an attempt to reduce this amount. During a 10-day trial period, the kilograms of scrap produced each day were: 25.0, 21.9, 23.5, 25.2, 22.0, 23.0, 24.5, 25.0, 26.1, 22.8. From the nature of the modification, no change in day-to-day variability of the amount of scrap will result. The normal distribution will apply. A

Statistical Inferences for the Mean

first glance at the figures suggests that the modification is effective in reducing the scrap level. Does a significance test confirm this at the 1% level?

4. The standard deviation of a particular dimension on a machine part is known to be 0.0053 inches. Four parts coming off the production line are measured, giving readings of 2.747, 2.740, 2.750 and 2.749 inches. The population mean is supposed to be 2.740 inches. The normal distribution applies.
   a) Is the sample mean significantly larger than 2.740 inches at the 1% level of significance?
   b) What is the probability of a Type II error (i.e., of accepting the null hypothesis of part (a) when in fact the true mean is 2.752 inches)? Assume the standard deviation remains unchanged.

5. The outlet stream of a continuous chemical reactor is sampled every thirty minutes and titrated. Extensive records of normal operation show the concentration of component A in this stream is approximately normally distributed with mean 41.2 g/L and standard deviation 0.90 g/L.
   a) What is the probability that the concentration of component A in this stream will be more than 42.3 g/L?
   b) Five determinations of concentration of component A are made. If the mean of these five concentrations is more than 42.3 g/L, action is taken. What is the level of significance associated with this test?
   c) State the null hypothesis and the alternative hypothesis that fit the test described in part (b).
   d) The test in part (b) is applied. Now suppose the true mean has changed to 43.5 g/L with no change in standard deviation. What is the probability of a Type II error?

6. A manufacturer produces a special alloy steel with an average tensile strength of 25,800 psi. The standard deviation of the tensile strength is 300 psi. Strengths are approximately normally distributed. A change in the composition of the alloy is tried in an attempt to increase its strength. A sample consisting of eight specimens of the new composition is tested. Unless an *increase* in the strength is significant at the 1% level, the manufacturer will return to the old composition. Standard deviation is not affected.
   a) If the mean strength of the sample of eight items is 26,100 psi, should the manufacturer continue with the new composition?
   b) What is the minimum mean strength that will justify continuing with the new composition?
   c) How large would the true mean strength of the new composition (i.e., a new population mean) have to be to make the odds 9 to 1 in favor of obtaining a sample mean at least as big as the one specified in part (b)?

7. Noise levels in the cabs of a large number of new farm tractors were measured ten years ago and were found to vary about a mean value of 76.5 decibels (db)

with a variance of 72.43 db². A researcher conducted a survey of this year's new tractors to determine whether or not tractor cab manufacturers have been successful in developing quieter cabs. In her final report, the researcher stated that the mean noise level in the cabs she studied was 74.5 db, and she concluded that there was only 12% probability of getting results at least this far different if there was no real reduction in noise level. Calculate the number of cabs that the researcher must have surveyed in order to have drawn this conclusion.

8. Jack Spratt is in charge of quality control of the concrete poured during the construction of a certain building. He has specimens of concrete tested to determine whether the concrete strength is within the specifications; these call for a mean concrete strength of no less than 30 MPa. It is known that the strength of such specimens of concrete will have a standard deviation of 3.8 MPa and that the normal distribution will apply. Mr. Spratt is authorized to order the removal of concrete which does not meet specifications. Since the general contractor is a burly sort, Mr. Spratt would like to avoid removing the concrete when the action is not justified. Therefore, the probability of rejecting the concrete when it actually meets the specification should be no more than 1%. What size sample should Mr. Spratt use if a sample mean 10% less than the specified mean strength will cause rejection of the concrete pour? State the null hypothesis and alternative hypothesis.

9. A scale for weighing bags of product either weighs correctly or slips out of adjustment so that it reads high by a constant 5 kg. The scale is used to weigh samples of 20 bags of product. The bags are intended to have a mean weight of 35 kg each, and the population standard deviation remains constant at 6 kg. The bagging machine is checked when the scale indicates that the mean weight of the bags is significantly higher than expected at the 5% level of significance.
   a) What is the maximum sample mean that will not trigger a bagging machine check?
   b) Using the value from part (a), what is the probability that the bagging machine will not be checked when it has slipped out of adjustment?
   c) At what cutoff value and level of significance will the probability of an undetected slippage equal the probability of an unnecessary machine check?

10. A manufacturer of fluorescent lamps claims (1) that his lamps have an average luminous flux of 3,600 lm at rated voltage and frequency and (2) that 90% of all lamps produced by an automatic process have a luminous flux higher than 3,300 lm. The luminous flux of the lamps follows a normal distribution. What standard deviation is implied by the manufacturer's claim? Assume that this standard deviation does not change. A random sample of 10 lamps is tested and gives a sample mean of 3,470 lm. At the 5% level of significance can we conclude that the mean luminous flux is significantly less than what the manufacturer claims? State your null hypothesis and alternative hypothesis.

Statistical Inferences for the Mean

9.1.2 Confidence Interval

We saw in the previous section that when the normal distribution applies, the rejection region for a sample mean at the 5% level of significance in two tails is the union of $z < -1.96$ and $z > +1.96$. In the rejection region sample means are far enough away from the assumed population mean that only 5% of sample means would fall there by chance.

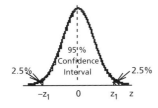

Figure 9.5: Confidence Interval

We can look at those same numbers from another point of view. If the population mean is μ and the normal distribution applies, the *probability* that a random sample mean will fall by chance in the region between $z_1 = -1.96$ and $z_1 = +1.96$ is $100\% - 5\% = 95\%$. Therefore, we can have 95% confidence that a random sample mean will fall in that interval. That is shown in Figure 9.5.

This is called the *95% confidence interval*. The *level of confidence* for the interval is 95%. Similarly, we might find a 98% confidence interval or some other interval for a stated level of confidence.

Example 9.3

Data taken over a long period of time have established that the standard deviation of percentage iron in an iron analysis is 0.12, and that is not expected to change. A representative, well-mixed ore sample is analyzed six times, so the sample size is 6. If the true iron content is 32.60 percent iron, if there is no systematic error, and if the normal distribution applies, what is the 95% confidence interval for sample means?

Answer: For the 95% confidence level, $z_1 = 1.96$.

Then the confidence interval for *sample means* is from

$$32.60 - 1.96\left(\frac{0.12}{\sqrt{6}}\right) \text{ to } 32.60 + 1.96\left(\frac{0.12}{\sqrt{6}}\right),$$

or 32.50 to 32.70 percent iron.

But the problem which we face in practice is usually not to find a confidence interval for sample means. Much more frequently we need to find a confidence interval for the *population mean*. The sample mean is known from measurements, and the population mean is the uncertain value for which we need an estimate. We already know that the sample mean gives a point estimate for the population mean, but now we need an *interval* estimate. That interval should correspond to a stated level of confidence that the interval contains the true population mean if the assumptions are satisfied. The assumptions are that the normal distribution applies, there is no systematic error, the sample is random and can therefore be considered representative, and the standard deviation of the population is known. Then the known sample

Chapter 9

mean is at the center of the confidence interval for the population mean. The sketch shown in Figure 9.5 still applies.

Example 9.4

As in Example 9.3, the standard deviation of percentage iron in analyses is 0.12, the sample size is 6, and the normal distribution applies with no systematic error. The sample mean is determined from measurements to be 32.56 percent iron. Find the 95% confidence interval for the true mean iron content of the population from which the sample came.

Answer: For 95% confidence level the interval is still from $z = -1.96$ to $z = +1.96$. $\bar{x} = 32.56$. Then the interval estimate for the population mean with 95% confidence is from $32.56 - 1.96\left(\dfrac{0.12}{\sqrt{6}}\right)$ to $32.56 + 1.96\left(\dfrac{0.12}{\sqrt{6}}\right)$, or from 32.46 to 32.66 percent iron. See Figure 9.6. This sort of result is often shown as 32.56 ± 0.10, or within ± 0.10 of the sample mean.

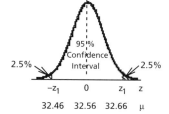

**Figure 9.6:
95% Confidence Interval**

Example 9.5

A large population is normally distributed with a standard deviation of 0.12. A random sample will be taken from this population and the sample mean will be calculated. We require at least 98% confidence that the true population mean will be within ± 0.05 of the sample mean, assuming there is no systematic error. What sample size is required?

Answer:

For 98% confidence $1 - \Phi(z_1) = 0.01$ and $\Phi(-z_1) = 0.01$. From Table A1 or from Excel function NORMSINV(Φ) we find that $z_1 = 2.33$. We have also that $z_1 = \dfrac{\bar{x}_1 - \mu}{\sigma_{\bar{x}}}$ and $\sigma_{\bar{x}} = \dfrac{\sigma}{\sqrt{n}}$, so $z_1 = \dfrac{\bar{x}_1 - \mu}{\left(\sigma/\sqrt{n}\right)}$.

Substituting for $\bar{x}_1 - \mu = 0.05$ (which will also give $\mu - \bar{x}_2 = 0.05$ with $-z_1 = \dfrac{\bar{x}_2 - \mu}{\sigma_{\bar{x}}}$), and substituting $\sigma = 0.12$ and $z_1 = 2.33$, we obtain

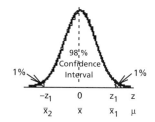

**Figure 9.7:
Confidence Interval**

Statistical Inferences for the Mean

$$2.33 = \frac{0.05}{\left(\dfrac{0.12}{\sqrt{n}}\right)}$$

$$\frac{0.12}{\sqrt{n}} = \frac{0.05}{2.33}$$

$$\sqrt{n} = \frac{(0.12)(2.33)}{0.05} = 5.59$$

and $\quad n = (5.59)^2 = 31.3$.

The sample size must be at least this large, and it must be an integer. The minimum sample size to give at least 98% confidence that the population mean will be within ±0.05 of the sample mean therefore is 32. The 98% confidence interval will then be $\bar{x} \pm 0.05$.

Remember that calculations for both test of hypothesis and confidence limits by the methods which have been discussed so far have three requirements:

1. The sample must be random and representative.
2. The distribution of the variable must be a normal distribution, at least to a good approximation. However, the Central Limit Theorem is helpful here. If the sample contains enough observations, the sample mean will be normally distributed (to whatever approximation is required) even though the original observations were not.
3. The standard deviation of the observations must be known reliably, probably from previous information.

These requirements are often not completely met. For example, probability distributions in the tails, far from the population mean, are often not exactly according to the normal distribution. Therefore, inferences may not be quite as reliable as they seem: a "98%" confidence interval may actually be a 96% confidence interval, and so on.

The next section will consider a calculation where the standard deviation of the observations is not known reliably before the experiment, so it must be estimated from a sample of moderate size.

Problems

1. A cocoa packaging machine fills bags so that the bag contents have a standard deviation of 3.5 g. Weights of contents of bags are normally distributed.
    a) If a random sample of 20 bags gives a mean of 102.0 g, what are the 99% confidence limits for the mean weight of the population (i.e., all bags)?

Chapter 9

b) What sample size (how many bags) would have to be taken so that a person would be 95% confident that the population mean was not smaller than the sample mean minus 1 g?

2. The diameters of shafts made by a particular process are approximately normally distributed with a standard deviation of 0.0120 cm. When all settings are correct, the mean diameter is 3.200 cm.
   a) If the settings are correct and random samples contain six specimens each, what proportion of the sample means will be smaller than 3.190 cm?
   b) How large should the sample be to give 98% confidence that the sample mean is within ±0.0080 cm of the true mean of all diameters of shafts produced under current conditions?

3. Carbon composition resistors with mean resistance 560 Ω and coefficient of variation 10% are produced by a factory. They are sampled each hour in the quality control lab. What sample size would be required so that there is 95% probability that the mean resistance of the sample lies within 10 Ω of 560 Ω if the population mean has not changed?

4. We want to estimate the mean distance traveled to work by employees of a large manufacturing firm. Past studies indicate that the standard deviation of these distances is 2.0 km and that the distances follow a normal distribution. How many employees should be chosen at random and polled if the estimated mean distance is to be within 0.1 km of the true mean with a confidence level of 95%?

5. A batch processor dispenses a mean volume of approximately 0.80 m^3 of grain with a standard deviation of 0.05 m^3. The volumes of batches are normally distributed. A test engineer wishes to check the calibration of the processor.
   a) How many batches would have to be measured for the engineer to be 90% confident that the mean volume from the sample is between 0.99 and 1.01 times the true population mean?
   b) If a sample of 50 batches of grain is measured, with what confidence can the claim be made that the sample mean volume is within 1% of the true mean volume of the population, if the true mean volume is approximately 0.80 m^3?
   c) If a sample of 200 batches of grain is measured, the engineer can be 90% confident that the sample mean is within what percentage of the true population mean? Again assume that the population mean is approximately 0.80 m^3.

6. Company A produces tires. The mean distance to wearout of these tires is 108,000 km, and the standard deviation of the wearout distances is 15,000 km.
   a) A distributor who is about to buy those tires wishes to test a random sample of them. What number of tires would have to be tested so there is 98% probability that the mean wearout distance for the sample is within 5% of the population mean?

Statistical Inferences for the Mean

b) For sample sizes of 4, 8, 16, 32, 64 and 128, calculate the probability with which the distributor can claim that the mean wearout distance for the sample is within 5% of the population mean.

c) If the manufacturer's claim is correct, how many tires from a shipment of 100 tires are expected to wear out in less than 120,000 km?

7. Carbon resistors of mean resistance approximately 560 ohms are produced in a certain factory. The standard deviation is 28 ohms, and resistances are normally distributed.

a) What confidence level is associated with a single resistor falling within ±10 ohms of the population mean?

b) How large a random sample is required to give 95% level of confidence that the sample mean is within ±10 ohms of the population mean?

8. a) The diameter of a certain shaft is normally distributed with a mean expected to be 2.79 cm and a standard deviation of 0.01 cm. The specification limits are 2.77 ± 0.03. If 1000 shafts are produced, how many can we expect will be unacceptable? If the sample mean for 1000 shafts is 2.786 cm, what are the 99% confidence limits for the true mean diameter?

b) What is the probability that a single diameter measurement will deviate from the true mean by at least ±0.02 cm?

c) Estimate the required size of the random sample in future measurements so that the 95% confidence interval for the population mean will not be wider than from the sample mean less 0.01 cm, to the sample mean plus 0.01 cm.

9. A small plant bags a blend of three types of fertilizer to meet the needs of a particular group of farmers. The different types of fertilizer are fed through three different machines. Each bag is supposed to contain 18.00 kg from machine 1, 7.00 kg from machine 2, and 5.00 kg from machine 3. It is found that the actual amounts are normally distributed about these means with the following standard deviations:

Machine	Standard Deviation
1	0.19 kg
2	0.07 kg
3	0.04 kg

a) What are the variance and coefficient of variation of the amount of fertilizer in a bag?

b) What percentage of the bags contain less than 29.50 kg?

c) How many bags must be sampled to establish with 99% confidence that the true mean is within ±0.5% of the sample mean, regardless of what the population mean is?

10. Portland cement is packed in bags of nominal weight 80 pounds. The actual mean weight of a bag is found to be 80.2 pounds with a coefficient of variation of 1.2%. A normal distribution applies. Railway flat cars are loaded with enough

bags to make up a *nominal* load of 60 tons on each car. A train is made up of fifty flat cars.
  a) What is the mean weight of a car load?
  b) What is the standard deviation of the weight of a car load?
  c) What are the 95% confidence limits for the weight of a train load of Portland cement?

11. The Soapy Suds Corporation owns a machine which fills boxes of laundry soap. The mean weight is 51 ounces per box, with a standard deviation of 1.10 ounces.
  a) Why might we expect the distribution of weights of boxes of soap to be approximately normal?
  b) Assuming the distribution is normal, what fraction of the boxes differ from the mean weight by more than 1.50 ounces?
  c) What sample size (number of boxes) must a quality control officer test so that he is 90% confident that the mean of the sample is within 0.500 ounces of the true population mean?

12. Fertilizer is sold in bags. The standard deviation of the content of a bag is 0.43 kg. Weights of fertilizer in bags are normally distributed. 40 bags are piled on a pallet and weighed.
  a) If the net weight of fertilizer in the 40 bags is 826 kg, (i) what percentage of the bags are expected to contain less than 20.00 kg each? (ii) find the 10th percentile (or smallest decile) of weight of fertilizer in a bag.
  b) How many pallets, carrying 40 bags each, will have to be weighed so that there is at least 96% probability that the true mean weight of fertilizer in a bag is known within 0.05 kg?

13. Insulators produced by a factory have a breakdown voltage distribution that can be approximated by a normal distribution. The coefficient of variation is 5%.
  a) What is the smallest sample size that will ensure probability of 90% that the sample mean measured is between 0.98 times the population mean and 1.02 times the population mean?
  b) If the sample size is 40, what is now the confidence level associated with the sample mean lying between 0.98 times the population mean and 1.02 times the population mean?

14. Two products A and B are added together to form a mixture. A carton of the mixed product has a mean weight of 100.0 kg and a standard deviation of 1.2 kg. The mean weight of Product A in each carton is 14.0 kg and the standard deviation is 0.6 kg. Weights of both Product A and Product B follow normal distributions.
  a) What are the mean weight and standard deviation of Product B in each carton?
  b) What is the probability that the weight of Product B in a carton chosen at random will be at least 5% lower than its specified mean weight?
  c) We take a random sample from a consignment of 1,000 cartons for a construction project. How big should the sample be to ensure with 95%

Statistical Inferences for the Mean

confidence that the population mean carton weight is within ±1% of the mean weight of a sample?

15. A coffee machine is adjusted to provide a population mean of 110 ml of coffee per cup and a standard deviation of 5 ml. The volume of coffee per cup is assumed to have a normal distribution. The machine is checked periodically by sampling 12 cups of coffee. If the mean volume, $\bar{x}$, of those 12 cups in ml falls in the interval $(110 - 2\sigma_{\bar{x}}) \leq \bar{x} \leq (110 + 2\sigma_{\bar{x}})$, no adjustment is made. Otherwise, the machine is adjusted.
   a) If a 12-cup test gives a mean volume of 107.0 ml, what should be done?
   b) What fraction of the total number of 12-cup tests would lead to an adjustment being made, even if the machine had not changed from its original correct setting?
   c) How many cups should be sampled randomly so there is 99% confidence that the mean volume of the sample will lie within ±2 ml of 110 ml when the machine is correctly adjusted?

16. Capacitors are manufactured on a production line. It is known that their capacitances have a coefficient of variation of 2.3%.
   a) What is the probability that the capacitance of a capacitor will be between 0.990μ and 1.010μ if μ is the mean capacitance of the population? State any assumption.
   b) We want to make the 99% confidence interval for the sample mean of the capacitances to be no larger than from 0.990μ to 1.010μ, where μ is the population mean. What is the minimum sample size?

17. A company received 200 electrical components that were claimed to have a mean life of 500 hours. Assume the distribution of component lives was normal. A sample of 25 components was selected randomly without replacement. It was decided to give that sample a special test that would allow the component life to be estimated accurately but nondestructively.
   a) What is the maximum value the standard deviation of the population could have if the sample mean was to be within ±10% of the population mean with a 95% confidence level?
   b) If the coefficient of variation of the population was 2% and the sample mean was found to be 487.0 hours, what conclusions can be made about the claims of the manufacturer? Use the 5% level of significance.

18. Insulators produced by a factory have a breakdown voltage distribution that can be approximated by a normal distribution. The coefficient of variation is 5%.
   a) What is the smallest sample size that will ensure a probability of 90% that the sample mean measured is within 2% of the population mean? The samples are taken randomly.
   b) If economic reasons dictate that the sample size should be 40, what is now the confidence level associated with the sample mean lying within 2% of the population mean?

c) What is the probability that an insulator will have a breakdown voltage 10% higher than the mean breakdown voltage?

9.2 Inferences for the Mean when Variance Is Estimated from a Sample

In most cases the variance or standard deviation must be estimated from a sample. Even if we have a reliable figure for variance or standard deviation from previous observations, it is often hard to be certain that the variance hasn't changed. What is the variance now? We can estimate it from the same sample as we use to estimate the mean of the population.

But if variance is estimated from a sample of moderate size, that estimate is also subject to random error related to the size of the sample. The larger the sample, the more reliable the estimate of variance becomes. The quantitative relation is expressed in terms of the *degrees of freedom* of the sample. The degrees of freedom refer to the number of pieces of independent information used to estimate the variance. The sample mean, $\bar{x}$, was calculated from n independent quantities, x_i. But the deviations from the mean, $x_i - \bar{x}$, are not all independent because, as we saw in Chapter 3, $\sum_{i=1}^{n}(x_i - \bar{x}) = 0$. The number of *independent* deviations from the mean is not n but $(n-1)$. Then the number of independent pieces of information on which the variance or standard deviation is based is $(n-1)$. We can check this by considering the case of a sample consisting of only one item, x_1, so n = 1. This gives a rough indication of the mean ($\mu \approx \bar{x} = x_1$), but no information at all about the variance of the population. For this case $s^2 = \dfrac{(x_1 - \bar{x})^2}{(n-1)} = \dfrac{0}{0}$, which is mathematically indeterminate. That agrees with the statement that the estimate of the variance for a sample of n items has $(n-1)$ degrees of freedom because in this case n = 1, so $(n-1)$ is equal to zero.

A sample of size n gives an estimate of variance, $s^2 = \dfrac{\sum_{i=1}^{n}(x_i - \bar{x})^2}{(n-1)}$. In words this estimate is the sum of squares of the deviations from the sample mean, divided by the number of degrees of freedom. In abbreviated form the relation is $s^2 = SS / df$, where SS is the sum of squares of deviations from the sample mean (or the equivalent), and df is the number of degrees of freedom (often represented by the Greek letter nu, ν). We will see in Chapter 14 that variance from a regression line is given by a similar relation, although both the sum of squares of deviations and the number of degrees of freedom are evaluated differently.

If we have only a limited number of degrees of freedom to use, the resulting estimate of variance, s^2, is less reliable than if an infinite number of degrees of freedom were available. This must be taken into account when we make statistical inferences about the mean.

Statistical Inferences for the Mean

This puzzle was considered in the early years of the twentieth century by W.S. Gosset, a chemist working for a brewery in Dublin. He had a practical problem: how to make valid inferences about the contents of beer on the basis of rather small samples. He worked out the mathematical solution to this problem. That gave a new probability distribution, a distribution *related* to the normal distribution but taking into account the number of degrees of freedom. He called this new distribution the *t-distribution*. He realized that other people would find the *t*-distribution useful, so he wanted to publish it in a scientific journal. But here was a difficulty of a different kind: the company for which he worked did not allow employees to publish in the open literature. He was sure that publication would not harm the company. He decided to publish using a pen-name, "Student." Thus the *t*-distribution is often called Student's *t*-distribution, and a test of hypothesis based on it is often called Student's *t*-test.

The independent variable of the normal distribution applied to sample means is $z = \dfrac{\bar{x} - \mu}{\sigma_{\bar{x}}} = \dfrac{\bar{x} - \mu}{\left(\dfrac{\sigma}{\sqrt{n}}\right)}$. If we don't know σ, we estimate it from a sample by the estimated standard deviation, s. Then instead of z we have the variable t, which for this case is equal to $\dfrac{\bar{x} - \mu}{\left(\dfrac{s}{\sqrt{n}}\right)}$. Probability according to the *t*-distribution is then a function of two independent variables, t and the number of degrees of freedom, which in this case is $(n - 1)$.

Figure 9.8 shows the probability density functions of *t*-distributions as functions of *t* for various numbers of degrees of freedom. The general shape is similar to the shape of the normal distribution: symmetrical and roughly bell-shaped. However, smaller numbers of degrees of freedom give lower and wider distributions (the total area under each curve must correspond to a probability of one, so if a distribution is lower at the center it is also wider in the tails). The highest curve is for infinite degrees of freedom and is identical to the normal curve as a function of z. The lowest (and hence widest) curve of Figure 9.8 is for one degree of freedom. That makes sense, because only one degree of freedom would correspond to little reliability.

Tables of the *t*-distribution often give one-tail probabilities, that is $\Pr[t > t_1]$, where t_1 is a critical or limiting value. The corresponding areas are shown in Figure 9.8. For any particular number of degrees of freedom, the one-tail probability is 1 minus the cumulative probability, which is $\Phi(t_1) = \Pr[t < t_1]$. For infinite degrees of freedom the *t*-distribution becomes the normal distribution, so the one-tail probability becomes $1 - \Phi(z_1) = \Phi(-z_1)$.

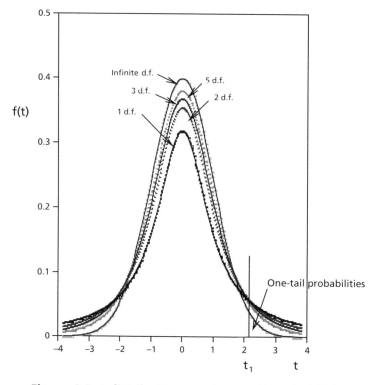

Figure 9.8: *t*-distributions and one-tail probabilities

Figure 9.9 shows the ratio of t (for the *t*-distribution) to z (for the normal distribution) corresponding to various values of the one-tail probability. This ratio shows the effect of limited numbers of degrees of freedom, as compared to infinite degrees of freedom for the normal distribution. Notice that the scales are logarithmic. For one degree of freedom the ratio of t to z is 2.4 for a one-tail probability of 0.10. Still for one degree of freedom (d.f.), that ratio rises to 103 for a one-tail probability of 0.001. Thus, we can see that at small numbers of degrees of freedom, the effect on calculations of estimating the variance from a sample of limited size can be large. On the other hand, the line for 30 degrees of freedom is not far from a ratio of 1. For 30 d.f. the ratio varies from 1.023 for a one-tail probability of 0.10, to 1.095 for a one-tail probability of 0.001. This indicates that for a sample size of 30 or more, the normal distribution is usually (but not always) a reasonable approximation to the *t*-distribution.

Table A2 in Appendix A gives values of t according to the *t*-distribution as functions of the one-tail probability (across the top in bold letters from 0.1 to 0.001) and the number of degrees of freedom, d.f. (along the lefthand and righthand sides in bold letters from 1 to 8). For example, for a one-tail probability of 0.025 and three degrees of freedom, the value of t is 3.182.

Statistical Inferences for the Mean

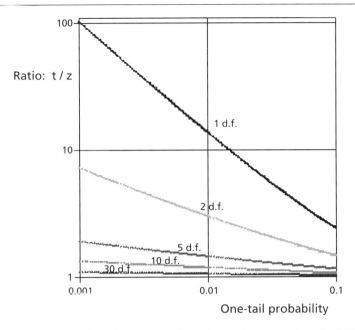

Figure 9.9: Ratio of *t* to *z* as function of one-tail probability

Alternatively, if a person has access to a computer with Excel (or an alternative) or a pocket calculator, values for the *t*-distribution can be obtained using Excel functions or their equivalents. The Excel function TINV gives the value of *t* for the desired *two-tail* probability and number of degrees of freedom. Since the *t*-distribution is symmetrical, the two-tail probability is just twice the one-tail probability. Thus, a one-tail probability of 0.025 corresponds to a two-tail probability of 0.05, and combining this with three degrees of freedom on Excel by entering =TINV(0.05,3) gives a value for *t* of 3.18244929 (to quote all the figures given). The function TDIST gives a one-tail or two-tail probability for a stated value of *t* and a stated number of degrees of freedom. The number of tails is specified by a third parameter, which can be 1 or 2. Thus, entering =TDIST(3.182,3,1) gives 0.02500857, and entering =TDIST(3.182,3,2) gives 0.05001714. Excel has some other related functions, but they are not needed during the learning process.

Various statistical inferences can be made using the *t*-distribution. They are not usually sensitive to small deviations of the underlying distribution from the normal distribution because of the Central Limit Theorem and because the larger value of *t* for small numbers of degrees of freedom reduces the effects of small deviations. In general, *if the variance is estimated from a sample*, statistical inferences should be made using the *t*-distribution rather than the normal distribution. If the number of degrees of freedom is large enough, the normal distribution can be used as an approximation to the *t*-distribution.

Let us now consider the inferences involving the *t*-distribution.

Chapter 9

9.2.1 Confidence Interval Using the t-distribution

Say we have a random sample of size n, from the measurements of which we calculate the sample mean, $\bar{x}$, and the estimate of variance, s^2. Then the estimated standard deviation is s, based on $(n-1)$ degrees of freedom. Because the variance or standard deviation is estimated from a sample, we must generally use the *t*-distribution rather than the normal distribution for calculations. From the number of degrees of freedom and the estimate of the standard deviation, we can calculate *t*, then use tables or computer functions to find a *confidence interval* for the population mean, μ, at a stated level of confidence. Once we find a value of *t*, the calculations are the same as using *z* with the normal distribution.

Example 9.6

A certain dimension is measured on four successive items coming off a production line. This sample gives $\bar{x} = 2.384$ and $s = 0.048$.

(a) On the basis of this sample, what is the 95% confidence interval for the population mean?

(b) If instead of estimating the standard deviation from a sample, we knew the true standard deviation was 0.048, what then would be the 95% confidence interval for the population mean?

Answer:

(a) The 95% confidence interval, two-sided assumed unless otherwise stated, corresponds to a one-tail probability of $(100-95)\%/2 = 2.5\%$. This is shown in Figure 9.10. The number of degrees of freedom is $4 - 1 = 3$. From Table A2 or the function TINV on Excel, the limiting value of *t* is $t_1 = 3.182$. Then, the 95% confidence interval for μ is

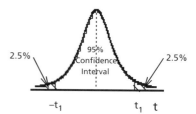

Figure 9.10: 95% Confidence Interval

$$\bar{x} \pm t_1 \left(\frac{s}{\sqrt{n}}\right) = 2.384 \pm (3.182)\left(\frac{0.048}{\sqrt{4}}\right) = 2.31 \text{ to } 2.46.$$

(b) If the standard deviation of the population were known reliably, we would find confidence intervals using the normal distribution. The 95% confidence interval extends from cumulative normal probability of 0.025 (at $z = -z_1$) to cumulative normal probability of 0.975 (at $z = +z_1$). From Table A1 or Excel function NORMSINV we find $z_1 = 1.96$. Then, if the standard deviation were known reliably to be 0.048, the 95% confidence interval for μ would be

$$\bar{x} \pm z_1 \left(\frac{\sigma}{\sqrt{n}}\right) = 2.384 \pm (1.96)\left(\frac{0.048}{\sqrt{4}}\right) = 2.34 \text{ to } 2.43.$$

Then the confidence interval would be appreciably narrower than in part (a).

Statistical Inferences for the Mean

9.2.2 Test of Significance: Comparing a Sample Mean to a Population Mean

For this case also, the calculation is very similar to the corresponding case using the normal distribution. The quantity t is calculated in nearly the same way as z, and then a probability is found from tables or the appropriate computer function taking the number of degrees of freedom into account. The null hypothesis, alternative hypothesis, and test statistic should be stated explicitly. A test of significance using the t-distribution is often called a *t-test*.

Example 9.7

The electrical resistances of components are measured as they are produced. A sample of six items gives a sample mean of 2.62 ohms and a sample standard deviation of 0.121 ohms. At what observed level of significance is this sample mean significantly different from a population mean of 2.80 ohms? Is there less than 2% probability of getting a sample mean this far away from 2.80 ohms or farther purely by chance when the population mean is 2.80 ohms?

Answer: $H_0: \mu = 2.80$

$H_a: \mu \neq 2.80$ (two-sided test)

The test statistic is $t = \dfrac{(\bar{x} - \mu)}{\left(\dfrac{s}{\sqrt{n}}\right)} = \dfrac{(\bar{x} - 2.80)}{\left(\dfrac{0.121}{\sqrt{6}}\right)}$.

Large values of $|t|$ indicate that H_0 is unlikely to be correct.

With $\bar{x} = 2.62$, $t_{observed} = \dfrac{(2.62 - 2.80)}{\left(\dfrac{0.121}{\sqrt{6}}\right)} = \dfrac{-0.18}{0.0494} = -3.64$

See Figure 9.11.

From Table A2 with $6 - 1 = 5$ degrees of freedom, for a one-tail probability of 0.01, $t_1 = 3.365$, and for a one-tail probability of 0.005, $t_1 = 4.032$. The observed value of $|t|$ is between 3.365 and 4.032 (remember that the distribution is symmetrical). Then the one-tail probability is between 0.005 and 0.01, and the sample mean is significantly different from a population mean of 2.80 ohms at a *two-sided* observed level of significance between 0.01 and 0.02.

If a computer with Excel (or some alternatives) is available, the observed level of significance can be found more exactly. The two-tail probability is given by

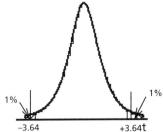

Figure 9.11: Level of Significance

entering =TDIST(3.64,5,2), giving 0.0149. Then the sample mean is significantly different from a population mean of 2.80 ohms at a two-sided observed level of significance of 0.0149 or 0.015.

There *is* less than a 2% probability of getting a sample mean this far from 2.80 ohms or farther purely by chance when the population mean is 2.80 ohms.

9.2.3 Comparison of Sample Means Using Unpaired Samples

In this case we have samples for each of two conditions. The question becomes: are the two sample means significantly different from one another, or could both plausibly come from the same population? We will have an estimate from each sample of the variance or standard deviation of the population, so these two estimates will have to be combined in a logical way. The two random samples will have been chosen separately and independently of one another, so the two sample means will be independent estimates. This test of significance is often called an *unpaired t-test*.

The two estimates of variance must be compatible with one another. This should be checked by the variance ratio test to be introduced in the next chapter. According to Walpole and Myers (see section 15.2 for reference) larger departures from equality of the variances can be tolerated if the two samples are of equal size ($n_1 = n_2$). The samples should be of equal size if that is feasible. In the examples and problems of the rest of this chapter, we assume that the estimates of variance are compatible with one another.

The estimates of variance, s_1^2 and s_2^2, are combined or pooled to give a combined estimate of variance, s_c^2. Say the estimate s_1^2 is based on $(n_1 - 1)$ degrees of freedom, and the estimate s_2^2 is based on $(n_2 - 1)$ degrees of freedom. Remember that the greater the number of degrees of freedom, the more reliable we expect the estimate to be. It can be shown theoretically that the separate estimates of variance should be weighted by their numbers of degrees of freedom before they are averaged. Then

$$s_c^2 = \frac{s_1^2(n_1-1) + s_2^2(n_2-1)}{(n_1-1)+(n_2-1)} \tag{9.1}$$

This is called the combined or pooled estimate of variance.

Since the product of the sample estimate of variance and number of degrees of freedom is the sum of squares of deviations from the sample mean, equation 9.1 can also be shown as the sum of squares of deviations for the first sample, plus the sum of squares of deviations for the second sample, then divided by the sum of the degrees of freedom.

The combined estimate of variance is based on more information than either of the two individual estimates, so it is reasonable that it has more degrees of freedom, $(n_1 - 1) + (n_2 - 1)$. Notice that this is the denominator of equation 9.1.

Using this combined estimate of variance, the estimated variance of the first

Statistical Inferences for the Mean

sample mean is $\dfrac{s_c^2}{n_1}$, and the estimated variance of the second sample mean is $\dfrac{s_c^2}{n_2}$. As we saw in section 8.1, the variance of the difference between two independent quantities is the sum of the variances of the separate quantities. Then

$$s^2_{(\bar{x}_1-\bar{x}_2)} = s_c^2\left(\frac{1}{n_1}+\frac{1}{n_2}\right) \tag{9.2}$$

Another notation is often used, letting $y = \bar{x}_1 - \bar{x}_2$, and then

$$s_y^2 = s_c^2\left(\frac{1}{n_1}+\frac{1}{n_2}\right) \tag{9.2a}$$

The null hypothesis is that both samples could have come from the same population, so $\mu_1 = \mu_2$, or $\mu_1 - \mu_2 = 0$, or $\mu_y = 0$. The alternative hypothesis may be either that $\mu_1 \neq \mu_2$ (a two-tailed test), or that $\mu_1 > \mu_2$ or $\mu_1 < \mu_2$ (a one-tailed test). In the notation for $y = \bar{x}_1 - \bar{x}_2$, the alternative hypothesis would be either $\mu_y \neq 0$ (a two-tailed test), or else $\mu_y > 0$ or $\mu_y < 0$ (a one-tailed test).

Example 9.8

Two methods of determining the nickel content of steel are compared using four determinations by each method. The results are:

For method 1: $\bar{x}_1 = 3.2850$, $s_1 = 0.00774$ (from 3 degrees of freedom)

For method 2: $\bar{x}_2 = 3.2580$, $s_2 = 0.00960$ (from 3 degrees of freedom)

Assuming that the two estimates of variance are compatible, is the difference in means statistically significant at the 5% level of significance?

Answer:

H_0: Both samples could have come from the same population, so $\mu_1 - \mu_2 = 0$
  (or in the other notation, $\mu_y = 0$).

H_a: $\mu_1 - \mu_2 \neq 0$ (or else $\mu_y \neq 0$) (a two-tailed test)

The test statistic is $t = \dfrac{\bar{x}_1 - \bar{x}_2}{s_{(\bar{x}_1-\bar{x}_2)}}$ $\left(\text{or else } t = \dfrac{y}{s_y}\right)$.

Large values of $|t|$ tend to make H_0 unlikely.

$$s_c^2 = \frac{s_1^2(n_1-1)+s_2^2(n_2-1)}{(n_1-1)+(n_2-1)} = \frac{(0.00774)^2(3)+(0.00960)^2(3)}{3+3}$$

$$= 76.03 \times 10^{-6}$$

(This value would correspond to $s_c = 0.00872$, but it is not necessary to make that calculation.)

Chapter 9

$$s^2_{(\bar{x}_1-\bar{x}_2)} = s_c^2 \left(\frac{1}{n_1}+\frac{1}{n_2}\right) = \left(76.03\times 10^{-6}\right)\left(\frac{1}{4}+\frac{1}{4}\right)$$

$$= 38.02 \times 10^{-6}$$

Then $s_{(\bar{x}_1-\bar{x}_2)} = s_y = \sqrt{38.02\times 10^{-6}} = 6.17\times 10^{-3}$, based on 3 + 3 = 6 d.f.

$$t = \frac{3.285-3.258}{6.17\times 10^{-3}} = 4.38$$

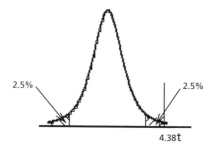

Figure 9.12:
Level of Significance

From Table A2 or TINV from Excel, for 5% level of significance in two tails and 6 degrees of freedom, $t_{\text{critical}} = 2.447$. Alternatively, the observed level of significance or p-value is given from Excel by TDIST(4.38,6,2) = 0.00467.

Since $t > t_{\text{critical}}$ (or since 0.00467 < 5% or even < 0.5%), the difference *is* statistically significant at the 5% level of significance.

This two-sample t-test or unpaired t-test is used very frequently in a planned experiment to see whether a change in the experimental conditions has any statistically significant effect on the product or result of a process. Comparing the two sample means gives a direct comparison between the two sets of conditions. However, we have to be as sure as we can that other significant factors are not affecting the result because they are changing at the same time. We can minimize the effect of *interfering factors* (which are sometimes called "lurking variables") by randomizing the choice of samples for different treatments and the order in which samples are taken and/or analyzed. Randomizing has been discussed briefly in Chapter 8 and will be considered more fully in Chapter 11.

If an interfering factor is known or suspected to be present and has an appreciable effect, the unpaired or two-sample t-test may not be the best plan. An alternative experiment may be a better choice.

Illustration: Suppose we want to compare rates of evaporation from a standard evaporation pan and from a pan using an experimental design. Both types are used to measure rates of evaporation at a weather station. The question we want to answer is, does the new type give *significantly* higher results than the standard type? We know that evaporation will be different on successive days because of changing weather conditions, but different weather conditions should not have an appreciable effect on any difference in rates of evaporation between the two types of pan. We want to focus on a comparison of evaporation rates between the two types, so for present purposes the variation from day to day is not of prime interest. If we use the comparison of

Statistical Inferences for the Mean

sample means by the t-test with unpaired samples, variation of evaporation from day to day will be an *interfering* factor. If we neglect randomizing, the effect of daily variation may be confused with the effect of type of evaporator; then the results of the tests may be quite misleading. If we make random choices of which type of evaporator should be used on a particular day, say by drawing lots, the effect of varying weather conditions will be minimized. But still the variation in evaporation due to varying conditions from day to day will give larger variance within populations for both heat treatments. Then the estimates of variance will very likely be inflated. This will give smaller values of *t*, so it will be difficult to show that one type is significantly better than another. The reader should compare the next two examples, Examples 9.9 and 9.10.

Example 9.9

Daily evaporation rates were measured on 20 successive days. Which of two types of evaporation pan would be used on a particular day was decided by tossing a coin. The mean daily evaporation for the 10 days on which Pan A was used was 19.10 mm, and the mean evaporation on the 10 days on which Pan B was used was 17.24 mm. The variance estimated from the sample from Pan A is 7.72 mm^2, and the variance estimated from the sample from Pan B is 5.36 mm^2. Assuming that these two estimates of variance are compatible, does the experimental evaporation pan, Pan A, give significantly higher evaporation rates than the standard pan, Pan B, at the 1% level of significance?

Answer: $H_0: \mu_A - \mu_B = 0$

$H_a: \mu_A - \mu_B > 0$ (one-tailed test, because the question to be answered is whether the experimental design gives *higher* evaporation rates than the standard pan)

Test statistic: $t = \dfrac{\bar{x}_A - \bar{x}_B}{s_{\text{diff}}}$

Large values of t make H_0 less likely.

$\bar{x}_A = 19.10,$ $\bar{x}_B = 17.24$

$s_A^2 = 7.72,$ $s_B^2 = 5.36$

$n_A = 10,$ $n_B = 10$

$df_A = 10 - 1 = 9,$ $df_B = 10 - 1 = 9$

$s_c^2 = \dfrac{(10-1)(7.72) + (10-1)(5.36)}{(10-1) + (10-1)} = 6.54$

$s_{\text{diff}}^2 = s_c^2 \left(\dfrac{1}{n_A} + \dfrac{1}{n_B} \right) = (6.54)\left(\dfrac{1}{10} + \dfrac{1}{10} \right)$

$= 1.308$

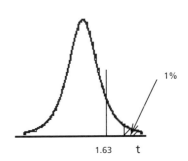

Figure 9.13: t-distribution for unpaired test

Chapter 9

$$s_{\text{diff}} = \sqrt{1.308} = 1.144$$

$$t_{\text{calculated}} = \frac{\bar{x}_A - \bar{x}_B}{s_{\text{diff}}} = \frac{19.10 - 17.24}{1.144}$$

$$= \frac{1.86}{1.144} = 1.626$$

This value of t must be compared with a limiting or critical value of t for $9 + 9 = 18$ degrees of freedom at the 1% level of significance for one tail. See Figure 9.13. According to Table A2 or the Excel function TINV, $t_{\text{critical}} = 2.552$. Since $t_{\text{calculated}} < t_{\text{critical}}$, the difference in rates of evaporation by these data is not significant at this level of significance. Alternatively, the observed level of significance is given from Excel or alternatives by TDIST(1.626,18,1) = 0.0607. Since 0.0607 > 1%, again the difference in evaporation rates is not significant at the 1% level.

However, since the effect of varying atmospheric conditions on the evaporation rates was appreciable, a different experiment involving paired samples might well show a significant difference. That type of experiment will be discussed next.

9.2.4 Comparison of Paired Samples

Although randomizing can effectively minimize incorrect definite conclusions due to interfering factors in the comparison of two sample means using unpaired samples, the interfering factors will still inflate the pooled estimate of variance of test results. This larger estimate of variance will make calculated values of t smaller, so it will be difficult to show that one treatment is *significantly* better than another. Thus, real effects may be missed. A more sensitive test is desirable.

In some cases it is possible to pair the measurements. One member of each matched pair comes from one value or characteristic of a variable or design, and the other member of each pair comes from the other characteristic, but everything else is nearly the same (as closely as possible) for the two members of the pair. For example, we might have one member of each pair from an experimental type of equipment and the other member from a standard type. Aside from the variable used to form the pairs, factors which might have appreciable effects must be kept as constant as possible. We try to match the two items forming a pair. Randomization should still be used to minimize interference from other factors. Then the difference between the members of a pair becomes the important variable, which will be examined by a test of significance using the t-distribution. Is the mean difference significantly different from zero? This technique blocks out the effect of interfering variables. It is called a *paired t-test* or a *t*-test using a matched pair.

Example 9.10

We decide to run a test using an experimental evaporation pan and a standard evaporation pan over ten successive days. The two types are set up side-by-side so

Statistical Inferences for the Mean

that atmospheric conditions should be the same. A coin is tossed to decide which evaporation pan is on the lefthand side and which on the righthand side on any particular day. The measured daily evaporations are as follows:

Pair or Day No.	1	2	3	4	5	6	7	8	9	10
Evaporation, mm:										
Pan A	9.1	4.6	14.0	16.9	11.4	10.7	27.4	22.8	42.8	29.4
Pan B	6.7	3.1	13.8	16.6	12.3	6.5	24.2	20.1	41.9	27.7
$d = \Delta_{evap}$, A − B	2.4	1.5	0.2	0.3	−0.9	4.2	3.2	2.7	0.9	1.7

Does the experimental Pan A give significantly higher evaporation than the standard Pan B at the 1% level of significance?

Answer: H_0: no real difference between the two methods, so $\mu_d = 0$.

H_a: $\mu_d > 0$ (one-tailed test)

The test statistic is $t = \dfrac{\bar{d} - 0}{s_{\bar{d}}}$.

Large enough values of t would show that H_0 is unlikely to be correct.

$n = 10,\ \bar{d} = \dfrac{\sum d}{n} = \dfrac{16.2}{10} = 1.62,$

$s_d = \sqrt{\dfrac{\sum d_i^2 - n(\bar{d})^2}{n-1}} =$

$\sqrt{\dfrac{(2.4^2 + 1.5^2 + 0.2^2 + \cdots + 1.7^2) - (10)(1.62)^2}{10-1}}$

$= 1.548$ (using calculator)

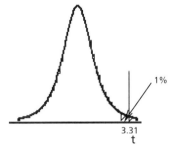

Figure 9.14: t-distribution for Paired t-test

Then $s_{\bar{d}} = \dfrac{1.548}{\sqrt{10}} = 0.4896$ and

$t_{calculated} = \dfrac{\bar{d} - 0}{s_{\bar{d}}} = \dfrac{1.62}{0.4896} = 3.31$

For 9 d.f., 1% single tail area, Table A2 or the Excel function TINV gives $t_{critical} = 2.821$. See Figure 9.14. Since $t_{calculated} > t_{critical}$, there is evidence at the 1% level of significance that the experimental treatment does give significantly higher evaporation. The alternative approach is to find the observed level of significance or

Chapter 9

p-value from Excel or alternatives. TDIST(3.31,9,1) = 0.0046. Since this is less than 1%, again we find the difference is significant at the 1% level of significance.

Notice that the paired t-test compares the mean difference of pairs to an assumed population mean difference of zero. From that point on, the calculation becomes the same as comparing a sample mean to a population mean as in section 9.2.2 above.

Notice also that the number of degrees of freedom is only half as great for the paired t-test as for the corresponding unpaired t-test, so if the variable (or variables) kept constant in forming the pairs has little effect, the unpaired t-test may actually be more sensitive.

A variation of the paired t-test asks whether the difference within pairs is more than a stated non-zero quantity.

However, a note of caution must be sounded at this point. The paired t-test *assumes* that if the interfering factor is kept constant within each pair, the difference in response will not be affected by the value of the interfering factor. This means that the effect on the response of the interfering factor and of the factor of interest must be purely *additive*. If the variable of interest and the interfering variable can interact to complicate the effect on the response, the paired t-test will not be as sensitive as we may think. For example, suppose we are studying the strengths of metal rods made of chromium-steel alloys of varying composition and want to see whether one heat treatment gives greater strength than another heat treatment. But for some compositions the strength is more sensitive to heat treatment than for some other compositions. (See the discussion of interaction in section 11.3.) In that case, even if we use a properly randomized, paired t-test, the interaction between heat treatment and composition will tend to inflate the estimate of random error and so make the test less sensitive than it should be. If that is the situation, rather than unpaired t-test or paired t-test, we should use a factorial design, which will be discussed in section 11.3.

Problems

1. Benzene in the air workers breathe can cause cancer. It is very important for the benzene content of air in a particular plant to be not more than 1.00 ppm. Samples are taken to check the benzene content of the air. 25 specimens of air from one location in the plant gave a mean content of 0.760 ppm, and the standard deviation of benzene content was estimated on the basis of the sample to be 0.45 ppm. Benzene contents in this case are found to be normally distributed.
   a) Is there evidence at the 1% level of significance that the true mean benzene content is less than or equal to 1.00 ppm?
   b) Find the 95% confidence interval for the true mean benzene content.
2. High sulfur content in steel is very undesirable, giving corrosion problems among other disadvantages. If the sulfur content becomes too high, steps have to

Statistical Inferences for the Mean

be taken. Five successive independent specimens in a steel-making process give values of percentage sulfur of 0.0307, 0.0324, 0.0314, 0.0311 and 0.0307. Do these data give evidence at the 5% level of significance that the true mean percentage sulfur is above 0.0300? What is the 90% two-sided confidence interval for the mean percentage sulfur in the steel?

3. The diameter of a mechanical component is normally distributed with a mean of approximately 28 cm. A standard deviation is found from the samples to be 0.25 cm. If we require a sample big enough so that there is at least 95% probability that the sample mean diameter ($\bar{x}$) is within 0.08 cm. of the true mean diameter (μ), what is the minimum sample size?

4. a) 25 standard reinforcing bars were tested in tension and found to have a mean yield strength of 31,500 psi with a sample variance of 25 x 10^4 psi^2. Another sample of 15 bars composed of a new alloy gave a mean and coefficient of variation of 32,000 psi and 2.0% respectively. Yield strengths follow a normal distribution. At the 1% level of significance, does the new alloy give an increased yield strength?
   b) If the industry-wide norm (so a population value) for yield strength of reinforcing bars is 31,600 psi, does the new alloy result in a significantly higher mean yield strength than the industry standard? Use the 5% level of significance.
   c) Find the 90% confidence interval for the true mean strength of the new alloy.

5. The mean height of 61 males from the same state was 68.2 inches with an estimated standard deviation of 2.5 inches, while 61 males from another state had a mean height of 67.5 inches with an estimated standard deviation of 2.8 inches. The heights are normally distributed. Test the hypothesis that males from the first state are taller than males from the second state.
   a) Use a level of significance of 5%.
   b) Use a level of significance of 10%.

6. On last year's final examination in statistics, the marks of two different sections had the numbers, means and standard deviations shown in the table below:

n	$\bar{x}$	s_x
41	64.3	15.6
51	59.5	17.2

The marks were normally distributed.
   a) Were the means in the two sections significantly different? Use the 5% level of significance.
   b) The overall average of all the students in the last 10 years of statistics final examinations was 61.7 with a standard deviation of 16.8. Was the section average for the 41 students shown above significantly higher than the overall average? Use the 5% level of significance.

Chapter 9

7. Two chemical processes for manufacturing the same product are being compared under the same conditions. Yield from Process A gives an average value of 96.2 from six runs, and the estimated standard deviation of yield is 2.75. Yield from Process B gives an average value of 93.3 from seven runs, and the estimated standard deviation is 3.35. Yields follow a normal distribution. Is the difference between the mean yields statistically significant? Use the 5% level of significance, and show rejection regions for the difference of mean yields on a sketch.

8. Two companies produce resistors with a nominal resistance of 4000 ohms. Resistors from company A give a sample of size 9 with sample mean 4025 ohms and estimated standard deviation 42.6 ohms. A shipment from company B gives a sample of size 13 with sample mean 3980 ohms and estimated standard deviation 30.6 ohms. Resistances are approximately normally distributed.
   a) At 5% level of significance, is there a difference in the mean values of the resistors produced by the two companies?
   b) Is either shipment significantly different from the nominal resistance of 4000 ohms? Use .05 level of significance.

9. Two different types of evaporation pans are used for measuring evaporation at a weather station. The evaporation for each pan for 6 different days is as follows:

Day No.	Evaporation (mm)	
	Pan A	Pan B
1	9	11
2	42	41
3	28	29
4	16	16
5	11	13
6	11	12

   At the 5% level of significance, is there a significant difference in the evaporation recorded by the two pans? Interaction between type of pan and weather variation from day to day can be neglected.

10. A new composition for car tires has been developed and is being compared with an older composition. Ten tires are manufactured from the new composition, and ten are manufactured from the old composition. One tire of the new composition and one of the old composition are placed on the front wheels of each of ten cars. Which composition goes on the lefthand or righthand wheel is determined randomly. The wheels are properly aligned. Each car is driven 60,000 km under a variety of driving conditions. Then the wear on each tire is measured. The results are:

Car No.	1	2	3	4	5	6	7	8	9	10
Wear of New Composition	2.4	1.3	4.2	3.8	2.8	4.7	3.2	4.8	3.8	2.9
Wear of Old Composition	2.7	1.9	4.3	4.2	3.0	4.8	3.8	5.3	3.7	3.1

Statistical Inferences for the Mean

Do the results show at the 1% level of significance that the new composition gives significantly less wear than the old composition? Interaction between the tire composition and the car can be neglected.

11. Nine specimens of unalloyed steel were taken and each was halved, one half being sent for analysis to a laboratory at the University of Antarctica and the other half to a laboratory at the University of Arctica. The determinations of percentage carbon content were as follows:

Specimen No.	1	2	3	4	5	6	7	8	9
University of Antarctica	0.22	0.11	0.46	0.32	0.27	0.19	0.08	0.12	0.18
University of Arctica	0.20	0.10	0.39	0.34	0.23	0.14	0.13	0.08	0.16

Test for a difference in determinations between the two laboratories at the 0.05 level of significance. Neglect any possibility of interaction.

12. Two flow meters, A and B, are used to measure the flow rate of brine in a potash processing plant. The two meters are identical in design and calibration and are mounted on two adjacent pipes, A on pipe 1 and B on pipe 2. On a certain day, the following flow rates (in m^3/sec) were observed at 10-minute intervals from 1:00 p.m. to 2:00 p.m.

	Meter A	Meter B
1:00 p.m.	1.7	2.0
1:10 p.m.	1.6	1.8
1:20 p.m.	1.5	1.6
1:30 p.m.	1.4	1.3
1:40 p.m.	1.5	1.6
1:50 p.m.	1.6	1.7
2:00 p.m.	1.7	1.9

Is the flow in pipe 2 significantly different from the flow in pipe 1 at the 5% level of significance?

13. The visibility of two traffic paints, A and B, was tested, each at 8 different locations. The measures of visibility were taken after exposure to weather and traffic during the period January 1 to July 1. The results were as follows:

Paint A		Paint B	
Location	Visibility	Location	Visibility
1A	7	1B	8
2A	7	2B	10
3A	8	3B	8
4A	5	4B	5
5A	5	5B	3

Chapter 9

6A	6	6B	9
7A	6	7B	8
8A	4	8B	5

Test the hypothesis that the mean visibility of paint A is less than that of paint B at the 5% level of significance under the following conditions:
a) if both paints were tested simultaneously under identical conditions, the signs in paint A and paint B being erected adjacent to one another. Neglect any possibility of interaction.
b) if the A locations are on the west side and the B locations are on the east side of the city.

14. A water quality lab tests for the bacterial count in drinking water in a certain northern city.
    a) A test is made of a claim in the literature that the time to equilibrium in bacterial growth is greater in northerly climates, the standard deviation remaining unaffected. The mean time in southerly cities has been found, from many measurements, to equal 24.1 hours with a standard deviation of 2.3 hours. The northern lab tests 21 water specimens and finds the mean time to equilibrium bacterial growth is 25.4 hours, with an estimated standard deviation of 2.2 hours, which is not significantly different from the standard deviation of 2.3 hours quoted above. Does this data bear out the claim in the literature about the increase in mean time to equilibrium, at the 5% level of significance?
    b) Two salesmen turn up at the laboratory one week, each claiming that the additive he is selling will decrease the time to equilibrium bacterial growth, compared to the other salesman's product. The laboratory decides to check out the claims and tests 6 specimens of water, half of each treated with each of the two products. You should neglect any possibility of interaction. What does the following data indicate as to the salesmen's claims (at the 5% level of significance)?

	Time to Equilibrium, hours	
Water Sample no.	Additive 1	Additive 2
1	23.8	24.5
2	34.1	34.4
3	22.1	23.2
4	15.3	16.7
5	31.8	31.8
6	22.5	22.9

15. a) 41 cars equipped with standard carburetors were tested for gas usage and yielded an average of 8.1 km/litre with a standard deviation of 1.2 km/l. 21 of these cars were then chosen randomly, fitted with special carburetors and

Statistical Inferences for the Mean

tested, yielding an average of 8.8 km/l with a standard deviation of 0.9 km/l. At the 5 percent level of significance, does the new carburetor decrease gas usage?

b) Does the following group of data bear out the same result? Neglect any possibility of interaction between the type of carburetor and other characteristics of the cars.

Car No.	Standard Carburetor	New Carburetor
1	7.6	8.2
2	7.9	7.8
3	6.5	8.1
4	5.6	8.6
5	7.3	9.5

Supplementary Problems

Students may need practice in deciding whether a particular problem can be done using the normal distribution or requires the t-distribution. The following problem set contains both types.

1. The lives of Glowbrite light bulbs made by Glownuff Inc. have a mean of 1000 hours and standard deviation 160 hours.
   a) Assuming a normal distribution for the sample means, find the probability that 25 bulbs will have a mean life of less than 920 hours.
   b) The Consumers Association demands that the mean life of samples of 25 bulbs be not below 920 hours with 99.9% confidence. What is the maximum permissible standard deviation (for $\mu = 1000$ hours)?
   c) The manufacturer has instituted a sampling program to maintain quality control. He intends that there be no more than 5% probability that the true mean bulb life is more than 20 hours different from the sample mean. What sample size should he use, assuming the standard deviation is still 160 hours?

2. Electrical resistors made by a particular factory have a coefficient of variation of 0.28% with a normal distribution of resistances.
   a) Find the 99% confidence interval for the mean of samples of size five if the population mean is 10.00 ohms.
   b) How many observations must a sample contain to give at least 99.5% probability that the sample mean is within 0.30% of the population mean?

3. Slaked lime is added to the furnace of an electric power station to reduce the production of SO_2 (a major cause of acid rain). Extensive previous data showed that a standard method of adding slaked lime reduced SO_2 emission by an average percentage of 31.0 with a standard deviation of 4.70. A test on a new method gives mean percentage removed of 33.5 based on a sample of size 15 with no change in the standard deviation. Is there evidence at the 1% level of

Chapter 9

significance that the new method gives higher removal of SO_2 than the standard method? A normal distribution is followed.

4. a) The manufacturer of the Energy-saver furnace claims a mean energy efficiency of at least 0.83. A sample of 21 Energy-saver furnaces gives a sample mean of 0.81 and sample standard deviation of 0.060. Data show approximately a normal distribution. Test whether the manufacturer's claim can be rejected at the 5% level of significance.
    b) It is known that the industry-standard furnace has a mean energy efficiency of 0.78 and a standard deviation of 0.055. Use the sample mean for Energy-saver furnaces to test whether these furnaces have a significantly higher efficiency than the industry standard at the 5% level of significance.

5. The mean yield stress of a certain plastic is specified to be 30.0 psi. The standard deviation is known to be 1.20 psi. A normal distribution is followed.
    a) If the population mean is 30.0 psi, what is the 95% confidence interval for the mean yield stress of 9 specimens?
    b) A sample of 9 specimens shows a mean of 27.4 psi. Is this sample mean significantly different from the specified mean value? Use the 5% level of significance?
    c) Is the sample mean from part (b) significantly larger than 26.3 psi at 1% level of significance?

6. The standard deviation of a particular dimension on a machine part is known to be 0.0053 inches. A normal distribution is followed. Four parts coming off the production line are measured, giving readings of 2.747 in, 2.739 in, 2.750 in, and 2.749 in. Is the sample mean significantly larger than 2.740 inches at the 1% level of significance? What is the probability of accepting the null hypothesis if the true mean is 2.752 in. and the standard deviation remains unchanged? (Notice that this would be a Type II error.)

7. Specimens of soil were obtained from a site both before and after compaction. Tests on 10 pre-compaction specimens gave a mean porosity of 0.413 and a standard deviation of 0.0324. Tests on 20 post-compaction specimens gave a mean porosity of 0.340 and a standard deviation of 0.0469. These standard deviations are not significantly different. Porosity follows a normal distribution.
    a) At the 5% level of significance, did the compaction correspond to a significant reduction in mean porosity?
    b) At the 5% level of significance, is the reduction in mean porosity significantly less than the desired reduction of 0.1?

8. Three machines are used to pack different colored crystals in a bath salt mixture. The machines are set for machines 1 and 2 to each add 500 grams of salts and machine 3 to add 750 grams. It has been found that the variation around the set point is normally distributed in each case with the following dispersions:

Statistical Inferences for the Mean

Machine	Standard Deviation
1	20 grams
2	10 grams
3	25 grams

a) What is the mean weight of a package of bath salts?
b) If packages of bath salts with weight less than 1.65 kg have to be repacked, what percentage of the day's output would fall into this category?
c) It is decided to sample the final output to estimate the mean weight of the packages. How big a sample must be taken to estimate with 99% confidence that the true mean lies between 99% and 101% of the sample mean?

9. Two different kinds of cereal designated A and B are combined to form a new product called Brand X. The cereal types are weighed independently and mixed automatically before being packed in a plastic bag which weighs 10 grams. The weighing machines are set so that μ_A = 1000 grams and μ_B = 500 grams. The weights are normally distributed, and the coefficient of variation in each case is 10%.
   a) What is the mean total weight of a bag of Brand X?
   b) What is the probability that a bag of Brand X will contain less than 950 grams of Cereal A and more than 450 grams of Cereal B?
   c) What is the probability that a bag of Brand X will contain exactly 1400 grams?
   d) What is the probability that a bag of Brand X will contain less than 1400 grams?
   e) How many bags must be weighed to ensure with 95% confidence that the true mean weight of a bag lies within 30 grams of the sample mean?

CHAPTER **10**

Statistical Inferences for Variance and Proportion

For this chapter the reader needs a good knowledge of Chapter 9. For section 10.1 a solid understanding of sections 3.1 and 3.2 is needed, while section 10.2 requires a good knowledge of section 5.3.

The general approach developed in Chapter 9 for tests of hypothesis and confidence intervals for means carries over to similar inferences for variances and proportions. The concepts of null hypothesis, alternative hypothesis, level of significance, confidence levels and confidence intervals can be applied directly.

10.1 Inferences for Variance

Is a sample variance significantly larger than a population variance? Or is one sample variance significantly larger than another, indicating that one population is more variable than another? Those are the sorts of question we are trying to answer when we compare two variances. To obtain answers we will introduce two more probability distributions, the chi-squared distribution and the F-distribution. Mathematically, the F-distribution is related to the ratio of two chi-squared distributions. We will use the chi-squared distribution in section 10.1.1 to compare a sample variance with a population variance, and we will use the F-distribution in section 10.1.2 to compare two sample variances. We will see in part (d) of section 10.1.2 that the F-distribution can be used also to compare a sample variance with a population variance. Therefore, at this time the reader can omit section 10.1.1, and so the chi-squared probability distribution, if that seems desirable. We will need the chi-squared distribution later when we come to Chapter 13, where we will encounter the chi-squared test for frequency distributions.

10.1.1 Comparing a Sample Variance with a Population Variance

Say we are trying to make the production from a particular process less variable, so more uniform. To assess whether we have been successful we might take a sample from current production and compare its sample variance with the population variance established under previous conditions. Is the new estimate of variance significantly smaller than the previous variance? If it is, we have an indication that the production has become less variable, so there is some evidence of success.

We would test the trial assumption that the new sample variance and the previous population variance differ only because of chance. Specifically, the null hypothesis is that the new population variance is equal to the previous population variance. The

Statistical Inferences for Variance and Proportion

alternative hypothesis would be that the new population variance is smaller than the previous one, so we have a one-sided test. Is the new sample variance so much smaller than the previous population variance that the null hypothesis is very unlikely? The size of the sample would, of course, affect the answer.

(a) Chi-squared Probability Distribution

If the sample is from a normal distribution, the probability distribution which applies to the variances in this situation is the **chi-squared distribution**. The chi-squared distribution and the normal distribution are related mathematically. Chi is a Greek letter, χ, which is pronounced "kigh," like high. A relationship can be derived among χ^2, σ^2, s^2, and the number of degrees of freedom on which s^2 is based, $(n-1)$. This relationship is

$$\chi^2 = \frac{(n-1)s^2}{\sigma^2} \tag{10.1}$$

The density function of the χ^2 distribution is unsymmetrical, and its shape depends on the number of degrees of freedom. Probability density functions for three different numbers of degrees of freedom are shown in Figure 10.1. As the number of degrees of freedom increases, the density function becomes more symmetrical as a function of χ^2. For any particular number of degrees of freedom, the mean of the distribution is equal to the number of degrees of freedom.

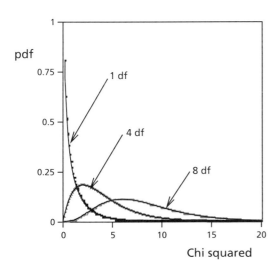

Figure 10.1: Shapes of Probability Density Functions for Some Chi-squared Distributions

Table A3 in Appendix A gives values of χ^2 corresponding to some values of the upper-tail probability.

If a computer with Excel or some alternatives is available, values can be found from the computer instead of from tables. Probabilities corresponding to values of χ^2 can be found from the Excel function CHIDIST. The arguments to be used with this function are the value of χ^2 and the number of degrees of freedom. The function then

Chapter 10

returns the upper-tail probability. For example, for $\chi^2 = 18.49$ at 30 degrees of freedom, we type in a cell for a work sheet the formula CHIDIST(18.49,30), or else we can paste in the function, CHIDIST(,), then type in the arguments and choose the OK button. The result is 0.95005, the probability of obtaining a value of χ^2 greater than 18.49 completely by chance.

If we have a value of the upper-tail probability and the number of degrees of freedom, we use the Excel function CHIINV to find the value of χ^2. Again, the function can be chosen using the Formula menu or it can be typed into a cell. For an upper-tail probability of 0.95 and 30 degrees of freedom, CHIINV(0.95,30) gives 18.4927.

We will use the χ^2 distribution in this chapter to compare a sample variance with a population variance. In Chapter 13 we will use this same distribution for an entirely different purpose, to compare two or more frequency distributions.

(b) Test of Significance for Variances

Let us look at an example.

Example 10.1

The population standard deviation of strengths of steel bars produced by a large manufacturer is 2.95. In order to meet tighter specifications engineers are trying to reduce the variability of the process. A sample of 28 bars gives a sample standard deviation of 2.65. Assume that the strengths of steel bars are normally distributed. Is there evidence at the 5% level of significance that the standard deviation has decreased?

Answer: H_0: $\sigma^2 = (2.95)^2 = 8.70$

H_a: $\sigma^2 < 8.70$ (one-tailed test)

The test statistic will be $\chi^2 = \dfrac{(n-1)s^2}{\sigma^2}$.

If $\chi^2_{calculated}$ is sufficiently small, then H_0 is not likely to be true.

$$\chi^2_{calculated} = \dfrac{(n-1)s^2}{\sigma^2} = \dfrac{(28-1)(2.65)^2}{(2.95)^2} = 20.98$$

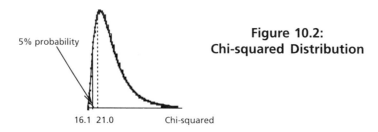

Figure 10.2: Chi-squared Distribution

Statistical Inferences for Variance and Proportion

From Table A3, for 5% probability in the lower tail (the lefthand tail) and therefore 95% probability in the upper tail, the one to the right, and with $28 - 1 = 27$ degrees of freedom, we find $\chi^2_{critical} = 16.15$. Then $\chi^2_{calculated} > \chi^2_{critical}$, so the calculated value does not fall in the cross-hatched tail for 5% probability. The population variance is not significantly less than 8.70, so the population standard deviation is not significantly less than 2.95. We do not have evidence at the 5% level of significance that the standard deviation of strengths of the steel bars has decreased.

An alternative method for solving this sort of problem using the F-distribution will be given in section 10.1.2(d).

(c) Confidence Intervals for Population Variance or Standard Deviation

If we have an estimate of the variance or standard deviation from a sample, we can determine a corresponding confidence interval for the variance or standard deviation for the population. Again, let's examine an example.

Example 10.2

A sample of 15 concrete cylinders was taken randomly from the production of a plant. The strength of each specimen was determined, giving a sample standard deviation of 215 kN/m². Find the 95% confidence interval (with equal probabilities in the two tails) for standard deviation of the strengths. Assume the strengths follow a normal distribution.

Answer: $s^2 = (215)^2 = 46{,}225$ based on $15 - 1 = 14$ degrees of freedom.

The relevant statistic to be found from tables or Excel is $\chi^2 = \dfrac{(n-1)s^2}{\sigma^2}$.

Then the confidence limits will be found from $\sigma^2 = \dfrac{(n-1)s^2}{\chi^2}$ using values of χ^2 at cumulative probabilities of 0.025 and 0.975 for 14 d.f.

The limiting values of χ^2 can be found from either Table A3 or Excel. From Table A3 for 14 d.f. the limiting values of χ^2 are 5.63 at a cumulative probability of 0.025 (so upper-tail area of 0.975) and 26.12 at a cumulative probability of 0.975 (so upper-tail area of 0.025). The same numbers (expressed in more figures) are found from CHIINV(0.025,14) and CHIINV(0.975,14). Limiting values are shown on Figure 10.3. The

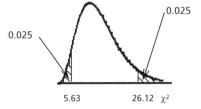

Figure 10.3: Confidence limits for Chi-squared Distribution

corresponding limits on σ^2 are $\dfrac{(14)(46225)}{5.63} = 115{,}000$ and $\dfrac{(14)(46225)}{26.12} = 24{,}800$.

The limits on σ are the square roots of these numbers, 339 and 157. Then, the 95% confidence interval for standard deviation is from 157 to 339 kN/m².

Chapter 10

10.1.2 Comparing Two Sample Variances

Say we have two sample variances. Is one sample variance significantly different from (or else larger than) the other? Or, on the other hand, is it reasonable to say that both sample variances might have come from the same population? The appropriate test of hypothesis is the **F-test** or **Variance-ratio test**. We calculate the ratio of the two sample variances:

$$F = \frac{s_1^2}{s_2^2} \qquad (10.2)$$

where s_1^2 is the estimate of population variance on the basis of sample 1, and s_2^2 is the estimate of population variance on the basis of sample 2. In this book we will put the larger estimate of variance in the numerator and call it s_1^2 so that the quantity F is **larger** than 1.

(a) Probability Distribution for Variance Ratio

A critical or limiting value of F is obtained from tables or Excel. These theoretical values must be related to the ratio of one χ^2 function to another. In fact, the theoretical statistic F is defined as the ratio of two independent chi-squared random variables, each divided by its number of degrees of freedom, but we don't need to go into the details here. Remember that we assumed that the sample came from a normal distribution in order to make the chi-squared distribution applicable to the variances, and the same assumption is required to make the F-distribution applicable here.

The shape of the F-distribution is always unsymmetrical, skewed to the right. The shape depends on the numbers of degrees of freedom in the sample variances in both the numerator and the denominator. Figure 10.4 shows the shapes of two F-distributions.

The probability that $F > f_1$ depends on the number of degrees of freedom in the numerator and the number of degrees of freedom in the denominator, as well as the value of f_1. To show all the combinations of parameters that might be needed in practical calculations would require a very extensive table. The usual practice is to show in a table only a limited selection of values. Table A4 in Appendix A is in two parts. For various combinations of degrees of freedom for variance in the numerator, df_1, and degrees of freedom for variance in the denominator, df_2,

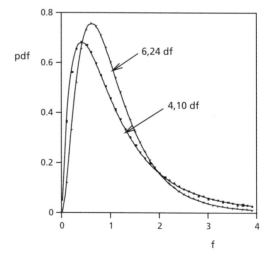

Figure 10.4: Shapes of Two F-distributions with Various Degrees of Freedom in Numerator and Denominator

Statistical Inferences for Variance and Proportion

values of F which will give an upper-tail probability of 0.05 are shown on the first page of Table A4. For various combinations of df_1 and df_2, values of F which will give an upper-tail probability of 0.01 are shown on the second page of Table A4. If combinations of df1 and df2 that are not shown on Table A4 are needed, interpolation is required.

If a computer is available with Excel or some alternative, it can be used to find probabilities corresponding to any applicable value of F, or else values of F corresponding to any applicable probability. These would both be for the required combination of degrees of freedom in the numerator and degrees of freedom in the denominator. The Excel function FDIST gives the probability distribution for F. The arguments to be used with this function are the value of F, the number of degrees of freedom for variance in the numerator, and the number of degrees of freedom for variance in the denominator. Then Excel will give the corresponding upper-tail probability, that is, $\Pr[F > f_1]$. Similarly, the Excel function FINV gives the value of F for stated upper-tail probability. If we enter FINV(upper-tail probability, degrees of freedom for variance in the numerator, degrees of freedom for variance in the denominator), Excel will give the corresponding value of F.

(b) Test of Significance: the F-test or Variance-ratio Test

Now we compare a calculated value of F to a chosen or critical value of F. Is the calculated value so large that it is very unlikely that it could have occurred by chance? The samples must have been chosen randomly and independently.

We make the null hypothesis that the difference between the two estimates of variance is entirely due to chance, so $\sigma_1^2 = \sigma_2^2$. The alternative hypothesis is either that $\sigma_1^2 \neq \sigma_2^2$ for a two-sided test, or else that $\sigma_1^2 > \sigma_2^2$ for a one-sided test. Because we put the larger estimate of variance in the numerator, $s_1^2 > s_2^2$, we have no reason to consider the possibility that $\sigma_1^2 < \sigma_2^2$.

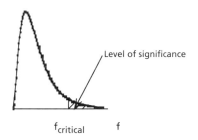

Figure 10.5:
Level of Significance
for a one-sided F-test

If the variance ratio, F, is too large, then there is little probability that the null hypothesis is true. Specifically, the probability of obtaining this large a value of F or larger purely by chance, when the null hypothesis is true, is equal to the observed level of significance. Such a probability must also depend on the numbers of degrees of freedom on which each estimate of variance is based. These are df_1 degrees of freedom for the larger estimate of variance and so for the numerator, and df_2 degrees of freedom for the smaller estimate of variance and so for the denominator.

For the 5% level of significance, the limiting value of F for a one-sided test must be such that $\Pr[F > f_{\text{critical}}] = 0.05$, and similarly for other levels of significance.

Chapter 10

For a two-sided F-test, but set up so that $f_{calculated} > 1$, the same values of $f_{critical}$ apply for levels of significance twice as great to allow for both tails. For example, $f_{critical}$ for a two-sided test at 2% level of significance is the same as $f_{critical}$ for a one-sided test at 1% level of significance.

Example 10.3

Two additives to Portland cement are being tested for their effect on the strength of concrete. 21 batches were made with Additive A, and their strengths showed standard deviation $s_A = 41.3$. 16 batches were made with the same percentage of Additive B, and their strengths showed standard deviation $s_B = 26.2$. Assume that the strengths of concrete follow a normal distribution. Is there evidence at the 1% level of significance that the concrete made with Additive A is more variable than concrete made with Additive B?

Answer: $H_0: \sigma_A^2 = \sigma_B^2$

$H_a: \sigma_A^2 > \sigma_B^2$ (one-tailed test)

The test statistic will be $F = \dfrac{s_A^2}{s_B^2}$. Large values of $f_{calculated}$ will indicate that the null hypothesis is not likely to be true.

$f_{calculated} = \dfrac{s_A^2}{s_B^2} = \dfrac{41.3^2}{26.2^2} = 2.485$ based on 20 degrees of freedom for the numerator and 15 degrees of freedom for the denominator.

From the second part of Table A4, for 1% level of significance with $df1 = 20$ and $df2 = 15$, $f_{critical} = 3.37$. Alternatively, from the function FINV in Excel, FINV(0.01,20,15) gives $f_{critical} = 3.37189476$.

Since $f_{calculated} < f_{critical}$, the difference is not significant at the 1% level of significance. Then at this level of significance there is not sufficient evidence to say that the strength of concrete made with Additive A is more variable than the strength of concrete made with Additive B.

Example 10.4

Using the same figures as in Example 10.3, is there evidence at the 10% level of significance that concrete made with Additive A and concrete made with Additive B have different variabilities? Again, assume that the strengths of concrete follow a normal distribution.

Answer: $H_0: \sigma_A^2 = \sigma_B^2$

$H_a: \sigma_B^2 \neq \sigma_B^2$ (two-tailed test)

Statistical Inferences for Variance and Proportion

The test statistic will be $F = \dfrac{s_A^2}{s_B^2}$. Large values of $f_{\text{calculated}}$ will indicate that H_0 is unlikely to be true. As before, $f_{\text{calculated}} = \dfrac{s_A^2}{s_B^2} = \dfrac{41.3^2}{26.2^2} = 2.485$ based on 20 degrees of freedom for the numerator and 15 degrees of freedom for the denominator.

From the first part of Table A4, for 5% upper-tail area, corresponding to 10 % level of significance for a two-tailed test, with $df1 = 20$ and $df2 = 15$, $f_{\text{limit}} = 2.33$. Alternatively, from the function FINV in Excel, FINV(0.01,20,15) gives 2.32753194.

Since $f_{\text{calculated}} > f_{\text{limit}}$, there *is* evidence at the 10% level of significance that concrete made with Additive A and concrete made with Additive B have different variabilities.

Besides comparisons in which the major objective is to see whether one set of data is significantly more variable or has different variability than another set, the F-test is used for two main purposes:

1. To see whether two estimates of variance can be combined or pooled to compare means by an unpaired *t*-test. In this case the *F*-test would be two-tailed. Usually, if the variances are not significantly different at (let us say) the 10% level of significance, they can be combined to give a better estimate of variance to use in the *t*-test.

2. To compare two estimates of variance from different types of data as part of the **analysis of variance**, which will be considered more fully in Chapter 12. In some cases the total variation of data from an experiment can be broken down into two estimated variances, say the variance within groups and the variance between groups. The variance within groups comes from repeated measurements at the same condition and so gives an estimate of the variance due to experimental error. The variance between groups arises from different treatments or different conditions as well as from experimental error. The question to be answered is, is the variance between groups significantly larger than the variance within groups? If so, that is an indication that the variation of treatments or conditions has an effect on the results. A critical level of significance must be stated. This is a one-tailed *F*-test.

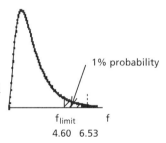

**Figure 10.6:
Test of Significance**

Example 10.5

In the results from an experiment the estimated variance within groups (WG), based on 27 degrees of freedom, is 233, while the estimated variance between the groups (BG), based on 3 degrees of freedom, is 1521. Is there evidence at the 1% level of significance that the difference in conditions between the groups has an effect on the results? The data have been plotted on normal probability paper, showing reasonable agreement with normal distributions.

Chapter 10

Answer: $H_0: \sigma_{WG}^2 = \sigma_{BG}^2$

$H_a: \sigma_{BG}^2 > \sigma_{WG}^2$ (one-tailed test)

The test statistic will be $F = \dfrac{s_{BG}^2}{s_{WG}^2}$, in that order because $s_{BG}^2 > s_{WG}^2$.

If $f_{calculated}$ is sufficiently large, H_0 will not be plausible.

$f_{calculated} = \dfrac{s_{BG}^2}{s_{WG}^2} = \dfrac{1521}{233} = 6.53$. Another name for $f_{critical}$ is f_{limit}. For 1% level of significance in a one-tailed test, with 3 degrees of freedom in the numerator and 27 degrees of freedom in the denominator, the second part of Table A4 gives $f_{limit} = 4.60$. Alternatively, from the function FINV in Excel, FINV(0.01,3,27) gives 4.60090632.

Since $f_{calculated} > f_{limit}$, there is evidence at the 1% level of significance that the difference in conditions between the groups has an effect on the results.

(c) Confidence Interval for Ratio of Sample Variances

A point estimate of the ratio of two population variances is given by the corresponding ratio of two sample variances, $\dfrac{s_1^2}{s_2^2}$. It is quite feasible to derive a confidence interval for the ratio of the population variances, $\dfrac{\sigma_1^2}{\sigma_2^2}$, as long as the samples were taken randomly from normal distributions. However, practical applications of this technique by engineers are hard to find, so these confidence intervals will not be discussed further here. If they should be needed, the reader is referred to books by Walpole and Myers and by Vardeman (references in section 15.2). On the other hand, confidence intervals for population variances (rather than their ratios) are very useful and have been discussed in section 10.1.1(c).

(d) Using the Variance Ratio to Compare a Sample Variance with a Population Variance

In section 10.1.1 we saw that the chi-squared distribution can be used to compare a sample variance with a population variance. An alternative method of making this comparison uses the F-distribution. If one of the variances is a population variance, its number of degrees of freedom will be infinite. In this section Example 10.1 will be solved by this alternative method.

Example 10.1 (Alternative solution)

The population standard deviation of strengths of steel bars produced by a large manufacturer is 2.95. In order to meet tighter specifications engineers are trying to reduce the variability of the process. A sample of 28 bars gives a sample standard deviation of 2.65. Assume that the strengths of steel bars are normally distributed. Is there evidence at the 5% level of significance that the standard deviation has decreased?

Statistical Inferences for Variance and Proportion

Answer: $H_0: \sigma^2 = (2.95)^2 = 8.70$

$H_a: \sigma^2 < 8.70$ (one-sided test)

The test statistic will be $F = \dfrac{s_1^2}{s_2^2} = \dfrac{\sigma^2}{s^2}$. If $F_{\text{calculated}}$ is sufficiently large, then H_0 is not likely to be true.

$$F_{\text{calculated}} = \dfrac{\sigma^2}{s^2} = \dfrac{(2.95)^2}{(2.65)^2} = 1.24$$

From Table A4 for 5% upper-tail probability, ∞ degrees of freedom in the numerator (df1) and 28 − 1 = 27 degrees of freedom in the denominator (df2), $F_{\text{limit}} = 1.67$. Then $F_{\text{calculated}} < F_{\text{limit}}$, so the calculated value is not significant at the 5% level of significance. Therefore, we do *not* have evidence at the 5% level of significance that the standard deviation has decreased.

Problems

1. A testing laboratory is trying to make its results more consistent by standardizing certain procedures. From a sample of size 28 the sample standard deviation by the revised procedure is found to be 1.74 units. Plotting concentrations on normal probability paper did not show any marked departure from a normal distribution. Is there evidence at the 5% level of significance that the sample standard deviation is significantly less than the former population standard deviation of 2.92 units?

2. It is known from long experience that, for a particular chemical compound, determinations made with a mass spectrometer have a variance of 0.24. An analyst who is new to the job makes a series of 28 determinations with the spectrometer and they give an unbiased estimate of variance of 0.32. Plotting the results on normal probability paper indicates that the data do not vary significantly from a normal distribution. Is the sample estimate of variance significantly larger than the variance based on long experience? Use a 5% level of significance.

3. Yield stresses for shear were measured in a random sample consisting of 28 soil specimens. Plotting the data on normal probability paper showed no apparent departure from normal distribution. The sample standard deviation was found to be 285 kN/m². Find the two-sided confidence limits (with equal probabilities in the two tails) for the standard deviation of the yield stress.

4. A sample consists of 21 specimens, each taken by a standard procedure from a different filter cake on an industrial filter. Moisture contents of the specimens were measured. Plotting the data on normal probability paper indicated negligible departure from normal distribution. The sample standard deviation of percentage moisture contents was found to be 3.21. Find the two-sided 90% confidence limits (with equal probabilities in the two tails) for the standard deviation of percentage moisture contents.

Chapter 10

5. The coefficients of thermal expansion of two alloys, A and B, are compared. Six random measurements are made for each alloy. For alloy A, the coefficients ($\times 10^6$) are 12.95, 14.05, 12.75, 12.10, 13.50 and 13.00. Coefficients ($\times 10^6$) for alloy B are 14.05, 15.35, 14.35, 15.15, 13 85 and 14.25. Assume the values for each alloy are normally distributed. Is the variance of coefficients for alloy A significantly different from the variance of coefficients for alloy B? Use the 10% level of significance.

6. The carbon dioxide concentration in the air within an energy-efficient house was measured once each month over an entire year. The measurements (in ppm) for January to December, respectively, were 650, 625, 480, 400, 325, 305, 310, 305, 490, 540, 695, and 600. Assume that these measurements follow a normal distribution. The concentration of carbon dioxide in an older house also was measured each month in the same year, but on a different day of the month than for the energy efficient house. The data for this house for January to December, respectively, were 505, 530, 430, 400, 300, 300, 305, 310, 320, 410, 520, and 540. At the 10% level of significance, is there a difference in the variability of carbon dioxide concentration between the two houses?

7. The standard way of measuring water suction in soil is by a tensiometer. A new instrument for measuring this parameter is an electrical resistivity probe. A purchaser is interested in the variability of the readings given by the new instrument. The purchaser put both instruments into a large tank of soil at ten different locations, both instruments side by side at each location, and obtained the following results.

Suction (in cm) Measured by	
Tensiometer	Electrical Resistivity Probe
355	365
305	300
360	375
330	360
345	340
315	320
375	385
350	380
330	330
350	390

a) Choose an appropriate level of significance and test for a significant difference in the variance of the two instruments.

b) It is known from extensive measurements that the variance of the tensiometer readings in a tank of soil like this should be 350 cm^2. Choose an appropriate level of significance and test whether the electrical resistivity probe gives a higher variability than expected.

Statistical Inferences for Variance and Proportion

8. A general contractor is considering purchasing lumber from one of two different suppliers. A sample of 12 boards is obtained from each supplier and the length of each board is measured. The estimated standard deviations from the samples are $s_1 = 0.13$ inch and $s_2 = 0.17$ inch, respectively. Assume the lengths follow a normal distribution. Does this data indicate the lengths of one supplier's boards are subject to less variability than those from the other supplier? Test using a level of significance equal to 0.02.

9. Wire of a certain type is supplied to an electrical retailer by each of two manufacturers, A and B. Users of the wire suggest that there is more variability (from specimen to specimen) in the resistance of the wire supplied by Company A than in that supplied by Company B. Random samples of wire from spools of the wire supplied by the two companies were taken. The resistances were measured with the following results:

Company	A	B
Number of Samples	13	21
Sum of Resistances	96.8	201.4
Sum of Squares of Resistances	732.30	1936.90

   Assume the resistances were normally distributed. Use the results of these samples to determine at the 5% level of significance whether or not there is evidence to support the suggestion of the users.

10. A study of wave action downstream of a dam spillway was carried out before and after a modification was made to the structure. The modification was intended to reduce wave action, which is indicated by variability in the depth of water. Depths of water were measured in meters. Before modification 41 measurements gave a sample standard deviation of 2.80. After modification 51 measurements gave a sample standard deviation of 1.49.
    a) Choosing an appropriate level of significance, determine if there is a significant reduction in variability in the water depth—i.e., a significant reduction in wave action.
    b) Is the pre-modification wave action at this site any different from that at another site where 51 measurements gave a sample variance of the depth of 2.65 m^2? Choose an appropriate level of significance.

11. In a random survey of gasoline stations in Saskatchewan and Alberta, the average prices per liter of unleaded regular gasoline and the corresponding standard deviations were as follows:

Province	Sample Size	Mean (Cents/liter)	Standard deviation (Cents/liter)
Alberta	14	68.8	1.1
Saskatchewan	9	70.7	0.8

Chapter 10

a) Using the 10% level of significance, test the claim that the price per liter of gasoline is equally variable in the two provinces.
b) At what level of significance can you conclude that the average gasoline price in Alberta is less than in Saskatchewan?

12. It was claimed by a sand filter salesman that the mean concentrations of solids after filtering are normally distributed and have an average value of .025 percent solids, and that 95% of recorded concentrations will not exceed .030 percent solids. In order to check the validity of this claim a sample of 21 measurements of solids concentration after filtering was taken. A mean value of .0265 percent solids and a sample standard deviation of 0.0042 percent solids were found.
    a) Is there reason at the 5% level of significance to suspect that the output is more variable than the salesman claims?
    b) Assuming the answer to part a) is no, is there reason to suspect that the filter is less efficient than the salesman claims at the 5% level of significance?

13. Six random determinations of sulfur content in steel at a particular point in a process gave the values 3.07, 3.11, 3.14, 3.24, 3.16, and 3.08. Assume the values are normally distributed. A previous study based on a sample of 21 random observations gave an estimate of variance of 1.51×10^{-3}. Is the variance significantly higher now? Use the 5% level of significance.

14. The following are the values, in millimeters, obtained by two engineers in ten successive measurements of the same dimension.

    Engineer A   10.06 10.00 9.94 10.10 9.90 10.04 9.98 10.02 9.96 10.00
    Engineer B   10.04 9.94 9.84 9.96 9.92 9.98 9.90 9.94 9.92 9.96
    a) At 10% level of significance, is one engineer more consistent in his measuring than the other?
    b) At 5% level of significance, is there a difference in the mean values obtained by the two engineers?

15. From a set of experimental results the sample estimate of the variance within groups, based on 40 degrees of freedom, is 312, and the sample estimate of the variance between groups, based on 5 degrees of freedom, is 987. At the 5% level of significance, can we say that the difference in conditions between groups has a significant effect? The data have been plotted on normal probability paper, showing reasonable agreement with a normal distribution.

16. Analysis of a set of experiments gives an estimated variance within groups, based on 20 degrees of freedom, of 4.55, and an estimated variance between groups, based on 4 degrees of freedom, of 21.3. Is there evidence to say at the 5% level of significance that the difference between groups is significant? When data are plotted on normal probability paper they show reasonable agreement with a normal distribution.

Statistical Inferences for Variance and Proportion

10.2 Inferences for Proportion

Let us consider a typical engineering problem involving inference for proportion, most often a problem from the area of quality control or quality assurance. Engineers in industry often need to find the proportion of rejected items among the units produced by a production line. We would attack such a problem by taking a random sample. We would examine a certain number of units, say n units, as they are produced. We would determine for that sample the number of rejected units, say x of them. Then the ratio of x to n gives an indication of the proportion of rejects in all the items produced under those conditions. In fact, this turns out to be an unbiased estimate of the proportion of rejects in that population, although it may be a very preliminary estimate. We will still need some indication of how precise the estimate is, and by taking a large enough sample we can make the estimate as precise as desired. Then we might find confidence limits for the proportion of rejects in the population.

If we later take a sample of a suitable size and find that the proportion of rejects in the sample is so large that the difference from the previous result is significant at a particular level of significance, that would be an indication that the proportion of rejects in the population has changed. As another possibility, we may make some modification of operating conditions and take a sample of suitable size. Analysis would indicate whether there is statistically significant evidence that the modification has reduced the proportion of rejects in the population.

The methods we used in Chapter 9 to find answers to similar questions for the mean (and in section 10.1 for the variance) can be applied to questions involving proportion without much modification, but now the binomial distribution will be appropriate instead of the normal distribution or t-distribution or F-distribution.

10.2.1 Proportion and the Binomial Distribution

We have seen in section 5.3(g) that if certain reasonable assumptions are satisfied, the proportion of rejects in a sample is governed by a form of the binomial distribution. If a random sample of *size n* is found to contain *x rejects*, then on the basis of that sample we would estimate the proportion of rejects in the relevant population to be $\hat{p} = \dfrac{x}{n}$. According to equation 5.13 the mathematical expectation of the sample proportion rejected is $\mu_{\hat{p}} = p$, where p is the true proportion of rejects in the population. According to equation 5.14, the variance of the proportion rejected in a random sample of that size is $\dfrac{p(1-p)}{n}$.

10.2.2 Test of Hypothesis for Proportion

If the number of defective items in a sample is too large, we have an indication that the proportion of defective items in the population has become unacceptable.

Chapter 10

(a) Direct Calculation from the Binomial Distribution

If the sample size and number of defective items in the sample are fairly small, we can calculate using the binomial distribution directly.

Example 10.7

Mechanical components are produced continuously in large numbers on a production line. When the machines are correctly adjusted, extensive data show that the proportion of defective components is 0.027. If the proportion of defectives in a sample of size 50 is so large that the result is significant at the 5% level, the production line will be stopped for adjustment.

a) What probability distribution applies?
b) What is the smallest proportion of defective items in a sample of 50 that will stop the production line?

Answer: (a) The binomial distribution applies because there are only two possible results, the probability of defective items is assumed constant, each result is independent of every other result, and the number of trials is fixed.

(b) The production line will be stopped if the proportion of rejected items in a sample of 50 is so large that the observed level of significance is 5% or less.

Null hypothesis, H_0: $p = 0.027$

Alternative hypothesis, H_a: $p > 0.027$ (one-tailed test)

The binomial distribution applies with $n = 50$ and $p = 0.027$, so

$$\Pr[X = x] = {}_{50}C_x \, (0.027)^x (0.973)^{(50-x)}$$

Let the limiting proportion of defective items to stop the production line be $\dfrac{x_{lim}}{n} = \dfrac{x_{lim}}{50}$. Then the cumulative probability of a proportion defective up to and including $\dfrac{x_{lim}}{50}$ must be no more than 5% when the true proportion defective is 0.027. That is, we choose the smallest value of $\hat{p}$ which will satisfy the requirement that $\Pr\left[\hat{p} \geq \dfrac{x_{lim}}{50}\right] \prec 0.05$ on condition that the true proportion defective is $p = 0.027$.

The probability that the sample will contain no rejects is

$$\Pr[\hat{p} = 0] = \Pr[X = 0] = (0.973)^{50} = 0.254$$

Similarly, $\Pr[\hat{p} = 0.02] = \Pr[X = 1] = (50)(0.027)^1 (0.973)^{49} = 0.353$

$$\Pr[\hat{p} = 0.04] = \Pr[X = 2] = \frac{(50)(49)}{2}(0.027)^2 (0.973)^{48} = 0.240$$

Statistical Inferences for Variance and Proportion

$$\Pr[\hat{p} = 0.06] = \Pr[X = 3] = \frac{(50)(49)(48)}{(3)(2)} (0.027)^3 (0.973)^{47} = 0.107$$

$$\Pr[\hat{p} = 0.08] = \Pr[X = 4] = \frac{(50)(49)(48)(47)}{(4)(3)(2)} (0.027)^4 (0.973)^{46} = 0.035$$

$$\Pr[\hat{p} = 0.10] = \Pr[X = 5] = \frac{(50)(49)(48)(47)(46)}{(5)(4)(3)(2)} (0.027)^5 (0.973)^{45} = 0.009$$

Probabilities are decreasing rapidly, and the total probability to this point is 0.998 (to three figures), so the critical number of rejected items at the 5% level of significance has been reached or exceeded. To see just where the boundary for that level of significance is located, we calculate successive cumulative probabilities:

$\Pr[\hat{p} \geq 0.02] = 1 - \Pr[\hat{p} = 0] = 1 - 0.254 = 0.746.$

$\Pr[\hat{p} \geq 0.04] = 1 - \Pr[\hat{p} = 0] - \Pr[\hat{p} = 0.02]$
$= 1 - 0.254 - 0.353 = 0.392$

$\Pr[\hat{p} \geq 0.06] = 1 - \Pr[\hat{p} = 0] - \Pr[\hat{p} = 0.02] - \Pr[\hat{p} = 0.04]$
$= 1 - 0.254 - 0.353 - 0.240 = 0.152$

$\Pr[\hat{p} \geq 0.08] = 1 - \Pr[\hat{p} = 0] - \Pr[\hat{p} = 0.02] - \Pr[\hat{p} = 0.04] - \Pr[\hat{p} = 0.06]$
$= 1 - 0.254 - 0.353 - 0.240 - 0.107 = 0.046$

Since this last result is less than 0.05, and $\Pr[\hat{p} \geq 0.08]$ corresponds to $\Pr[X \geq 4]$, 4 or more defective items in a sample of 50 will be significant at the 5% level of significance. Then the smallest proportion of defective items in a sample of 50 items which will stop the production line will be 0.08.

Example 10.8

This is a continuation of Example 10.7. Now the true probability that any one component is defective has increased to 0.045. What is the probability of a Type II error?

Answer: Remember that a Type II error is accepting a null hypothesis when in fact the null hypothesis is incorrect.

Then $\Pr[\text{Type II error}] = \Pr[\text{observed level of significance} > 5\% \mid H_0 \text{ is not true}]$

In this specific case $\Pr[\text{Type II error} \mid p = 0.045] =$

$\Pr[\text{fewer than 4 defective items in a sample of 50} \mid p = 0.045]$

The binomial distribution still applies, but now

$\Pr[X = x] = {}_{50}C_x (0.045)^x (0.955)^{(50-x)}$

Then $\Pr[X = 0] = (0.955)^{50} = 0.100$

$\Pr[X = 1] = (50)(0.045)^1 (0.955)^{49} = 0.236$

Chapter 10

$$\Pr[X = 2] = \frac{(50)(49)}{2} (0.045)^2 (0.955)^{48} = 0.272$$

$$\Pr[X = 3] = \frac{(50)(49)(48)}{(3)(2)} (0.045)^3 (0.955)^{47} = 0.205$$

Then $\Pr[\text{Type II error} \mid p = 0.045] = \Pr[X \leq 3 \mid p = 0.045]$

$$= 0.100 + 0.236 + 0.272 + 0.205$$

$$= 0.813$$

Thus, if the probability of a defective item has increased to 0.045, the probability that the production line will *not* be stopped for adjustment is 0.813, so the fair odds are more than 4 to 1 that the increased likelihood of defectives will not be detected by any one sample. In almost all practical cases we would require a larger probability of detecting such a large increase in the likelihood of a defective item, so we would probably need to increase the sample size.

As the sample size increases, calculations using the binomial distribution directly become time-consuming, so an alternative method of calculation becomes very desirable. The normal approximation to the binomial distribution can be used if the probability of a defective item or an item of another specific class is close enough to 0.5 and the sample size is large enough. (See the discussion of the rough rule in section 7.6.) Remember that if p, the probability that any single item comes within a particular class, is close to 0 or 1, a larger value of np or $n(1 - p)$ will be required. Engineers often need confidence intervals for quality control problems in which p, the probability of a defective item, is small in relation to 1. In that case very large samples are required before the normal approximation provides satisfactory results. See Example 7.8.

(b) Calculation Using Excel

If a computer with Excel or alternative software is available, another possibility is to use computer calculations. The use of the function BINOMDIST has been discussed in section 5.3(f). It can be used to calculate the individual terms or the cumulative distribution function of the binomial distribution. It requires four parameters: the number of "successes" in a fixed number of trials, the number of trials, the probability of "success" on each trial, and either TRUE to direct the program to calculate cumulative probabilities or FALSE to direct the program to calculate individual probabilities according to the binomial distribution.

Example 10.9

Electrical components are manufactured continuously on a production line. Extensive data show that when all machines are correctly adjusted, a fraction 0.026 of the components are defective. However, some settings tend to vary as production contin-

Statistical Inferences for Variance and Proportion

ues, so the fraction of defective components may increase. A sample of 420 components is taken at regular intervals, and the number of defective components in the sample is counted. If there are more than 16 defective components in the sample of 420, the production line will be stopped and adjustments will be made.

(a) State the null hypothesis and alternative hypothesis in terms of p.

(b) What is the observed level of significance if the number of defective components is just large enough to stop the production line?

(c) Suppose the probability that a component will be defective has increased to 0.040. Then what is the probability of a Type II error?

Answer: a) H_0: $p = 0.026$

H_a: $p > 0.026$ (one-tailed test)

The binomial distribution applies with $n = 420$ and $p = 0.026$.

b) The production line will be stopped if a sample of 420 components contains more than 16 defective items. Then the observed level of significance will be the probability of finding more than 16 defective items in a sample of size 420.

MS Excel can be used to find the observed level of significance. It will be 1 minus the cumulative probability of finding 16 or fewer defective components in a sample of size 420 if the null hypothesis is correct. That will be given by Excel if we enter the expression =1 – BINOMDIST (16,420,0.026,TRUE), where BINOMDIST is an Excel function giving probabilities for the binomial distribution, 16 is the number of defective items, 420 is the sample size, 0.026 is the probability that any one component will be defective, and "TRUE" indicates that we want a cumulative probability. That gives an observed level of significance of 0.0507 or 0.051 or 5.1%.

This is a more accurate result than an answer obtained using the normal approximation to the binomial distribution.

c) Now we want to find the probability of a Type II error when the probability of a defective component on any one trial has increased to 0.040. If we obtain 16 or fewer defective components in a sample consisting of 420 components, we will have no reason to stop the production line.

Pr [16 or fewer defective components in a sample of size 420 | $p = 0.040$] will be given by entering the expression =BINOMDIST(16,420,0.040,TRUE) in Excel. We find that the probability of a Type II error is 0.486 or 48.6%. The probability of detecting an increase in the proportion defective from 0.026 to 0.040 by this scheme of sampling is not much more than 50%. That situation is almost certainly unacceptable. We can reduce the probability of a Type II error by making the sample larger.

Chapter 10

10.2.3 Confidence Interval for Proportion

Unless the sample size is very small, it is not practical to find confidence intervals for proportion by calculations of individual probabilities directly from the binomial distribution. We need to use either a normal approximation or a computer solution.

A computer solution with Excel (except for rather small sample sizes) involves using the function BINOMDIST to obtain cumulative probabilities. Then the goal-seeking algorithm can be used to find the upper limit or the lower limit of the appropriate confidence interval for the proportion p, say the probability that any one item will be defective.

Example 10.10

Mechanical components are being produced continuously. A quality control program for the mechanical components requires a close estimate of the proportion defective in production when all settings are correct. 1020 components are examined under these conditions, and 27 of the 1020 items are found to be defective.

(a) Find a point estimate of the proportion defective.

(b) Find a 95% two-sided confidence interval.

(c) Find an upper limit giving 95% level of confidence that the true proportion defective is less than this limiting value.

Use Excel in parts (b) and (c).

Answer: a) The point estimate of the proportion defective is just $\frac{27}{1020} = 0.0265$.

b) If the probability distribution is not symmetrical, various two-sided confidence intervals can be defined. We will use the confidence interval with equal tails, that is, one in which the probability of a value above the upper limit is equal to the probability of a value below the lower limit. For this problem that would mean 2.5% probability that the proportion defective is above the upper boundary of the confidence interval and 2.5% probability that it is below the lower limit.

These limits can be found using the goal-seeking method on the Formula menu or Tools menu of Excel. At the upper limit we seek a proportion p_{upper} (or p_u) such that the probability of finding 27 or fewer defective items in a sample of size 1020 is 2.5%. In the work sheet shown in Table 10.1 the function =BINOMDIST(27,1020,p_u,true) was entered in cell B10. The cell B9 was selected and named p_u using "Define Name" on the Formula menu. Then cell B10 was selected, and from the Formula or Tools menu "Goal Seek" was chosen. In the "Set Cell" box, the reference B10 appeared. In the "To Value" box the quantity 0.025 was entered. In the "By Changing Cell" box the name p_u was entered, referring to cell B9. Then the OK button was chosen. Then Excel began a numerical algorithm to change the value of p_u in such a way that the goal, 0.025, was approached by the content of the cell B10. The goal can not be attained exactly: the process is terminated by the algorithm when the content of that cell comes within a preset difference from the goal. In the present example the final content of cell B10 was 0.0244 when the value of p_u was 0.0383. The accuracy of the upper confidence limit was checked by entering values close to the given quantity in cells A22:A25. The array function =BINOMDIST(26,1020,A22:A25,true) was entered in cells B22:B25. In this case the value 0.0383 was found to be correct to four decimal places, or three significant figures. The binomial distribution for this situation is shown in Figure 10.7.

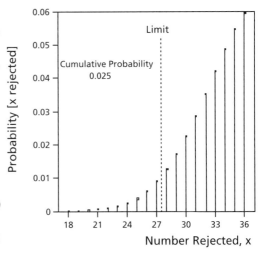

Figure 10.7:
Binomial Distribution at Upper Limit of 95% Confidence Interval, p_{upper} = 0.0383

Similarly, at the lower confidence limit we seek a proportion p_l such that the probability of finding 27 or more defective items in a sample of size 1020 is 2.5.%. But the available function finds a cumulative probability that the number of defective items will be *less* than, or equal to, a limiting number. That limiting number must now be 26 rather than 27 because the binomial distribution is discrete; Pr [R ≥ 27] = 1 − Pr [R ≤ 26]. The binomial distribution for this relationship is shown in Figure 10.8.

The function BINOMDIST(26,1020,p_l,true) was entered in cell B15. The cell B14 was defined as p_l. Then "Goal Seek" was chosen. The reference B15 was placed in the "Set Cell" box, and the quantity 0.975 was entered in the "To Value" box. The name p_l, which refers then to cell B14, was entered in the "By Changing Cell" box. The OK button was chosen to start the algorithm of changing the content of cell B14 so that the content of cell B15 approached the goal of 0.975. The final content of cell B15 was 0.9749 when the content of cell B14 was 0.0175. Checking indicated that this gave a correct answer to four decimal places.

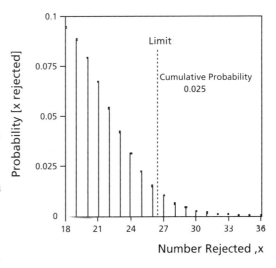

Figure 10.8:
Binomial Distribution at Lower Limit of 95% Confidence Interval, p_{lower} = 0.0175

Chapter 10

Then the 95% two-sided confidence interval is from 0.0175 to 0.0383.

The work sheet is shown in Table 10.1.

Table 10.1: Work Sheet for Example 10.10

	A	B	C	D
1	Confidence Interval for Proportion	**Formula Menu: Goal Seek**		
2				
3	Sample Size = n	1020		
4	Number rejected = x	27		
5	Point Estimate, p_hat = x/n	0.02647059		
6	1 − p_hat =	0.97352941		
7				
8	**Pr[R<=27	p=p_u] -> 0.025**	Set cell B10 to value 0.025 by changing B9		
9	Upper boundary of interval, p_u =	0.0383459		
10	Pr[R<=27	p=p_u]	0.02440541		
11	[Binomdist(27,1020,p_u,true)=]			
12				
13	**Pr[R>=27	p=p_l] ->0.025 or**	**Pr[R<=26	p=p_l] -> 0.975**		Set cell B15 by
14	Lower boundary of interval, p_l =	0.01752218		changing B14
15	Pr[R<=26	p=p_l]	0.97489228		
16	[Binomdist(26,1020,p_l,true)=]			
17				
18	Then 95% confidence interval seems to be			
19	from	0.0175	to	0.0383
20				
21	Check Upper Confidence Limit:	CumProb		
22	0.038	0.02771796	Binomdist(27,1020,A22:A25,true)	
23	0.039	0.01908482		
24	0.0382	0.02575748		
25	0.0383	0.02482385		
26	Check Lower Confidence Limit:	CumProb		
27	0.017	0.98196003	Binomdist(26,1020,A27:A30,true)	
28	0.018	0.96669979		
29	0.0176	0.97367698		

30		0.0175	0.97523058		
31					
32	Then the true limits are from 0.0175 to 0.0383.				

c) A one-sided confidence interval corresponding to Pr $[0 < P \leq p_{upper,2}]$ can be found in the same way as the upper limit for part (b) was found. That gives a 95% one-sided confidence interval of 0 to 0.0363.

10.2.4 Extension

(a) Comparison of Two Sample Proportions

In discussing hypothesis testing for proportion in section 10.2.2 we have assumed that we know without appreciable error the proportion of the defective components when all machines are correctly adjusted. This would require a very large sample, which is often not available. In many cases we must take into account both the variance when all adjustments are correct and the variance in the case being tested. The variance of the sample mean proportion at correct adjustment must be added to the variance of the sample mean proportion being tested, giving the variance of the difference. The only simple calculation available in such a case involves a normal approximation to the binomial distribution.

(b) Sample Size for Required Level of Confidence

Similar to the way sample sizes to reduce standard errors of the mean to required values were found in Examples 7.3 and 7.4, we can find at least approximately the sample size needed to give a required level of confidence that a proportion is within stated limits. This can be found either by "goal-seeking" with Excel or by using a normal approximation to the binomial distribution. For these we need an assumed value of p, the probability of "success," so satisfactory results require a close estimate of p. That estimate is often obtained from a preliminary sample of the population.

Closing Comment

To make confidence intervals for proportion reasonably small often requires large sample sizes, particularly for small proportions such as proportion defective. If the property that makes the items defective can be measured fairly precisely, it will usually be more satisfactory to base quality control on that measurement rather than on the proportion defective.

On the other hand, proportion defective is often quoted as some indication of quality. If that is done, there should be some indication of the confidence limits for proportion to see how reliable this indication is.

Chapter 10

Problems

1. A production line is producing electrical components. Under normal conditions 2.4% of the components are defective. To monitor production, a sample of 18 components is taken each hour. If the number of defectives becomes too high, the production line is stopped and adjustments are made. What distribution applies to the number of defectives in a sample? Write down specifically the null hypothesis and the alternative hypothesis. For 1% level of significance, what is the smallest number of defectives in the sample which should shut down the production line?

2. In problem 1 the probability that any one component will be defective has increased to 6.3%. Now what is the probability of a Type II error?

3. When a production line is properly adjusted, it is found that 4% of the mechanical components produced are defective. Occasionally settings go out of adjustment, and more defectives are produced. A sample of 12 components is examined and the number of defectives is counted. What distribution applies to the number of defectives? What are the null hypothesis and the alternative hypothesis? If the level of significance is set at 1%, how many defectives can be allowed in the sample before any action is taken?

4. In problem 3 adjustments have gone badly wrong so that 7.5% of the components are defective. Now what is the probability of a Type II error?

5. A continuous production line is producing electrical components. When all adjustments are correct, 3.2% of the components from the line are defective. A sample of 480 components is taken every few hours, and the number of defectives is counted. If there are more than 21 defectives in the sample, extensive adjustments will be made. Use the normal approximation to the binomial distribution, remembering the correction for continuity.
   a) State the null hypothesis and alternative hypothesis.
   b) What is the observed level of significance if there are just more than 21 defectives in the sample?
   c) If the probability that a component will be defective has increased to 6.0%, what is the probability of a Type II error?

6. Mechanical components are being produced continuously. When all adjustments are correct, 3.0% of the components from the production line are defective. A sample of 500 components is taken at regular intervals, and the number of defectives is counted. If the number of defectives is large enough to be significant at the 5% level of significance, the production line will be shut down for adjustment. Use the normal approximation to the binomial distribution.
   a) State the null hypothesis and Alternative Hypothesis.
   b) What is the minimum number of defectives in a sample which will result in a shut-down?
   c) If the probability that a component will be defective has increased to 0.060, what is the probability of a Type II error?

Statistical Inferences for Variance and Proportion

Computer Problems

C7. A continuous production line is producing electrical components. When all adjustments are correct, 3.2% of the components from the line are defective. A sample of 480 components is taken every four hours, and the number of defectives is counted. If there are more than 21 defectives in the sample, extensive adjustments will be made.

Use Excel. This is the same problem as number 5, except that that problem was done using a normal approximation.

a) State the null hypothesis and alternative hypothesis.
b) What is the observed level of significance if there are 22 defectives in the sample?
c) If the probability that a component will be defective has increased to 6.0%, what is the probability of a Type II error?

C8. Mechanical components are being produced continuously. When all adjustments are correct, 3.0% of the components from the production line are defective. A sample of 500 components is taken at regular intervals, and the number of defectives is counted. If the number of defectives is large enough to be significant at the 5% level of significance, the production line will be shut down for adjustment.

Use Excel. This is the same problem as number 6, except that that problem was done using the normal approximation to the binomial distribution.

a) State the null hypothesis and alternative hypothesis.
b) What is the minimum number of defectives in a sample which will result in a shut-down?
c) If the probability that a component will be defective has increased to 0.060, what is the probability of a Type II error?

C9. Mechanical components are being produced continuously. When all settings are correct and checked frequently, a sample of 1800 components contains 44 items which are rejected.

a) Find a point estimate of the proportion of the components which are rejected.
b) Find the two-sided 90% confidence interval with equal probability in the two tails.

CHAPTER 11

Introduction to Design of Experiments

This chapter is largely independent of previous chapters, although some previous vocabulary is used here.

Professional engineers in industry or in research positions are very frequently responsible for devising experiments to answer practical problems. There are many pitfalls in the design of experiments, and on the other hand there are well-tried methods which can be used to plan experiments that will give the engineer the maximum information and often more reliable information for a particular amount of effort. Thus, we need to consider some of the more important factors involved in the design of experiments. Complete discussion of design of experiments will be beyond the scope of this book, so the contents of this chapter will be introductory in nature.

We have seen in section 9.2.4 that more information can be gained in some cases by designing experiments to use the paired t-test rather than the unpaired *t*-test. In many other cases there is a similar advantage in designing experiments carefully.

There are complications in many experiments in industry (and also in many research programs) that are not found in most undergraduate engineering laboratories. First, several different factors may be present and may affect the results of the experiments but are not readily controlled. It may be that some factors affect the results but are not of prime interest: they are *interfering* factors, or lurking factors. Often these interfering factors can not be controlled at all, or perhaps they can be controlled only at considerable expense. Very frequently, not all the factors act independently of one another. That is, some of the factors interact in the sense that a higher value of one factor makes the results either more or less sensitive to another factor. We have to consider these complicating factors in planning the set of experiments.

There are several expressions that are key to understanding the design of experiments. Among the most important are:

    Factorial Design
    Interaction
    Replication
    Randomization
    Blocking

Introduction to Design of Experiments

We will see the meaning of these key words and begin to see how to use them in later sections.

11.1 Experimentation vs. Use of Routine Operating Data

Rather than design a special experiment to answer questions concerning the effects of varying operating conditions, some engineers choose to change operating conditions entirely on the basis of routine data recorded during normal operations. Routine production data often provide useful clues to desirable changes in operating conditions, but those clues are usually ambiguous. That is because in normal operation often *more than one governing factor* is changing, and not in any planned pattern. Often some changes in operating conditions are needed to adjust for changes in inputs or conditions beyond the operator's control. Some factors may change uncontrollably. The operator may or may not know how they are changing. If he or she knows what factors have changed, it may be extremely desirable to make compensating changes in other variables. For example, the composition of material fed to a unit may change because of modifications in operations in upstream units or because of changes in the feed to the entire plant. The composition of crude oil to a refinery often changes, for instance, with increase or decrease of rates of flow from the individual fields or wells, which give petroleum of different compositions. In some cases considerable time is required before steady, reliable data are obtained after any change in operating conditions, so another change may be made, consciously or unconsciously, before the full effects of the first change are felt.

If more than one factor changes during routine operation, it becomes very difficult to say whether the changes in results are due to one factor or to another, or perhaps to some combination of the two. Two or more factors may change in such a way as to reinforce one another or cancel one another out. The results become ambiguous. In general, it is much better to use *planned experiments* in which changes are chosen carefully.

An exception to this statement is when all the factors affecting a result and the mathematical form of the function are well known without question, the mathematical forms of the different factors are different from one another, but the values of the coefficients of the relations are not known. In that case, data from routine operations can give satisfactory results.

11.2 Scale of Experimentation

Experiments should be done on as small a scale as will give the desired information. Managers in charge of full-scale industrial production units are frequently reluctant to allow any experimentation with the operating conditions in their units. This is because experiments might result in production of off-specification products, or the rate of production might be reduced. Either of these might result in very appreciable financial penalties. Production managers will probably be more willing to perform

Chapter 11

experiments if conditions are changed only moderately, especially if experiments can be done when the plant is not operating at full capacity. A technique of making a series of small planned changes in operating conditions is known as *evolutionary operation*, or EVOP. The changes at each step can be made small enough so that serious consequences are very unlikely. After each step, results to that point are evaluated in order to decide the most logical next step. For further information see the book by Box, Hunter, and Hunter, shown in the List of Selected References in section 15.2.

Sometimes experiments to give the desired information can be done on a laboratory scale at very moderate cost. In other cases the information can be gained from a pilot plant which is much smaller than full industrial scale, but with characteristics very similar to full-scale operation. In still other cases, there is no substitute for experiments at full scale, and the costs are justified by the improved technique of operation.

11.3 One-factor-at-a-time vs. Factorial Design

What sort of experimental design should be adopted? One approach is to set up standard operating conditions for all factors and then to vary conditions from the standard set, one factor at a time. An optimum value of one factor might be found by trying the effects of several values of that factor. Then attention would shift to a different factor. This is a plan that has been used frequently, but in general it is not a good choice at all.

It would be a reasonable plan if all the factors operated independently of one another, although even then it would not be an efficient method for obtaining information. If the factors operated independently, the results of changing two factors

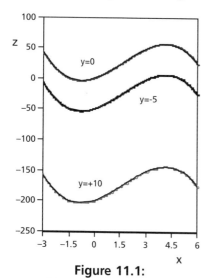

**Figure 11.1:
Profiles without Interaction**

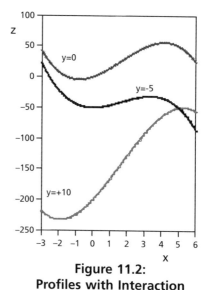

**Figure 11.2:
Profiles with Interaction**

together would be just the sum of the effects of changing each factor separately. Figure 11.1 shows profiles for such a situation. Each profile represents the variation of the response, z, as a function of one factor, x, at a constant value of the other factor, y. If the factors are completely independent and so have no interaction, the profiles of the response variable all have the same shape. The profiles for different values of y differ from one another only by a constant quantity, as can be seen in Figure 11.1. In that case it would be reasonable to perform measurements of the response at various values of x with a constant value of y, and then at various values of y with a constant value of x. If we knew that x and y affected the response independently of one another, that set of measurements would give a complete description of the response over the ranges of x and y used. But that is a very uncommon situation in practice.

Very frequently some of the factors *interact*. That is, changing factor A makes the process more or less sensitive to change in factor B. This is shown in Figure 11.2, where an interaction term is added to the variables shown in Figure 11.1. Now the profiles do not have the same shape, so measurements are required for various combinations of the variables.

For example, the effect of increasing temperature may be greater at higher pressure than at lower pressure. If there were no interaction the simplest mathematical model of the relation would be

$$R_i = K_0 + K_1 P + K_2 T + \varepsilon_i \tag{11.1}$$

where R_i is the response (the dependent variable) for test i,

K_0 is a constant,

P is pressure,

K_1 is the constant coefficient of pressure,

T is temperature,

K_2 is the constant coefficient of temperature,

and ε_i is the error for test i.

This is similar to the profiles of Figure 11.1, except that Figure 11.1 does not include errors of measurement.

When interaction is present the simplest corresponding mathematical model would be

$$R_i = K_0 + K_1 P + K_2 T + K_3 PT + \varepsilon_i \tag{11.2}$$

where K_3 is the constant coefficient of the product of temperature and pressure. Then the term $K_3 PT$ represents the interaction. In this case the interaction is second-order because it involves two independent variables, temperature and pressure. If it involved three independent variables it would be a third-order interaction, and so on.

Chapter 11

Second-order interactions are very common, third-order interactions are less common, and fourth order (and higher order) interactions are much less common.

Consider an example from fluid mechanics. The drag force on a solid cylinder moving through a fluid such as air or water varies with the factors in a complex way. Under certain conditions the drag force is found to be proportional to the product of the density of the fluid and the square of the relative velocity between the cylinder and the fluid far from the cylinder, say $F_d = K\rho u^2$, omitting the effect of errors of measurement. This does not correspond to equation 11.1. If a density increase of 1 kg m^{-3} at a relative velocity of 0.1 m s^{-1} increased the drag force by 1 N, the same density increase at a relative velocity of 0.2 m s^{-1} would increase the drag force by 4 N. Then in such a case density and relative velocity interact. In this case the interacting relationship could be changed to a non-interacting relationship by a change of variables, taking logarithms of the measurements, but there are other relationships involving interaction which can not be simplified by any change of variable.

Interaction is found very frequently, and its possibility must always be considered. However, the one-factor-at-a-time design would not give us *any precise information about the interaction*, and results from that plan of experimentation might be extremely misleading. In order to determine the effects of interaction, we must compare the effects of increasing one variable at different values of a second variable.

What is an alternative to changing one factor at a time? Often the best alternative is to conduct tests at all possible combinations of the operating factors. Let's say we decide to do tests at three different values (levels) of the first factor and two different levels of the second factor. Then measurements at *each* of the three levels of the first factor would be done at each level of the second factor, so at (3)(2) = 6 different combinations of levels of the two factors. This is called a **factorial design**. Then suitable algebraic manipulation of the data can be used to separate the results of changes in the various factors from one another. The techniques of analysis of variance (which will be introduced in chapter 12) and multilinear regression (which will be introduced in Chapter 14) are often used to analyze the data.

Now let us look at an example of factorial design.

Example 11.1

Figure 11.3 shows a case where three factors are important: temperature, pressure, and flow rate. We choose to operate each one at two different levels, a low level and a high level. That will require $2^3 = 8$ different experiments for a complete factorial design. If the number of factors increases, the required number of runs goes up exponentially.

Introduction to Design of Experiments

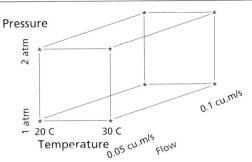

Figure 11.3: Factorial Design

For the three factors, each at two levels, measurements would be taken at the following conditions:

	Pressure	Temperature	Flow Rate
1.	1 atm	20° C	0.05 m^{3}s^{-1}
2.	1 atm	20° C	0.1 m^{3}s^{-1}
3.	1 atm	30° C	0.05 m^{3}s^{-1}
4.	1 atm	30° C	0.1 m^{3}s^{-1}
5.	2 atm	20° C	0.05 m^{3}s^{-1}
6.	2 atm	20° C	0.1 m^{3}s^{-1}
7.	2 atm	30° C	0.05 m^{3}s^{-1}
8.	2 atm	30° C	0.1 m^{3}s^{-1}

Each of these conditions is marked by an asterisk in Figure 11.3.

A possible set of results (for one flow rate) is illustrated in Figure 11.4. The interaction between temperature and pressure is shown by the result that increasing pressure increases the response considerably more at the higher temperature than at the lower temperature.

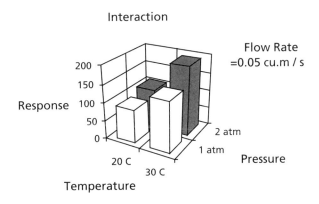

Figure 11.4: An Illustration of Interaction

277

Chapter 11

In the early stages of industrial experimentation it is usually best to choose only two levels for each factor varied in the factorial design. On the basis of results from the first set of experiments, further experiments can be designed logically and may well involve more than two levels for some factors. If a complete factorial design is used, experiments would be done at every possible combination of the chosen levels for each factor.

In general we should not try to lay out the whole program of experiments at the very beginning. The knowledge gained in early trials should be used in designing later trials. At the beginning we may not know the ranges of variables that will be of chief interest. Furthermore, before we can decide logically how many repetitions or replications of a measurement are needed, we require some information about the variance corresponding to errors of measurement, and that will often not be available until data are obtained from preliminary experiments. Early objectives of the study may be modified in the light of later results. In summary, the experimental design should usually be *sequential* or *evolutionary* in nature.

In some cases the number of experiments required for a complete factorial design may not be practical or desirable. Then some other design, such as a fractional factorial design (to be discussed briefly in section 11.6), may be a good choice. These alternative designs do not give as much information as the corresponding full factorial design, so care is required in considering the relative advantages and disadvantages. For example, the results of a particular alternative design may indicate either that certain factors of the experiment have important effects on the results, or else that certain interactions among the factors are of major importance. Which explanation applies may not be at all clear. In some instances one of the possible explanations is very unlikely, so the other explanation is the only reasonable one. Then the alternative design would be a logical choice. But assumptions always need to be recognized and analyzed. *Never* adopt an alternative experimental design without examining the assumptions on which it is based.

Everything we know about a process or the theory behind it should be used, both in planning the experiment and in evaluating the results. The results of previous experiments, whether at bench scale, pilot scale, or industrial scale, should be carefully considered. Theoretical knowledge and previous experience must be taken into account. At the same time, the possibility of effects that have not been encountered or considered before must not be neglected.

These are some of the basic considerations, but several other factors must be kept in mind, particularly replication of trials and strategies to prevent bias due to interfering factors.

11.4 Replication

Replication means doing each trial more than once at the same set of conditions. In some preliminary exploratory experiments each experiment is done just twice (two replications). This gives only a very rough indication of how reproducible the results are for each set of conditions, but it allows a large number of factors to be investigated fairly quickly and economically. We will see later that in some cases no replication is used, particularly in some types of preliminary experiments.

Usually some (perhaps many) of the factors studied in preliminary experiments will have negligible effect and so can be eliminated from further tests. As we zoom in on the factors of greatest importance, larger numbers of replications are often used. Besides giving better estimates of reproducibility, further replication reduces the standard error of the mean and tends to make the distribution of means closer to the normal distribution. We have already seen in section 8.3 that the mean of repeated results for the same condition has a standard deviation that becomes smaller as the sample size (or number of replications) increases. Thus, larger numbers of replications give more reliable results. Furthermore, we have seen in section 8.4 that as the sample size increases, the distribution of sample means comes closer to the normal distribution. This stems from the Central Limit Theorem, and it justifies use of such tests as the t-test and the F-test.

11.5 Bias Due to Interfering Factors

Very frequently in industrial experiments an interfering factor is present that will bias the results, giving systematic error unless we take suitable precautions. Such interfering factors are sometimes called "lurking variables" because they can suddenly assault the conclusions of the unwary experimenter. We are often unaware of interfering factors, and they are present more often than we may suspect.

(a) Some Examples of Interfering Factors

We will consider several examples. First, the temperature of the surrounding air may affect the temperature of a measurement, and so the results of that measurement. This is particularly so if measurements are performed outdoors. Variations of air temperature between summer and winter are so great that they are unlikely to be neglected, but temperature variations from day to day or from hour to hour may be overlooked. Atmospheric temperature has some tendency to persist: if the outside air temperature is above average today, the air temperature tomorrow is also likely to be above average. But at some point the weather pattern changes, so air temperature may be higher than average today and tomorrow, but below average a week from today. Thus, taking some measurements today and others of the same type a week later could bias the results. If we do experiments with one set of conditions today and similar experiments with a different set of conditions a week later, the differences in results may in fact be due to the change in air temperature rather than to the intended difference in conditions. Shorter-term variations in air temperature could also cause bias, since the temperature of outside air varies during the day. There may be a systematic difference between results taken at 9 A.M. and results taken at 1:30 P.M. We have used temperature variation as an example of an interfering factor, but of course this factor can be taken into account by suitable temperature measurements. Other interfering factors are not so easily measured or controlled.

An unknown trace contaminant may affect the results of an experiment. If the feed to the experimental equipment comes from a large surge tank with continual flow in and out and good mixing, higher than average concentration of a contaminant is likely to persist over an interval of time. This might mean that high results today are likely to be associated with high results tomorrow, and low results today are likely to be associated with low results tomorrow, but the situation might be quite different a week later. Thus, tests today may show a systematic difference, or bias, from tests a week from today.

Another instance involves tests on a machine that is subject to wear. Wear on the machine occurs slowly and gradually, so the effect of wear may be much the same today as tomorrow, but it may be quite different in a month's time (or a week's time, depending on the rate of wear). Thus, wear might be an uncontrolled variable that introduces bias.

(b) Preventing Bias by Randomization

One remedy for systematic error in measurements is to make **random choices** of the assignment of material to different experiments and of the order in which experiments are done. This ensures that the interfering factors affecting the results are, to a good approximation, independent of the intended changes in experimental conditions. We may say that the interfering factors are "averaged out." Then, the biases are minimized and usually become negligible. The random choices might be made by flipping a coin, but more often they are made using tables of random numbers or using random numbers generated by computer software. Random numbers can be obtained on Excel from the function RAND, and that procedure will be discussed briefly in section 11.5(c).

Very often we don't know enough about the factors affecting a measurement to be sure that there is no correlation of results with time. Therefore, if we don't take precautions, some of the intentional changes in operating factors may by chance coincide with (and become confused with) some accidental changes in other operating factors over which we may have no control. Only by randomizing can we ensure that the factors affecting the results are statistically independent of one another. Randomization should *always* be used if interfering factors may be present.

However, in some cases randomization may not be practical because of difficulties in adjusting conditions over a wide range in a reasonable time. Then some alternative scheme may be required; at the very least the possibility of bias should be recognized clearly and some scheme should be developed to minimize the effects of interfering factors. Wherever possible, randomization *must* be used to deal with possible interfering factors.

Example 11.1 (continued)

Now let's add randomization to the experimental design begun in Example 11.1. Let each test be done twice in random order. The order of performing the experiments has been randomized using random numbers from computer software with the following results:

Order		Conditions	
1	1 atm	30°C	0.05 m^3 s^{-1}
2	2 atm	20°C	0.05 m^3 s^{-1}
3	1 atm	20°C	0.05 m^3 s^{-1}
4	1 atm	20°C	0.1 m^3 s^{-1}
5	2 atm	30°C	0.05 m^3 s^{-1}
6	1 atm	30°C	0.1 m^3 s^{-1}
7	2 atm	20°C	0.1 m^3 s^{-1}
8	2 atm	30°C	0.1 m^3 s^{-1}
9	1 atm	20°C	0.05 m^3 s^{-1}
10	2 atm	20°C	0.05 m^3 s^{-1}
11	1 atm	30°C	0.05 m^3 s^{-1}
12	1 atm	20°C	0.1 m^3 s^{-1}
13	2 atm	30°C	0.05 m^3 s^{-1}
14	1 atm	30°C	0.1 m^3 s^{-1}
15	2 atm	20°C	0.1 m^3 s^{-1}
16	2 atm	30°C	0.1 m^3 s^{-1}

Example 11.2

A stirred liquid-phase reactor produces a polymer used (in small concentrations) to increase rates of filtration. A pilot plant has been built to investigate this process. The factors being studied are temperature, concentration of reactant A, concentration of reactant B, and stirring rate. Each factor will be studied at two levels in a factorial design, and each combination of conditions will be repeated to give two replications.

a) How many tests will be required?

b) List all tests.

c) What order of tests should be used?

Chapter 11

Answer:

a) Number of tests = $(2)(2^4) = 32$.

b) The tests are shown in Table 11.1 below:

Table 11.1: List of Tests for Example 11.2

Temperature	Concentration of A	Concentration of B	Stirring Rate	Number of Tests
Low	Low	Low	Low	2
High	Low	Low	Low	2
Low	High	Low	Low	2
Low	Low	High	Low	2
Low	Low	Low	High	2
High	High	Low	Low	2
High	Low	High	Low	2
High	Low	Low	High	2
Low	High	High	Low	2
Low	High	Low	High	2
Low	Low	High	High	2
High	High	High	Low	2
Low	High	High	High	2
High	Low	High	High	2
High	High	Low	High	2
High	High	High	High	2
			Total:	**32**

c) The order in which tests are performed should be determined using random numbers from a table or computer software.

Example 11.3

A mechanical engineer has decided to test a novel heat exchanger in an oil refinery. The major result will be the amount of heating produced in a petroleum stream which varies in composition. Tests will be done at two compositions and three flow rates. To get sufficient precision each combination of composition and flow rate will be tested five times. A factorial design will be used.

a) How many tests will be required?
b) List all tests.
c) How will the order of testing be determined?

Introduction to Design of Experiments

Answer:

a) (2)(3)(5) = 30 tests will be required.

b) The tests will be:

Composition	Flow Rate	Number of Tests
Low	Low	5
Low	Middle	5
Low	High	5
High	Low	5
High	Middle	5
High	High	5
	Total	30

c) Order of testing will be determined by random numbers from a table or from computer software.

Example 11.4

Previous studies in a pilot plant have been used to set the operating conditions (temperature and pressure) in an industrial reactor. However, some of the conditions in the full scale industrial equipment are not quite the same, so the plant engineer has decided to perform tests in the industrial plant. The plant manager is afraid that changes in operating conditions may produce off-specification product, so only small changes in conditions will be allowed at each stage of experimentation. If results from the first stage appear encouraging, further stages of experimentation can be done. (This is a form of evolutionary operation, or EVOP.) A simple factorial pattern will be used: temperature settings will be increased and decreased from normal by 2°C, and pressure will be increased and decreased from normal by 0.05 atm. The plant engineer has calculated that to get sufficient precision with these small changes, eight tests will be required at each set of conditions. The normal temperature is 125° C, and normal pressure is 1.80 atm.

a) How many tests will be required in the first stage of experimentation?
b) List the tests.
c) How will the order of testing be determined?

Answer:

a) (8)(2)(2) = 32 tests will be required.

b) The tests will be as follows:

Temperature	Pressure	Number Of Tests
123°C	1.75 atm	8
123°C	1.85 atm	8
127°C	1.75 atm	8
127°C	1.85 atm	8
	Total	32

Chapter 11

c) Order of testing will be determined by random numbers from a table or from computer software.

(c) Obtaining Random Numbers Using Excel

Excel can be used to generate random numbers to randomize experimental designs. The function = RAND( ) will return a random number greater than or equal to zero and less than one. We obtain a new number every time the function is entered or the work sheet is recalculated. If we want a random number between 0 and 10, we multiply RAND( ) by 10. But that will often give a number with a fraction.

If we want an integer, we can apply the function = INT(number), which rounds the number down, not up, to the nearest integer; e.g., INT(7.8) equals 7. We can nest the functions inside one another, so INT(RAND( )*10) will give a random integer in the inclusive interval from 0 to 9. If we want a random integer in the inclusive interval from 1 to 10 , we can use INT(RAND( )*10) + 1. Similarly, if we want a random integer in the inclusive interval from 1 to 8, that will be given by the function INT(RAND( )*8) +1, and so on for other choices. If we want a whole sequence of random numbers we can use an array function.

Example 11.5

To obtain a sequence of thirty random integers in the inclusive interval from 1 to 6, cells A1 to A30 were selected, and the formula =INT(RAND( ) *6) + 1 was entered as an array formula. The results were as follows in row form:

```
2  1  5  6  2  1  2  3  3  2  2  5  6  6  3
1  2  5  1  1  2  2  3  5  5  2  1  2  2  1
```

(Notice that this could be considered a numerical simulation of a discrete random variable in which the integers from 1 to 6 inclusive are equally likely.)

Now suppose we assign the numbers 1 to 6 to six different engineering measurements. We want a random order of these six measurements. The result of Example 11.5 would give us that random order if we use each digit the first time it appears and discard all repetitions. If by chance the thirty random digits do not contain at least five of the six digits from 1 to 6, we can repeat the whole operation. But the probability of that is small.

A complication is that changing other parts of the work sheet causes the random number generator to re-calculate, giving a new set of random integers. To avoid that, we can convert the contents of cells A1 to A30 to constant values. (See references or the Help function on Excel to see how to do that.)

Example 11.5 (continued)

Every time a cell in column A contained a repetition of one of the integers from 1 to 6, an x was placed in the corresponding row of column B until all the integers in the required interval had appeared or the list of integers was exhausted. Then the unrepeated integers (in order) were entered into column C. The results (as a row instead of a column) were as follows:

    2   1   5   6   3   [4]

Notice that, as it happened, 4 did not appear among the thirty integers. However, since 4 is the only missing integer, it must be the last of the six.

This, then, would give a random order of performing the six engineering measurements.

(d) Preventing Bias by Blocking

Blocking means dividing the complete experiment into groups or blocks in which the various interfering factors (especially uncontrolled factors) can be assumed to be more homogeneous. Comparisons are made using the various factors involved in each block. Blocking is used to *increase the precision* of an experiment by reducing the effect of interfering factors. Results from "block" experiments are applicable to a wider range of conditions than if experiments were limited to a *single* uniform set of conditions. For example, technicians may perform tests in somewhat different ways, so we might want to remove the differences due to different technicians in exploring the effect of using raw material from different sources.

A paired t-test is an example of an experimental design using blocking. In section 9.2.4 we examined the comparison of samples using paired data. Two different treatments (evaporator pans) were investigated over several days. Then the day was the blocking factor. The randomization was of the relative positions of the pans. The measured evaporation was the response. The difference between daily amounts of evaporation from the paired pans was taken as the variate in order to eliminate the effects of day-to-day variation in atmospheric conditions. Example 9.10 illustrated this procedure.

The paired t-test involved two factors or treatments, but blocking can be extended to include more than two treatments. Randomization is used to protect us from unknown sources of bias by performing treatments in *random* order within each block. (Notice that randomization is still required within the block.)

A block design does not give as much information as a complete factorial design, but it generally requires fewer tests, and the extra information from the factorial design may not be desired. Fewer tests are required because we do not usually repeat tests within blocks for the same experimental conditions. Error is estimated from variations which are left after variance between blocks and variance between experimental treatments have been removed. This assumes that the average effect of different treatments is the same in all the blocks. In other words, we assume there is no interaction between the effects of treatments and blocks. *Caution*: if there is any reason to think that treatments and blocks may not be independent and so may interact, we should include adequate replication so that that interaction can be

Chapter 11

checked. If interaction between treatment and block is present but not determined, the randomized blocking design will result in an inflation of the error estimate, so the test for significant effects becomes less sensitive.

Blocking should be used when we wish to eliminate the distortion caused by an interfering variable but are not very interested in determining the effects of that variable. If two factors are of comparable interest, a design blocking out one of the two factors should not be used. In that case, we should go to a complete factorial design.

Why is blocking required if randomization is used? If some factor is having an effect even though we may not know it, randomizing will tend to prevent us from coming to incorrect conclusions. However, the interfering effect will still increase the standard deviation or variance due to error. That is, the variance due to experimental error will be inflated by the variance due to the interfering factor; in other words, the randomized interference will add to the random "noise" of the measurements. If the error variance is larger, we are less likely to conclude that the effect of an experimental variable is statistically significant. In that case, we are less likely to be able to come to a definite conclusion. (This is essentially the same as the effect of a larger standard deviation in the t-test: if the standard deviation is larger, t will be smaller, so the difference is not so likely to be significant. See the comparison of Examples 9.9 and 9.10 in sections 9.2.3 and 9.2.4.) If we use blocking, we can both avoid incorrect conclusions and increase the probability of coming to results that are statistically significant. Therefore, the priority is to *block* all the interfering factors we can (so long as interaction is not appreciable), then to *randomize* in order to minimize the effects of factors we can't block.

Example 11.6

Example 9.10 was for a paired t-test. The evaporation from two types of evaporation pans placed side-by-side was compared over ten days. Any difference due to relative position was "averaged out" by randomizing the placement of the pans.

Now we wish to compare three evaporator pans: A, B, and C. The pans are placed side-by-side again, and their relative positions are decided randomly using random numbers. We know that evaporation from the pans will vary from one day to another with changing weather conditions, but that variation is not of prime interest in the current test. The day becomes the blocking variable. The resulting order is shown in terms of A, B, and C for the relative positions of the evaporator pans, and 1, 2, 3, 4, 5, 6, 7, 8, 9, 10 for the day. Then the order in which tests are done might be 1:BCA, 2:BAC, 3:CBA,4: ACB, 5:CAB,6: ABC,7: CBA, 8:ACB,9: ABC, 10:CAB.

Example 11.7

Let us modify Example 11.1 by adding blocking for effects associated with time. The principal factors are still temperature, pressure, and flow, but we suspect that some

Introduction to Design of Experiments

interfering factors may vary from one day to another but are very unlikely to interact appreciably with the principal factors. After considering the possible interfering factors and their time scales, we decide that variations within an eight-hour shift are likely to be negligible, but variations between Tuesday and Friday may well be appreciable. We can do eight trials on day shift each day. Then the eight trials on Tuesday will be one block, and the eight trials on Friday will be another block. The order of performing tests on each day will be randomized, again using random numbers from computer software.

Answer:

The orders of trials for Tuesday and for Wednesday are shown below.

Order for Tuesday	**Conditions**		
1	1 atm	30°C	0.1 m^3 s^{-1}
2	2 atm	30°C	0.1 m^3 s^{-1}
3	1 atm	30°C	0.05 m^3 s^{-1}
4	1 atm	20°C	0.05 m^3 s^{-1}
5	2 atm	20°C	0.05 m^3 s^{-1}
6	2 atm	30°C	0.05 m^3 s^{-1}
7	2 atm	20°C	0.1 m^3 s^{-1}
8	1 atm	20°C	0.1 m^3 s^{-1}

Order for Friday	**Conditions**		
1	1 atm	30°C	0.1 m^3 s^{-1}
2	1 atm	30°C	0.05 m^3 s^{-1}
3	1 atm	20°C	0.1 m^3 s^{-1}
4	1 atm	20°C	0.05 m^3 s^{-1}
5	2 atm	30°C	0.1 m^3 s^{-1}
6	2 atm	20°C	0.05 m^3 s^{-1}
7	2 atm	30°C	0.05 m^3 s^{-1}
8	2 atm	20°C	0.1 m^3 s^{-1}

We should note two points. First, there is no replication, so error is estimated from residuals left after the effects of temperature, pressure, and flow rate (and their interactions), and differences between blocks, have been accounted for. (If there is any interaction between the main effects (temperature, pressure, flow) and the blocking variable (time of run), that will inflate the error estimate.) Second, if a complete four-factor experimental design had been used with two replications, the required number of tests would have been $(2)(2^4) = 32$, whereas the block design requires 16 tests.

Chapter 11

Variations of blocking are used for specific situations. If the number of factors being examined is too large, not all of them can be included in each block. Then a plan called a *balanced incomplete block* design would be used. If more than one interfering factor needs to be blocked, then plans called *Latin squares* and *Graeco-Latin squares* would be considered.

The design of experiments that include blocking is discussed in more detail in books by Box, Hunter, and Hunter and by Montgomery (for references see section 15.2). There are considerations and pitfalls not discussed here.

Example 11.8

A civil engineer is planning an experiment to compare the levels of biological oxygen demand (B.O.D.) at three different points in a river. These are just upstream of a sewage plant, five kilometers downstream, and ten kilometers downstream. Assume that in each case samples will be taken in the middle of the stream (in practice samples would likely be taken at several positions across the stream and averaged). One set of samples will be taken at 6 a.m., another will be taken at 2 p.m., and a third will be taken at 10 p.m. The design will block the effect of time of day, as some interfering factors may be different at different times. However, the interaction of these interfering factors with location is expected to be negligible.

a) If there is no replication aside from blocking, how many tests are required?
b) List all tests.
c) Specify which set of tests constitute each block.
d) How should the order of tests be determined?

Answer:

a) Number of tests = (3)(3) = 9.
b) The tests are:

   6 a.m.: just upstream, 5 kilometers downstream, and 10 kilometers downstream.

   2 p.m.: just upstream, 5 kilometers downstream, and 10 kilometers downstream.

   10 p.m.: just upstream, 5 kilometers downstream, and 10 kilometers downstream.

c) The 6 a.m. tests make up one block, the 2 p.m. tests make up another block, and the 10 p.m. tests will make up a third block.
d) The order in which tests are performed should be determined using random numbers from a table or computer software.

11.6 Fractional Factorial Designs

As the number of different factors increases, the number of experiments required for a full factorial design increases exponentially. Even if we test each factor at only two levels, with only one measurement for each combination of conditions, a complete factorial design for n factors requires 2^n separate measurements. If there are five

Introduction to Design of Experiments

factors, that comes to thirty-two separate measurements. If there are six factors, sixty-four separate measurements are required.

In many cases nearly as much useful information can be obtained by doing only half or perhaps a quarter of the full factorial design. Certainly, somewhat less information is obtained, but by careful design of the experiment the omitted information is not likely to be important. This is referred to as a fractional factorial design or a fractional replication.

Fractional design will be illustrated for the simple case of three factors, each at only two levels, with no replication of measurements. In Example 11.1 we saw the complete factorial design for this set of conditions. The asterisk (*) marked in Figure 11.3 each of the $2^3 = 8$ combinations of conditions for the full factorial design. In Figure 11.5, on the other hand, only half of the 2^3 combinations of conditions are marked by asterisks, and only these measurements would be made for the fractional factorial design.

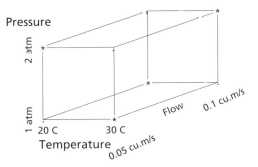

Figure 11.5: Fractional Factorial Design

Thus, measurements would be made at the following conditions:

1. 1 atm 20° C 0.1 m³ s⁻¹
2. 2 atm 20° C 0.05 m³ s⁻¹
3. 1 atm 30° C 0.05 m³ s⁻¹
4. 2 atm 30° C 0.1 m³ s⁻¹

Notice that in the first three sets of conditions two factors are at the lower value and one is at the higher value, but in the last set all three factors are at the higher value. Then half of the sets show each factor at its lower value, and half show each factor at its higher value. The order of performing the experiments would be randomized.

This half-fraction, three-factor design involving $\frac{1}{2}(2^3) = 4$ combinations of conditions is not really practical because it does not allow us to separate a main-factor effect from the second-order interaction of the other two effects. For example, the effect of pressure is confused (or confounded, as the statisticians say) with the interaction between temperature and flow rate, and these quantities cannot be sepa-

Chapter 11

rated. Since second-order interactions are found frequently, this is not a satisfactory situation.

The half-fraction design is more useful when the number of factors is larger. Consider the case where five factors are being investigated, so the full factorial design at two levels with no replication would require $2^5 = 32$ runs. A half-fraction factorial design would require $\frac{1}{2}(2^5) = 16$ runs. It will give essentially the same information as the full factorial design if either of two conditions is met: either (1) at least one factor has negligible effect on any of the results, so that its main effect and all its interactions with other variables are negligible; or else (2) all the three-factor and four-factor interactions are negligible, so that only the main effects and second-order interactions are appreciable. In exploratory studies we wish to see which factors are important, so we will often find that one or more factors have no appreciable effect. In that situation the first condition will be satisfied. Furthermore, the effects of three-factor and higher order interactions are usually negligible, so that the second condition would frequently be satisfied. There are some cases in which third-order interactions produce appreciable effects, but these cases are not common. Then if analysis of the half-factor factorial design indicates that we cannot neglect *any* of the factors, further investigation should be done.

Remember that the half-fraction design requires measurements at half of the combinations of conditions needed for a full factorial design. Then if the results of the half-fraction experimental designs are ambiguous, often the *other* half of a complete factorial design can be run later. The two halves are together equivalent to a full factorial, run in two blocks at different times. Then analyzing the two blocks together gives nearly as much information as a complete factorial design, provided that interfering factors do not change too much in the intervening time.

Problems

1. A mechanical engineer has designed a new electronic fuel injector. He is developing a plan for testing it. He will use a factorial design to investigate the effects of high, medium and low fuel flow, and high, medium and low fuel temperature. Four tests will be done at each combination of conditions.
   a) How many tests will be required?
   b) List them.

2. A civil engineer is performing tests on a screening device to remove coarser solids from storm overflow of untreated sewage. A stream is directed at a rotating collar screen, 7.5 feet in diameter, made of stainless steel mesh. The engineer intends to try three mesh sizes (150 mesh, 200 mesh and 230 mesh), two rotational speeds (30 and 60 R.P.M.), three flow rates (550 gpm, 900 gpm and 1450 gpm), and three time intervals between back washes (20, 40 and 60 seconds).
   a) How many tests will be required for a complete factorial design without replication?

b) List them.
c) How will the order of tests be determined?

3. A pilot plant investigation is concerned with three variables. These are temperature (160° C and 170° C), concentration of reactant (1.0 mol / L and 1.5 mol / L), and catalyst (Catalyst A and Catalyst B). The response variable is the percentage yield of the desired product. A factorial design will be used.
   a) If each combination of variables is tested twice (two replications), how many tests will be required?
   b) List them.
   c) How will the order of tests be determined?

4. A metal alloy was modified by adding small amounts of nickel and / or manganese. The breaking strength of each resulting alloy was measured. Tests were performed in the following order:
   1. 1.5% Ni, 0% Mn
   2. 3% Ni, 2% Mn
   3. 1.5% Ni, 1% Mn
   4. 1.5% Ni, 0% Mn
   5. 0% Ni, 1% Mn
   6. 3% Ni, 1% Mn
   7. 0% Ni, 2% Mn
   8. 1.5% Ni, 1% Mn
   9. 0% Ni, 0% Mn
   10. 3% Ni, 0% Mn
   11. 0% Ni, 1% Mn
   12. 1.5% Ni, 2% Mn
   13. 3% Ni, 1% Mn
   14. 1.5% Ni, 2% Mn
   15. 0% Ni, 0% Mn
   16. 3% Ni, 2% Mn
   17. 3% Ni, 0% Mn
   18. 0% Ni, 2% Mn

   a) How many factors are there? How many levels have been used for each factor? How many replications have been used (remember that this can be a fraction)?
   b) Then summarize the experimental design: factorial design or alternatives, characteristics.
   c) Verify that these characteristics would result in the number of test runs shown.

5. Four different methods of determining the concentration of a pollutant in parts per million are being compared. We suspect that two technicians obtain somewhat different results, so a randomized block design will be used. Each technician will run all four methods on different samples. Unknown to the

technicians, all samples will be taken from the same well stirred container. All determinations will be run in the same morning.
a) How many determinations are required?
b) List them.
c) How will the order of determinations be decided?

6. Tests are carried out to determine the effects of various factors on the percentage of a particular reactant which is reacted in a pilot-scale chemical reactor. The effects of feed rate, agitation rate, temperature, and concentrations of two reactants are determined. Test runs were performed in the following order:

	Feed Rate L/m	Agitation Rate RPM	Temperature °C	Conc. of A mol/L	Conc. of B mol/L
1.	15	120	150	0.5	1.0
2.	10	100	150	1.0	0.5
3.	15	100	150	1.0	1.0
4.	10	120	150	0.5	0.5
5.	15	120	160	0.5	0.5
6.	10	120	150	1.0	1.0
7.	15	100	160	1.0	0.5
8.	15	120	160	1.0	1.0
9.	10	100	150	0.5	1.0
10.	15	100	160	0.5	1.0
11.	15	100	150	0.5	0.5
12.	10	100	160	0.5	0.5
13.	10	100	160	1.0	1.0
14.	15	120	150	1.0	0.5
15.	10	120	160	0.5	1.0
16.	10	120	160	1.0	0.5

a) How many factors are there? How many levels have been used for each factor? How many replications have been used (remember that this can be a fraction)?
b) Then summarize the experimental design: factorial design or alternatives, characteristics.
c) Verify that these characteristics would result in the number of test runs shown.

Introduction to Design of Experiments

Computer Problems

C7. A program of testing the effects of temperature and pressure on a piece of equipment involves a total of eight runs, two at each of four combinations of temperature and pressure. Let us call these four combinations of conditions numbers 1, 2, 3, and 4. Use random numbers from Excel to find two random orders of four tests each in which the tests of conditions 1, 2, 3, and 4 might be conducted.

C8. An engineer is planning tests on a heat exchanger. Six different combinations of flow rates and fluid compositions will be used, and the engineer labels them as 1, 2, 3, 4, 5, 6. She will test each combination of conditions twice. Use random numbers from Excel to find a random order of performing the twelve tests.

C9. Simulate two samples of size ten from a binomial distribution with $n = 5$ and $p = 0.12$. Use the *Analysis Tools* command on Excel. Produce an output table with ten columns and two rows. Use the *Frequency* function to prepare a frequency table, which must be labeled clearly.

C10. Use the *Analysis Tools* command on Excel to simulate the results of a sampling scheme. The probability of a defect on any one item is 0.07, and each sample contains 12 items. Simulate the results of three samples. Use the *Frequency* function to prepare a frequency table, and label it clearly.

CHAPTER **12**

Introduction to Analysis of Variance

This chapter requires an understanding of the material in sections 3.1, 3.2, 3.4, and 10.1.2.

In Chapter 11 we have looked briefly at some of the principal ideas and techniques of designing experiments to solve industrial problems. Once the data have been obtained, how can we analyze them?

The analysis of data from designed experiments is based on the methods developed previously in this book. The data are summarized as means and variances, and graphical presentations are used, especially to check the assumptions of the methods. Confidence intervals and tests of hypothesis are used to infer results. But some techniques beyond those described previously are usually needed to complete the analysis.

The two main techniques used in analysis of data from factorial experiments, with and without blocking, are the *analysis of variance* and *multiple linear regression*. Analysis of variance will be introduced here. Multiple linear regression will be introduced in section 14.6.

Analysis of variance, or ANOVA, is used with both quantitative data and qualitative data, such as data categorizing products as good or defective, light or heavy, and so on. With both quantitative and qualitative data, the function of analysis of variance is to find whether each input has a significant effect on the system's response. Thus, analysis of variance is often used at an early stage in the analysis of quantitative data. Multiple linear regression is often used to obtain a quantitative relation between the inputs and the responses. But analysis of variance often has another function, which is to test the results of multiple linear regression for significance.

We saw in Chapter 8 that one of the most desirable properties of variance is that independent estimates of variance can be added together. This idea can be extended to separating the quantities leading to variance into various logical components. One component can be ascribed to differences resulting from various main effects such as varying pressure or temperature. Another component may be due to interactions between main effects. A third component may come from blocking, which has been discussed in section 11.5 (d). A final component may correspond to experimental error.

Introduction to Analysis of Variance

We have seen in section 9.2 that an estimate of variance is found by dividing the sum of squares of deviations from a mean by the number of degrees of freedom. We can partition both the total sum of squares and the total number of degrees of freedom into components corresponding to main effects, interactions, perhaps blocking, and experimental error. Then, for each of these components the sum of squares is divided by the corresponding number of degrees of freedom to give an estimate of the variance. Estimates of variance are often called "mean squares."

Then the *F*-test, which was discussed in section 10.1.2, is used to examine the various ratios of variances to see which ratios are statistically significant. Is a ratio of variances consistent with the hypothesis that the two population variances are equal, so that differences between them are due only to random chance variations? More specifically, the null hypothesis to be tested is usually that various factors make no difference to population means from different treatments or different levels of treatment. The F-test is used to see whether the null hypothesis can be accepted at a stated level of significance.

We will find that calculations for the analysis of variance using a calculator involve considerable labor, especially if the number of components investigated is fairly large. Almost always in practice, therefore, computer software is used to do the calculations more easily. The problems in this chapter can be solved using a calculator. If the reader chooses, he or she can use a computer spreadsheet with formulas involving basic operations. In some problems that will save considerable time. However, more complex, pre-programmed computer packages such as SAS or SPSS should not be used until the reader has the basic ideas firmly in mind. This is because these use "black-box" functions which require little thinking from someone who is learning.

12.1 One-way Analysis of Variance

Let us consider the simplest case, analyzing a randomized experiment in which only one factor is being investigated. Two or more replicates are used for each separate treatment or level of treatment, and there will be three or more treatments or levels. The null hypothesis will be that all treatments produce equal results, so that all population means for the various treatments are equal. The alternative hypothesis will be that at least two of the treatment means are not equal.

(a) Basic Relations

Say there are m different treatments or levels of treatment, and for each of these there are r different observations, so r replicates of each treatment. Let y_{ik} be the kth observation from the ith treatment. Let the mean observation for treatment i be $\bar{y}_i$, and let the mean of all N observations be $\bar{\bar{y}}$, where $N = (m)(r)$. Then

$$\bar{y}_i = \frac{\sum_{k=1}^{r} y_{ik}}{r} \qquad (12.1)$$

and
$$\bar{\bar{y}} = \frac{\sum_{i=1}^{m} \bar{y}_i}{m} = \frac{m \sum_{k=1}^{r} y_{ik}}{N} \tag{12.2}$$

The total sum of squares of the deviations from the mean of all the observations, abbreviated as *SST*, is

$$SST = \sum_{i=1}^{m} \sum_{k=1}^{r} (y_{ik} - \bar{\bar{y}})^2 \tag{12.3}$$

The treatment sum of squares of the deviations of the treatment means from the mean of all the observations, abbreviated as *SSA*, is

$$SSA = r \sum_{i=1}^{m} (\bar{y}_i - \bar{\bar{y}})^2 \tag{12.4}$$

The within-treatment or residual sum of squares of the deviations from the means within treatments is

$$SSR = \sum_{i=1}^{m} \sum_{k=1}^{r} (y_{ik} - \bar{y}_i)^2 \tag{12.5}$$

This residual sum of squares can give an estimate of the error.

It can be shown algebraically that

$$\sum_{i=1}^{m} \sum_{k=1}^{r} (y_{ik} - \bar{\bar{y}})^2 = r \sum_{i=1}^{m} (\bar{y}_i - \bar{\bar{y}})^2 + \sum_{i=1}^{m} \sum_{k=1}^{r_i} (y_{ik} - \bar{y}_i)^2 \tag{12.6}$$

or

$$SST = SSA + SSR \tag{12.7}$$

Thus, the total sum of squares is partitioned into two parts.

The degrees of freedom are partitioned similarly. The total number of degrees of freedom is $(N-1)$. The number of degrees of freedom between treatment means is the number of treatments minus one, or $(m-1)$. By subtraction, the residual number of degrees of freedom is $(N-1) - (m-1) = (N-m)$. This must be the number of degrees of freedom within treatments.

Then the estimate of the variance within treatments is

$$s_R^2 = \frac{SSR}{N-m} \tag{12.8}$$

This is often called the "within treatments" mean square. It is an estimate of error, giving an indication of the precision of the measurements.

The estimate of the variance obtained from differences of the treatment means (so between treatments) is

$$s_A^2 = \frac{SSA}{m-1} \tag{12.9}$$

Introduction to Analysis of Variance

This is often called the "between treatments" mean square.

Now the question becomes: are these two estimates of variance (or mean square deviations) compatible with one another? The specific null hypothesis is that the population means for different treatments or levels are equal. If that null hypothesis is true, the variability of the sample means will reflect the intrinsic variability of the individual measurements. (Of course the variance of the sample means is different from the variance of individual measurements, as we have seen in connection with the standard error of the mean, but that has already been taken into account.) If the population means are *not* equal, the true population variance between treatments will be *larger* than the true population variance within treatments. Is s_A^2, the estimate of the variance between means, significantly larger than s_R^2, the estimate of the variance within means? (Note that this is a one-tailed test.) But before the question can be addressed properly, we need to check that the necessary assumptions have been met.

(b) Assumptions

The first assumption is the mathematical model of the relationship we are investigating. Usually for a start the analysis of variance is based on the simplest mathematical model for each situation. In this section we are considering a single factor at various levels and with some replication. The simplest model for this case is

$$y_{ik} = \mu + \alpha_i + \varepsilon_{ik} \tag{12.10}$$

where y_{ik} is the kth observation from the the ith treatment (as before),

μ is the true overall mean (for the numbers of treatments and replicates used in this experiment),

α_i is the incremental effect of treatment i, such that $\alpha_i = \mu_i - \mu$,

μ_i is the true population mean for treatment i, and

ε_{ik} is the error for the kth observation from the ith treatment.

This mathematical model is the simplest for this situation, but if we find that it is not consistent with the data, we will have to modify it. For example, if we turn up evidence of some interfering factor or "lurking variable," a more elaborate model will be required; or, if the data do not fit the linear relation of equation 12.10, changes may be required to get a better fit.

Other important assumptions are that the observations for each treatment are at least approximately normally distributed, and that observations for all the treatments have the same population variance, σ^2, but the treatment means do not have to be the same. More specifically, the errors ε_{ik} must (to a reasonable approximation) be independently and identically distributed according to a normal distribution with mean zero and unknown but fixed variance σ^2. However, according to Box et al. (see section 15.2 for references) the analysis of variance as discussed in this chapter is not

Chapter 12

sensitive to moderate departures from a normal distribution or from equal population variances. In this sense the ANOVA method is said, like the *t*-test, to be robust.

If there are any biasing interfering factors, randomizing the order of taking and testing the sample will usually make the normal distribution approximately appropriate. However, there will still be an inflation of the error variance if biasing factors are present. Notice that if randomization has not been done properly in the situation where biases are present, the assumption of a normal distribution will not be appropriate. Any *outliers*, points with very large errors, may cause serious problems.

(c) Diagnostic Plots

The assumptions should be checked by various diagnostic plots of the residuals, which are the differences between the observations, y_{ik}, and $\hat{y}_{ik}$, the best estimates of the true values according to the mathematical model. Thus, the residuals are $(y_{ik} - \hat{y}_{ik})$. In the case of one-way analysis of variance, where only one factor is an input, the best estimates would be $\bar{y}_i$. The plots are meant to diagnose any major discrepancies between the assumptions and reality in the situation being studied. If there are any unexplained systematic variations of the residuals, the assumptions must be questioned skeptically.

The following plots should be examined carefully:

(1) a stem-and-leaf display (or equivalent, such as a dot diagram or normal probability plot) of all the residuals. Is this consistent with a normal distribution of mean zero and constant variance? Are there any outliers? If we have sufficient data, a similar plot should be shown for each treatment or level.

(2) a plot of residuals against estimated values ($\hat{y}_{ik}$, which is equal to $\bar{y}_i$ in this case). Is there any indication that variance becomes larger or smaller as $\hat{y}_{ik}$ increases?

(3) a plot of residuals against time sequence of measurement (and also time sequence of sampling if that is different). Is there any indication that errors are changing with time?

(4) a plot of residuals against any variable, such as laboratory temperature, which might conceivably affect the results (if such a plot seems useful). Are there any trends?

These plots are similar to those recommended in the books by Box, Hunter, and Hunter and by Montgomery (see references in section 15.2).

Each plot should be considered carefully. If plot (1) is not reasonably symmetrical and consistent with a normal distribution, some change of variable should be considered. If plot (1) shows one or more outliers, the corresponding numbers should be checked to see if some obvious mistake (such as an error of recording an observation) is present. However, in the absence of any obvious error the outlier should not

Introduction to Analysis of Variance

be discarded, although the assumption of an underlying normal distribution should be questioned. Careful examination of remaining outliers will often give useful information, clues to desirable changes to the assumed relationship.

If plot (2) indicates that variance is not constant with varying magnitude of estimated values, then the assumption of constant variance is apparently not satisfied. Then the mathematical model needs to be adjusted. For example, if the residuals tend to increase as $\hat{y}_{ik}$ increases, the percentage error may be approximately constant. This would imply that the mathematical model might be improved by replacing y_{ik} by $\log(y_{ik})$ in equation 12.10.

If plot (3) shows a systematic trend, there is some interfering factor which is a function of time. It might be a temperature variation, or possibly improvement in experimental technique as the experimenter learns to make measurements more exactly. If the order of testing has been properly randomized, the assumption of a normal distribution of errors will be approximately satisfied, but the estimated error will be inflated by any systematic interfering factor.

Any trends in plot (4) will require modification of the whole analysis.

Let us begin an example by calculating means for the various treatments and examining the diagnostic plots.

Example 12.1

Four specimens of soil were taken from each of three different locations in the same locality, and their shear strengths were measured. Data are shown below. Does the location affect the shear strength significantly? Use the 5% level of significance.

Sequence of testing	Location number	Shear strength N/m2
1	2	2940
2	2	2940
3	2	2940
4	3	3482.5
5	1	4000
6	1	4000
7	1	4000
8	3	3482.5
9	2	2940
10	3	3482.5
11	1	4000
12	3	3482.5

Chapter 12

Answer:

The simplest mathematical model for this case is given by equation 12.10:

$$y_{ik} = \mu + \alpha_i + \varepsilon_{ik}$$

The best estimates of the shear strength, $\hat{y}_{ik}$, are given by the means for the various locations.

First, we need to arrange the data by location (which is the "treatment" in this case) and calculate the treatment means and the overall mean. Then the residuals are calculated in lines 15 to 28 of the spreadsheet of Table 12.1 with results in lines 31 to 43.

Calculations of sums of squares are shown in lines 44 to 50, and estimates of variances or "mean squares" are shown in lines 51 and 52. Now the degrees of freedom should be partitioned. The total number of degrees of freedom is $(N - 1) = (3)(4) - 1 = 11$. Between treatment means we have $(m - 1) = 3 - 1 = 2$ degrees of freedom. The number of degrees of freedom within treatments by difference is then $(N - m) = 12 - 3 = 9$. An observed variance ratio is shown in line 53. Calculations can be done using either a pocket calculator or a spreadsheet: Table 12.1 shows a spreadsheet.

Table 12.1: Spreadsheet for Example 12.1, One-way Analysis of Variance

	A	B	C	D	E	F
15	Sorted Data:	y ik				
16	Location, i	Shear Strength	Sequence	Observ. no., k	Location Means, y i(bar)	
17	1	4010	5	1		
18	1	3550	6	2		
19	1	4350	7	3	SUM(B17:B20)/4=	
20	1	4090	11	4		4000
21	2	2970	1	1		
22	2	2320	2	2		
23	2	2910	3	3	SUM(B21:B24)/4=	
24	2	3560	9	4		2940
25	3	3650	4	1		
26	3	3470	8	2		
27	3	3650	10	3	SUM(B25:B28)/4=	
28	3	3160	12	4		3482.5

29 Overall Mean, y(barbar) = (E20+E24+E28)/3 = 3474.16667

30 Residuals: y ik - y i(bar)

Introduction to Analysis of Variance

31	Location,i		Residual			
32	1	B17:B20-E20	10	(Array formulas:		
33			-450	see Appendix B)		
34			350			
35			90			
36	2	B21:B24-E24	30			
37			-620			
38			-30			
39			620			
40	3	B25:B28-E28	167.5			
41			-12.5			
42			167.5			
43			-322.5			

44 **SSA, eqn** 12.4:

45 $\quad$ 4 *SUM(y i(bar)-y(bar bar))^2=

46 $\quad$ =4*((E20-E29)^2 +(E24-E29)^2+(E28-E29)^2= **2247616.67 SSA**

47 $\quad$ **SSR**, eqn 12.5 :

48 SUMi(SUMk(y ik -y i(bar))^2=

49 $\quad\quad$ =(10^2+450^2+350^2+90^2) +

50 $\quad\quad\quad$ +(167.5^2+12.5^2+167.5^2+322.5^2)= **1264075 SSR**

51 $\quad$ (s A)^2= $\quad$ SSA/(m-1)= $\quad$ E46/(3-1)= $\quad$ 1123808.33

52 $\quad$ (s R)^2= $\quad$ SSR/(N-m)= $\quad$ E50/(12-3)= $\quad$ 140452.778

53 **f obs** =(sA)^2/(s R)^2 = $\quad\quad$ D51/D52= $\quad$ **8.00132508**

Now we check the residuals. A stem-and-leaf display of all the residuals is shown in Table 12.2. The stem is the digit corresponding to hundreds, from −6 to +6.

Table 12.2: Stem-and-leaf display of residuals

Stem	Leaf	Frequency
−6	2	1
−5		0
−4	5	1
−3	2	1
−2		0
−1		0
−0	3 1	2
+0	1 3 9	3

Chapter 12

1	6 6	2
2		0
3	5	1
4		0
5		0
6	2	1

Considering the small number of data, Table 12.2 is consistent with a normal distribution of mean zero and constant variance. There is no indication of any outliers.

Plots of residuals against treatment means, $\hat{y}_{ik}$, and against time sequence of measurement are shown in Figure 12.1.

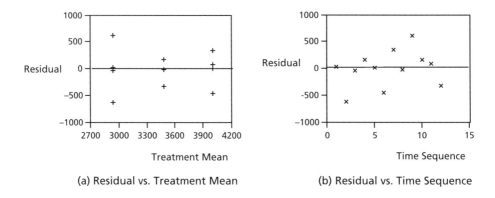

(a) Residual vs. Treatment Mean   (b) Residual vs. Time Sequence

Figure 12.1: Plots of Residuals

Neither plot of Figure 12.1 shows any significant pattern, so the assumptions appear to be satisfied. If calculations were being done with a calculator, the residuals would be checked before proceeding with calculations of sums of squares.

(d) Table for Analysis of Variance

Now we are ready to proceed to the analysis of variance, which we will discuss in general first, then apply it to Example 12.1. A table should be constructed like the one shown in general in Table 12.3 below.

Table 12.3: Table of One-way Analysis of Variance

Sources of Variation	Sums of Squares	Degrees of Freedom	Mean Squares	Variance Ratios
Between treatments	SSA	$(m-1)$	s_A^2	$f_{observed} = \dfrac{s_A^2}{s_R^2}$
Within treatments	SSR	$(N-m)$	s_R^2	
Total (about the grand mean, $\bar{\bar{y}}$)	SST	$(N-1)$		

Introduction to Analysis of Variance

The null hypothesis and the alternative hypothesis must be stated:

H_0: $\alpha_i = 0$ for all values of i (or $\mu_1 = \mu_2 = \mu_3 = ...$).

H_a: $\alpha_i > 0$ for at least one treatment.

If the null hypothesis is true, then $\sigma_A^2 = \sigma_R^2$, so the departure of $f_{observed}$ from 1 is due only to random fluctuations. If the null hypothesis is not true, the variance between treatments, σ_A^2, will be larger than the variance within treatments, σ_R^2, because at least one of the true treatment means will be different from the others.

The calculated variance ratio, $f_{observed} = \dfrac{s_A^2}{s_R^2}$, should be compared with critical values of the F-distribution for $(m-1)$ and $(N-m)$ degrees of freedom. This is the same comparison as was done in section 10.1.2. If $f_{observed} > f_{critical}$ at a particular level of significance, the test results are significant at that level.

Notice that this analysis of variance tests the null hypothesis that the means of all the treatment populations are equal. If we have only two treatments, this is equivalent to the t-test with the null hypothesis that the two population means are equal. This has been described in section 9.2.3. If we have more than two treatments, by the analysis of variance we examine all the treatment means together and so avoid problems of perhaps selecting subjectively the pairs of treatment means which are most favorable to a particular conclusion.

The table of analysis of variance for Example 12.1 is shown below in Table 12.4. This table summarizes the results of lines 44 to 53 in Table 12.1.

Example 12.1 (continued)

Table 12.4: Table of One-way Analysis of Variance for Soil Strengths

Sources of Variation	Sums of Squares	Degrees of Freedom	Mean Squares	Variance Ratios
Between treatments	2,247,616.7	2	1,123,808.3	$f_{observed} =$
Within treatments	1,264,075	9	140,452.78	$= \dfrac{1,123,808.3}{140,452.78}$
Total	3,511,691.7	11		$= 8.001$

The null hypothesis and alternative hypothesis are as follows:

H_0: $\mu_{Location\ 1} = \mu_{Location\ 2} = \mu_{Location\ 3} = \mu_{Location\ 4}$

H_a: At least one of the population means for a location is not equal to the others.

The observed value of f is compared to the limiting value of f for the corresponding degrees of freedom at the 5% level of significance from tables or software. We

Chapter 12

find f_{limit} or $f_{\text{critical}} = 4.26$, whereas f_{observed} from Table 12.4 is 8.001. Since $f_{\text{observed}} > f_{\text{limit}}$, the variance ratio is significant at the 5% level of significance.

Therefore we reject the null hypothesis and so accept the alternative hypothesis. Thus, at the 5% level of significance the location does affect the shear strength.

12.2 Two-way Analysis of Variance

Now the effects of more than one factor are considered in a factorial design. The possibility of interaction must be taken into account. This corresponds to the experimental designs used in Examples 11.2, 11.3, and 11.4, and, with modification, to the fractional factorial designs discussed briefly in section 11.6.

Let's consider the relations for a factorial design involving two separate factors. Later we can extend the analysis for larger numbers of factors.

Say there are m different treatments or levels for factor A, and n different treatments or levels for factor B. All possible combinations of the treatments of factor A and factor B will be investigated. Let us perform r replications of each combination of treatments (the number of replications could vary from one factor to another, but we will simplify a little here).

For this case the observations y must have three subscripts instead of two: i, j, and k for the ith treatment of factor A, the jth treatment of factor B, and the kth replication of that combination. Then an individual observation is represented by y_{ijk}. $\bar{y}_{ij}$ will be the mean observation for the (ij)th combination or cell, the mean of all the observations for the ith treatment of factor A and the jth treatment for factor B. $\bar{\bar{y}}_i$ will be the mean observation for the ith treatment of factor A (at all levels of factor B). $\bar{\bar{y}}_j$ will be the mean observation for the jth observation of factor B (at all levels of factor A). Then we have

$$\bar{y}_{ij} = \frac{\sum_{k=1}^{r} y_{ijk}}{r} \tag{12.11}$$

Averaging all the observations for the ith treatment of factor A gives

$$\bar{\bar{y}}_i = \frac{\sum_{j=1}^{n} \bar{y}_{ij}}{n} = \frac{\sum_{k=1}^{r}\sum_{j=1}^{n} y_{ijk}}{(r)(n)} \tag{12.12}$$

Similarly, averaging all the observations for the jth treatment of factor B gives

$$\bar{\bar{y}}_j = \frac{\sum_{i=1}^{m} \bar{y}_{ij}}{m} = \frac{\sum_{k=1}^{r}\sum_{i=1}^{m} y_{ijk}}{(r)(m)} \tag{12.13}$$

Introduction to Analysis of Variance

The grand average, $\overline{\overline{\overline{y}}}$, is given by

$$\overline{\overline{\overline{y}}} = \frac{\sum_{i=1}^{m} \overline{\overline{y}}_i}{m} = \frac{\sum_{j=1}^{n} \overline{\overline{y}}_j}{n} = \frac{\sum_{i=1}^{m}\sum_{j=1}^{n}\sum_{k=1}^{r} y_{ijk}}{mnr} \tag{12.14}$$

The *total sum of squares* of the deviations of individual observations from the mean of all the observations is

$$SST = \sum_{i=1}^{m}\sum_{j=1}^{n}\sum_{k=1}^{r} (y_{ijk} - \overline{\overline{\overline{y}}})^2 \tag{12.15}$$

For *factor A* the *treatment sum of squares* of the deviations of the treatment means from the mean of all the observations is

$$SSA = nr \sum_{i=1}^{m} (\overline{\overline{y}}_i - \overline{\overline{\overline{y}}})^2 \tag{12.16}$$

Similarly for *factor B* the *treatment sum of squares* is

$$SSB = mr \sum_{j=1}^{n} (\overline{\overline{y}}_j - \overline{\overline{\overline{y}}})^2 \tag{12.17}$$

But we have also the interaction between factors A and B. The *interaction sum of squares* for these two factors is

$$SS(AB) = r \sum_{i=1}^{m}\sum_{j=1}^{n} (\overline{y}_{ij} - \overline{\overline{y}}_i - \overline{\overline{y}}_j - \overline{\overline{\overline{y}}})^2 \tag{12.18}$$

The *residual sum of squares,* from which an estimate of error can be calculated, is

$$SSR = \sum_{i=1}^{m}\sum_{j=1}^{n}\sum_{k=1}^{r} (y_{ijk} - \overline{y}_{ij})^2 \tag{12.19}$$

It can be shown algebraically that

$$SST = SSA + SSB + SS(AB) + SSR \tag{12.20}$$

The total *number of degrees of freedom*, $N - 1 = (m)(n)(r) - 1$, is partitioned into the degrees of freedom for factor A, $(m - 1)$; the degrees of freedom for factor B, $(n - 1)$; the degrees of freedom for interaction, $(m - 1)(n - 1)$. The degrees of freedom within cells available for estimating error is the remaining number, $(m)(n)(r - 1)$.

Then the estimate of the variance obtained from the variability within cells is

$$s_R^2 = \frac{SSR}{(m)(n)(r-1)} \tag{12.21}$$

This is often called the *residual mean square* or sometimes the error mean square.

The estimate of the variance obtained from the variability of the treatment means for factor A is called *the mean square for Main Effect A* and is given by

$$s_A^2 = \frac{SSA}{m-1} \qquad (12.22)$$

The estimate of the variance obtained from the variability of means for factor B is called *the mean square for Main Effect B* and is given by

$$s_B^2 = \frac{SSB}{n-1} \qquad (12.23)$$

The estimate of the variance obtained from the interaction between effects A and B is

$$s_{(AB)}^2 = \frac{SS(AB)}{(m-1)(n-1)} \qquad (12.24)$$

This is called the *mean square for interaction between A and B*.

Once again, various assumptions must be examined before we can proceed to the variance-ratio test. The first assumption is the mathematical model. The simplest mathematical model for the case of two experimental factors with replication and interaction is

$$y_{ijk} = \mu + \alpha_i + \beta_j + (\alpha\beta)_{ij} + \varepsilon_{ijk} \qquad (12.25)$$

where y_{ijk} is the kth observation of the ith level of factor A and the jth level of factor B

μ is the true overall mean

α_i is the incremental effect of treatment i, such that $\alpha_i = \mu_i - \mu$

β_j is the incremental effect of treatment j, such that $\beta_j = \mu_j - \mu$

μ_i is the true population mean for the ith level of factor A

μ_j is the true population mean for the jth level of factor B

$(\alpha\beta)_{ij}$ is the interaction effect for the ith level of factor A and the jth level of factor B,

and ε_{ijk} is the error for the kth observation of the ith level of factor A and the jth level of factor B.

This is the simplest mathematical model, but again it may not be the most appropriate. If equation 12.25 applies, the best estimate of the true values would be

$$\hat{y}_{ijk} = \bar{\bar{\bar{y}}} + (\bar{\bar{y}}_i - \bar{\bar{\bar{y}}}) + (\bar{\bar{y}}_j - \bar{\bar{\bar{y}}}) + (\bar{y}_{ij} - \bar{\bar{y}}_i - \bar{\bar{y}}_j + \bar{\bar{\bar{y}}}) = \bar{y}_{ij} \qquad (12.26)$$

Then residuals are given by $(y_{ijk} - \hat{y}_{ijk})$.

Again, we are assuming that the errors ε_{ijk} are (to a good approximation) independently and identically distributed according to a normal distribution with mean zero and fixed but unknown variance σ^2.

Introduction to Analysis of Variance

These assumptions are checked by the same plots as were used in section 12.1 for the one-way analysis of variance. These were plots (1) to (4) of section (c) of that section.

If plot (2) indicates that the variance is not constant with varying estimates of the measured output, $\hat{y}_{ijk}$, some modification of the mathematical model is required. The book by Box, Hunter, and Hunter gives a full discussion and an example in their section 7.8.

If the plots give no significant indication that any of the assumptions are incorrect, we can go on to the analysis of variance for this case. The null hypotheses and alternative hypotheses are as follows:

H_0: $\alpha_i = 0$ for all values of i (or $\mu_1 = \mu_2 = \mu_3 = ... = \mu_a$);
H_a: $\alpha_i > 0$ for at least one treatment.

H_0: $\beta_j = 0$ for all values of j (or equal values of μ_j);
H_a: $\beta_j > 0$ for at least one treatment.

H_0: $(\alpha\beta)_{ij} = 0$ for all values of i and j;
H_a: $(\alpha\beta)_{ij} > 0$ for at least one cell.

As we discussed in section 12.1, if some of the alternative hypotheses are true, some of the corresponding true population variances will be larger than the true variance for error. Then the F-test is used to see whether the other estimates of variance are significantly larger than the estimate of variance corresponding to error (note again that this is a one-sided test).

A table should be constructed as in Table 12.5 below.

Table 12.5: Table of Analysis of Variance for a Factorial Design

Sources of Variation	Sums of Squares	Degrees of Freedom	Mean Squares	Variance Ratios
Main effect A	SSA	$(m-1)$	s_A^2	$f_{observed1} = \dfrac{s_A^2}{s_R^2}$
Main effect B	SSB	$(n-1)$	s_B^2	$f_{observed2} = \dfrac{s_B^2}{s_R^2}$
Interaction AB	SS(AB)	$(m-1)(n-1)$	$s_{(AB)}^2$	$f_{observed3} = \dfrac{s_{(AB)}^2}{s_R^2}$
Error	SSR	$(m)(n)(r-1)$	s_R^2	
Total	SST	$(N-1)$		

Chapter 12

Again, the observed ratios of variance can be compared to the tabulated values of F for the appropriate numbers of degrees of freedom and for various levels of significance according to the one-sided F-test.

If there are more than two factors in the factorial design, there will be further main effects in Table 12.7 and further interactions. Thus, if there are three factors, the main effects might be A, B, and C, and the corresponding interactions would be AB, AC, BC, and ABC.

If some main effects or interactions show no sign of being statistically significant, their sums of squares and degrees of freedom are sometimes combined with the error sum of squares and error degrees of freedom, respectively, to give improved estimates of the error mean squares.

Sometimes it is assumed that fourth-order (or perhaps third-order) and higher-order interactions are negligible. Then replications are sometimes omitted, so that the error mean squares are estimated entirely from the higher-order interactions.

An extension of this is used in analyzing fractional factorial designs. If the number of factors is large, some main effects and their interactions with other main effects are likely to have negligible significance. Then these main effects and interactions are used to estimate the error of measurements. This is discussed in Chapters 12 and 13 of the book by Box, Hunter, and Hunter. Since fractional factorial designs are used for exploratory investigations, often graphical analysis of the results is sufficient to show which variables require more detailed examination. This, also, is discussed in the book by Box, Hunter, and Hunter.

Example 12.2

A chemical process is being investigated in a pilot plant. The factors under study are, first the catalyst, Liquid Catalyst 1 or Liquid Catalyst 2, and then the concentration of each (1 gram/liter or 2 grams/liter). Two replicate runs are done for each combination of factors. The response, or dependent variable, is the percentage yield of the desired product. Results are shown in Table 12.6 below.

Table 12.6: Percentage Yields for Catalyst Study

	Yield	
Concentration	Catalyst 1	Catalyst 2
1g/L	49.3	47.4
	53.4	50.1
2g/L	63.6	49.7
	59.2	49.9

Introduction to Analysis of Variance

Is the yield significantly different for a different catalyst or concentration or combination of the two? Use the 5% level of significance.

Answer: The plots of the residuals were examined in the same way as for the previous example and showed no significant patterns. They will be omitted here for the sake of brevity.

Table 12.7: Spreadsheet for Example 12.2, Two-way Analysis of Variance

	A	B	C	D	E	F
3	Catalyst,i ->	1	2			
4	Concentration	Yield	, y ijk			
5	1g/L,j=1	49.3	47.4			
6		53.4	50.1			
7	2g/L,j=2	63.6	49.7			
8		59.2	49.9			
9	y ij(bar):	cat 1 (i=1)	cat 2 (i=2)		cat 1 (i=1)	cat 2 (i=2)
10	1g/L,j=1	(B5+B6)/2=	(C5+C6)/2=	1g/L, j=1	51.35	48.75
11	2g/L, j=2	(B7+B8)/2=	(C7+C8)/2=	2g/L, j=2	61.4	49.8
12	y i(barbar):	cat 1 (i=1)	cat 2 (i=2)		cat 1 (i=1)	cat 2 (i=2)
13		(E10+E11)/2=	(F10+F11)/2=	56.375	49.275	
14	y j(barbar):					
15	1g/L, j=1	(E10+F10)/2=	1g/L,j=1	50.05		
16	2g/L, j=2	(E11+F11)/2=	2g/L,j=2	55.6		
17	y (barbarbar):	(E13+F13)/2		52.825		
18	(check:)	(E15+E16)/2=	52.825			
19	Residuals	y ijk -y ij(bar)	i=1	i=2		
20		j=1	-2.05	-1.35	B5:C5-E10:F10	
21			2.05	1.35	B6:C6-E10:F10	
22		j=2	2.2	-0.1	B7:C7-E11:F11	
23			-2.2	0.1	B8:C8-E11:F11	
24	For catalyst, SSA=2*2*SUM((y i(barbar)-y(barbarbar))^2) =					
25	4*((E13-D17)^2+(F13-D17)^2)=			100.82	**SSA**	
26	For concentration, SSB=2*2*SUM((y j(barbar)-y(barbarbar))^2 =					
27	4*((E15-D17)^2+(E16-D17)^2)=			61.605	**SSB**	
28	For interaction, SS(AB)=2*SUMiSUMj((y ij(bar)-y i(barbar)-y j(barbar)+y(barbarbar))^2					
29	= 2*((E10-E13-E15+D17)^2+(E11-E13-E16+D17)^2+					
30	+(F10-F13-E15+D17)^2+(F11-F13-E16+D17)^2)=		40.5	**SS(AB)**		

Chapter 12

31	For residual,	SSR=SUMiSUMjSUMk(y ijk-y ij(bar))=					
32		C20^2+D20^2+C21^2+D21^2+C22^2+D22^2+C23^2+D23^2=					
33				21.75		**SSR**	
34	**df** total=(2)*(2)*(2)-1=			7 df:			
35	Catalyst, df(A)	= 2-1 =		1	A		
36	Conc., df(B)	= 2-1 =		1	B		
37	Interaction, df	df(AB)=(2-1)	*(2-1)=	1	AB		
38	df(error)=	(2)*(2)*(2-1)=		4	error		
39	Check:	D29-D30-D31-D32=		4	(check)		
40	s(A)^2:	SSA/df(A)=	E25/D35=	100.82			
41	s(B)^2:	SSB/df(B)=	E27/D36=	61.605			
42	s(AB)^2:	SS(AB)/df(AB)=	E30/D37=	40.5			
43	s(R)^2:	SSR/df(error)=	E33/D38=	5.4375			
44	f(obs,A)=	s(A)^2/s(R)^2=	D40/D43=	18.5416092			
45	f(obs,B)=	s(B)^2/s(R)^2=	D41/D43=	11.3296552			
46	f(obs,AB)=	s(AB)^2/s(R)^2=	D42/D43=	7.44827586			

Calculations are shown in the spreadsheet of Table 12.7. The mean yields for the cells were found by averaging the yields found for each set of conditions, as shown in lines 10 and 11. The mean yields for the catalysts (at all levels of concentration) are shown in line 13. The mean yields for concentrations of 1 g/L and 2 g/L (for both catalysts) are shown in lines 15 and 16. The overall mean (or grand average) is calculated in line 17. Residuals are calculated in lines 19 to 23 using array formulas.

The treatment sum of squares for catalyst, *SSA*, is calculated in lines 24 and 25. The treatment sum of squares for concentration, *SSB*, is calculated in lines 26 and 27. The interaction sum of squares between catalyst and concentration, *SS(AB)*, is calculated in lines 29 and 30. The residual sum of squares, *SSR*, is calculated in lines 31 to 33.

The total number of degrees of freedom is calculated in line 34 and then partitioned into degrees of freedom for catalyst, concentration, interaction, and the residual used for estimating error. These are shown in lines 35 to 39. Mean squares, estimates of variances, are calculated in lines 40 to 43. They are each found by dividing the corresponding sum of squares by the degrees of freedom. Observed variance ratios are calculated in lines 44 to 46.

Introduction to Analysis of Variance

The table of analysis of variance for this case is shown in Table 12.8.

Table 12.8: Table of Analysis of Variance for Study of Catalysts, Two-way Analysis of Variance

Sources of Variation	Sums of Squares	Degrees of Freedom	Mean Squares	Variance Ratios
Main effect, catalyst	61.605	1	61.605	$f_{observed1} = \dfrac{61.605}{5.4375} = 11.33$
Main effect, concentration	100.82	1	100.82	$f_{observed2} = \dfrac{100.82}{5.4375} = 18.54$
Interaction between catalyst and concentration	40.5	1	40.5	$f_{observed3} = \dfrac{40.5}{5.4375} = 7.45$
Error	21.75	4	5.4375	
Total	224.675	7		

The observed variance ratios in Table 12.8 were found by dividing the mean squares for catalyst, concentration, and interaction, respectively, by the mean square for error.

On the basis of the simplest mathematical model for this case,

$$y_{ijk} = \mu + \alpha_i + \beta_j + (\alpha\beta)_{ij} + \varepsilon_{ijk},$$

the null hypotheses and alternative hypotheses are as follows:

$H_{0,1}$:   The true effect of the catalyst is zero, as opposed to
$H_{a,1}$:   the true effect of the catalyst is not zero.

$H_{0,2}$:   The true effect of concentration is zero, versus
$H_{a,2}$:   the true effect of concentration is not equal.

$H_{0,3}$ The true effect of interaction between catalyst and concentration is zero, vs.
$H_{a,3}$ the true effect of interaction between catalyst and concentration is not zero.

If the alternative hypotheses are correct, the corresponding true variance ratios for the populations will be greater than 1.

Now we can apply the F-test. $f_{observed\,1}$ should be compared with $f_{critical}$ for a one-sided test with one degree of freedom and four degrees of freedom at the 5% level of significance; this is 7.71. $f_{observed\,2}$ should also be compared with 7.71, and so should $f_{observed\,3}$. Then we can reject the null hypotheses that the true population means for both catalysts are equal and the true population means for both concentrations are equal, both at the 5% level of significance, and accept the alternative hypotheses that

Chapter 12

the catalyst and the concentration make a difference. However, we do not have enough evidence at the 5% level of significance to reject the hypothesis that the mean result of the interaction between catalyst and concentration is zero. Because the value of $f_{observed}$ for interaction between catalyst and concentration is only a little smaller than the corresponding value of $f_{critical}$, we may well decide to collect some more data on this point.

Thus, we conclude (at the 5% level of significance) that the yield is affected by both the catalyst and the concentration, but we do not have enough evidence to conclude that the yield is affected by the interaction between catalyst and concentration.

Notice that the analysis discussed in this chapter allows us to conclude that certain factors have an effect, but it does not allow us to say quantitatively how the yield is changed by any particular level of a factor. In other words, we have not determined the functional relationship between the variables. For that, we would have to use a regression analysis, which will be discussed in Chapter 14.

Example 12.3

Concrete specimens are made using three different experimental additives. The purpose of the additives is to try to accelerate the gain of strength as the concrete sets. All specimens have the same mass ratio of additive to Portland cement, and the same mass ratio of aggregate to cement, but three different mass ratios of water to cement. Two replicate specimens are made for each of nine combinations of factors. All specimens are kept under standard conditions. After twenty-eight days the compression strengths of the specimens are measured. The results (in MPa) are shown in Table 12.9.

Table 12.9: Strengths of Concrete Specimens

Ratio, water to cement	Additives		
	#1	#2	#3
	Compressive Strengths		
0.45	40.7	42.5	40.4
	39.9	41.4	41.7
0.55	36	35.6	26.6
	26.3	30.7	28.2
0.65	24.7	30.6	21.9
	23.9	23.9	27.6

Do these data provide evidence that the additives or the water:cement ratios or interactions of the two affect the yield strength? Use the 5% level of significance.

Introduction to Analysis of Variance

Answer: Again the plots of the residuals were examined in the same way as for Example 12.1 but showed no significant patterns that would indicate that some of the assumptions were not valid. Again they are omitted for the sake of brevity. Calculations are shown in the spreadsheet, Table 12.10.

Table 12.10: Spreadsheet for Study of Additives to Concrete, Two-Way Analysis of Variance with Interaction

	A	B	C	D	E	F
1			Additives, i			
2	j, Ratio, w/c	i=1	i=2	i=3	Row sums	Totals, ratio j
3			y ijk			
4	j=1, 0.45	40.7	42.5	40.4	123.6	
5		39.9	41.4	41.7	123	246.6
6	j=2, 0.55	36	35.6	26.6	98.2	
7		26.3	30.7	28.2	85.2	183.4
8	j=3, 0.65	24.7	30.6	21.9	77.2	
9		23.9	23.9	27.6	75.4	152.6
10	Totals, addtv i	191.5	204.7	186.4		582.6
11	Cell means,	(B4+B5)/2, and copy:				Grand total, y
12	y ij(bar):	i=1	i=2	i=3		
13	j=1	40.3	41.95	41.05		
14	j=2	31.15	33.15	27.4		
15	j=3	24.3	27.25	24.75		
16	Residuals, y ijk-y ij(bar):					
17		i=1	i=2	i=3	Array formulas:	
18	j=1	0.4	0.55	-0.65	B4:B5-B13, and copy	
19		-0.4	-0.55	0.65		
20	j=2	4.85	2.45	-0.8	B6:B7-B14, and copy	
21		-4.85	-2.45	0.8		
22	j=3	0.4	3.35	-2.85	B8:B9-B15, and copy	
23		-0.4	-3.35	2.85		
24	Means, addtv i	B10/6, and copy				
25		i=1	i=2	i=3		
26	y i(barbar):	31.9166667	34.1166667	31.0666667		
27	Means, ratio j:	F5/6, etc.:				
28	j=1	41.1				

Chapter 12

29		j=2	30.5666667			
30		j=3	25.4333333			
31		Overall mean:	F10/18=	32.3666667	< y (barbarbar)	
32	For additive,		SSA=3*2*SUM((y i(barbar)-y(barbarbar))^2)			
33			6*((B26-C31)^2+(C26-C31)^2+(D26-C31)^2)		29.73	SSA
34	For w/c ratio,		SSB=3*2*SUM((y j(barbar)-y(barbarbar))^2			
35			6*((B28-C31)^2+(B29-C31)^2+(B30-C31)^2)=		765.493333	SSB
36	Interaction:		SS(AB)=2*SUMiSUMj((y ij(bar)-y i(barbar)-y j(barbar)+y(barbarbar))^2			
37			(B13-B26-B28+C31)^2, and copy:			
38			i=1	i=2	i=3	
39	j=1		0.1225	0.81	1.5625	
40	j=2		1.06777778	0.69444444	3.48444444	
41	j=3		0.46694444	0.00444444	0.38027778	SUMj	2*SUMj
42		SUMi	1.65722222	1.50888889	5.42722222	8.59333333	17.1866667
43	For residuals:		SSR=SUMiSUMjSUMk(y ijk-y ij(bar))^2			SS(AB) ^
44	(B18:D23)^2:		0.16	0.3025	0.4225	< Array formula
45			0.16	0.3025	0.4225	
46			23.5225	6.0025	0.64	
47			23.5225	6.0025	0.64	
48			0.16	11.2225	8.1225	
49			0.16	11.2225	8.1225	SUMj
50	SUMi		47.685	35.055	18.37	101.11	SSR
51	**Degrees of freedom**					
52	df total =			3*3*2 - 1=	17	**df:**
53	Addtves, df(A):			3-1=	2	**A**
54	w:c ratio,df(B):			3-1=	2	**B**
55	Interaction, df(AB)=			(3-1)*(3-1)=	4	**AB**
56	df(error)=			(3)*(3)*(2-1)=	9	**error**
57	Check:		D52-D53-D54-D55=		9	(check)
58	**Mean Squares:**					
59	s(A)^2:		SSA/df(A)=	E33/D53=	14.865	
60	s(B)^2:		SSB/df(B)=	E35/D54=	382.746667	
61	s(AB)^2:		SS(AB)/df(AB)=	F42/D55=	4.29666667	
62	s(R)^2:		SSR/df(error)=	E50/D56=	11.2344444	

Introduction to Analysis of Variance

63	f(obsrvd,A)=	s(A)^2/s(R)^2=	D59/D62=	1.323163	f(A)	
64	f(obsrvd,B)=	s(B)^2/s(R)^2=	D60/D62=	34.06903	f(B)	
65	f(obsrvd,AB)=	s(AB)^2/s(R)^2=	D61/D62=	0.382455	f(AB)	

The given data are shown in lines 1 to 9, totals for additive i are shown in line 10, and totals for water/cement ratio j are shown in column F. Cell means are calculated in cells B13:D15, as shown in cell B11. Then residuals are calculated in cells B18:D23. Line 26 shows means $\bar{\bar{y}}_i$ for additive i for all values of the w/c ratio j according to equation 12.12, and similarly cells B28:B30 show means $\bar{\bar{y}}_j$ for w/c ratio j for all values for additive i according to equation 12.13. The overall mean, $\bar{\bar{\bar{y}}}$, is calculated in cell C31.

The treatment sum of squares for factor A (additives), SSA, is calculated in lines 32 and 33. The treatment sum of squares for factor B, (w/c ratio), SSB, is calculated in lines 34 and 35. The treatment sum of squares for interaction, SS(AB), is calculated in lines 36 to 42 with the result in cell F42. The residual sum of squares, SSR, is calculated in lines 43 to 50 with the result in cell E50.

The degrees of freedom are calculated in lines 51 to 57. In line 57 we check that the number of degrees of freedom available for estimating error is the difference between the total degrees of freedom and the degrees of freedom allocated to A, B, and interaction AB.

Finally, mean squares for estimating variances for A, B, AB, and error are calculated in lines 58 to 62, and the observed variance ratios are calculate in lines 63 to 65.

The analysis of variance for this case is summarized in Table 12.11. Once again, the mean squares, or estimates of variance, are found by dividing the corresponding sums of squares and degrees of freedom.

Table 12.11: Analysis of Variance for Strength of Concrete

Sources of Variation	Sums of Squares	Degrees of Freedom	Mean Squares	Variance Ratios	
Additives	29.730	2	14.865	$f_{observed1} = \dfrac{14.865}{11.234}$	$= 1.32$
Water - cement ratio	765.493	2	382.747	$f_{observed2} = \dfrac{382.747}{11.234}$	$= 34.07$
Interaction between additives and water - cement ratio	17.187	4	4.297	$f_{observed3} = \dfrac{4.297}{11.234}$	$= 0.38$
Error	101.110	9	11.234		
Total	913.520	17			

Again the simplest mathematical model for this case is

$$y_{ijk} = \mu + \alpha_i + \beta_j + (\alpha\beta)_{ij} + \varepsilon_{ijk}$$

The corresponding null hypotheses and alternative hypotheses are as follows:

$H_{0,1}$:   The true effect of the additives is zero, as opposed to
$H_{a,1}$:   the true effect of the additives is not zero.

$H_{0,2}$:   The true effect of the water : cement ratio is zero, versus
$H_{a,2}$:   the true effect of the water : cement ratio is not zero.

$H_{0,3}$:   The true effect of interaction between additive water:cement ratio is zero, vs.
$H_{a,3}$:   the true effect of interaction between additive and water:cement ratio is not zero.

The observed variance ratios in Table 12.11 are compared with critical values of the F-distribution for corresponding numbers of degrees of freedom at the 5% level of significance. For two degrees of freedom in the numerator and nine degrees of freedom in the denominator, tables indicate that the critical or limiting variance ratio is 4.26. Thus $f_{observed,1}$ is not significantly different from 1, but $f_{observed,2}$ clearly is. Since the interaction mean square is smaller than the error mean square, there is no indication at all that the interaction has a significant effect.

We can conclude, then, that the data provide evidence at the 5% level of significance that the water:cement ratio affects the yield strength, but not that the additives or the interaction between additives and cement-water ratio affect the yield strength.

12.3 Analysis of Randomized Block Design

As we discussed in section 11.5 (d), blocking is used to eliminate the distortion caused by an interfering variable that is not of primary interest. In randomized block designs there is no replication within a block, and interactions between treatments and blocks are assumed to be negligible (subject to checking). In this section we will discuss a simple case in which there is only one treatment in two or more levels.

The nomenclature is a little simpler than in the previous section. y_{ij} is an observation of the i th level of factor A and the jth block. There are a different levels for factor A and b different blocks. m is the true overall mean. α_i is the incremental effect of treatment i, such that $\alpha_i = \mu_i - \mu$, where μ_i is the true population mean for the ith level of factor A. β_j is the incremental effect of block j, such that $\beta_j = \mu_j - \mu$, where μ_j is the true population mean for block j. The error ε_{ij} is the difference between y_{ij} and the corresponding true value.

The simplest mathematical model is

$$y_{ij} = \mu + \alpha_i + \beta_j + \varepsilon_{ij} \qquad (12.27)$$

The quantities μ, α_i, and β_j are estimated from the data. The best estimate of μ is the grand mean, the mean of all observations:

$$\bar{\bar{y}} = \frac{\sum_{i=1}^{a}\sum_{j=1}^{b} y_{ij}}{(a)(b)}$$

The best estimate of μ_i, the population mean for the ith level factor A, is the treatment mean,

$$\bar{y}_i = \frac{\sum_{j=1}^{b} y_{ij}}{b},$$

so the best estimate of α_i is $(\bar{y}_i - \bar{\bar{y}})$. Similarly, the best estimate of μ_j, the population mean for block j, is the block mean,

$$\bar{y}_j = \frac{\sum_{i=1}^{a} y_{ij}}{a},$$

so the best estimate of β_j is $(\bar{y}_j - \bar{\bar{y}})$. Then the best estimate of $(\mu + \alpha_i + \beta_j)$ is $[\bar{\bar{y}} + (\bar{y}_i - \bar{\bar{y}}) + (\bar{y}_j - \bar{\bar{y}})]$. Since for a block design there is no replication, the error ε_{ij} is estimated by the residual, $y_{ij} - [\bar{\bar{y}} + (\bar{y}_i - \bar{\bar{y}}) + (\bar{y}_j - \bar{\bar{y}})] = y_{ij} - \bar{y}_i - \bar{y}_j + \bar{\bar{y}}$. Remember that for a block design, interaction with the blocking variable is assumed to be absent.

The total sum of squares of the deviations of individual observations from the grand average is $SST = \sum_{i=1}^{a}\sum_{j=1}^{b}(y_{ij} - \bar{\bar{y}})^2$. The treatment sum of squares of the deviations of the treatment means from the mean of all the observations is $SSA = b\sum_{i=1}^{a}(\bar{y}_i - \bar{\bar{y}})^2$. The block sum of squares of the deviations of the block means from the mean of all the observations is $SSB = a\sum_{j=1}^{b}(\bar{y}_j - \bar{\bar{y}})^2$. The residual sum of squares is $SSR = \sum_{i=1}^{a}\sum_{j=1}^{b}(y_{ij} - \bar{y}_i - \bar{y}_j + \bar{\bar{y}})^2$. It can be shown algebraically that

$$SST = SSA + SSB + SSR \tag{12.28}$$

The total number of degrees of freedom, $N - 1 = (a)(b) - 1$, is partitioned similarly into the component degrees of freedom. The number of degrees of freedom between treatment means is the number of treatments minus one, or $(a - 1)$. The number of degrees of freedom between block means is the number of blocks minus one, or $(b - 1)$. The residual number of degrees of freedom is $(ab - 1) - (a - 1) - (b - 1) = ab - a - b + 1 = (a - 1)(b - 1)$.

Chapter 12

Once again, the assumptions should be checked by the same plots of residuals as were used in section 12.1. If, contrary to assumption, there is an interaction between treatments and the blocks, a plot of residuals versus expected values may show a curvilinear shape, that is, a systematic pattern that is not linear. If that occurs, a transformation of variable should be attempted (see the book by Montgomery). A full factorial design may be required.

If these plots give no indication of serious error, a table of analysis of variance should be prepared as in Table 12.12 below.

Table 12.12: Table of Analysis of Variance for a Randomized Block Design

Sources of Variation	Sums of Squares	Degrees of Freedom	Mean Squares	Variance Ratio
Between treatments	SSA	$(a-1)$	s_A^2	$f_{observed,1} = \dfrac{s_A^2}{s_R^2}$
Between blocks	SSB	$(b-1)$	s_B^2	
Residuals	SSR	$(a-1)(b-1)$	s_R^2	
Total (about the grand mean, $\bar{\bar{y}}$)	SST	$(N-1)$		

The null hypothesis and alternative hypothesis are similar to the ones we have seen before, and the observed variance ratios are compared as before to the tabulated values for the F-test. Then appropriate conclusions are drawn.

Once again, more than one factor may be present, and the table of analysis of variance can be modified accordingly.

Example 12.4

Three similar methods of determining the biological oxygen demand of a waste stream are compared. Two technicians who are experienced in this type of work are available, but there is some indication that they obtain different results. A randomized block design is used, in which the blocking factor is the technician. Preliminary examination of residuals shows no systematic trends or other indication of difficulty. Results in parts per million are shown in Table 12.13.

Table 12.13: Results of B.O.D. Study in parts per million

	Method 1	Method 2	Method 3
Technician 1	827	819	847
Technician 2	835	845	867

Is there evidence at the 5% level of significance that one or two methods of determination give higher results than the others?

Introduction to Analysis of Variance

Answer:

Table 12.14: Spreadsheet for Example 12.4, Randomized Block Design

	A	B	C	D	E	F
1			Methods, i			
2		i=1	i=2	i=3		
3	j, technician		y ij		Totals, ratio j	
4	j=1	827	819	847	2493	
5	j=2	835	845	867	2547	
6	Totals, method i	1662	1664	1714	5040	Grand Total
7		B6/2	C6/2	D6/2		
8	Means, method i	831	832	857	Overall Mean:	
9	Means, techn j	831	E4/3, j=1		y(bar,bar)=	840
10		849	E5/3, j=2			(E6/6)
11	Residuals, y ij-y i(bar)-y j(bar)+y(bar,bar):					
12		5	-4	-1	B4:D4-B$8:D$8-B9+F$9	
13		-5	4	1	(array formula), and copy	
14	SSA=2*SUM(y i(bar)-y(bar,bar))^2			=2*((B8-F9)^2+(C8-F9)^2+(D8-F9)^2)=		
15	**SSA=**	**868**				
16	SSB=3*SUM(y j(bar)-y(bar,bar)^2)			=3*((B9-F9)^2+(B10-F9)^2)=		
17	**SSB=**	**486**				
18	SSR=SUM(residual^2)		=B12^2+B13^2+C12^2+C13^2+D12^2+D13^2=			
19	**SSR=**	**84**				
20	df(A) = a-1 =		3-1=		2	
21	df(B) = b-1 =		2-1=		1	
22	df(resid) = (a-1)*(b-1) =		D20*D21=		2	
23	Mean Square, A=B15/D20=	434				
24	Mean Square, B=B17/D21=	486				
25	Mean Square, residual=	B19/D22=		42		
26	f obs, A =	D23/D25=		10.3333333		

The spreadsheet is shown in Table 12.14. The mean result for each technician (for all methods) was calculated in column E. The mean result for each method (for both technicians) was calculated in line 8. The overall mean (or grand average) was found in column F. Residuals were calculated in rows 12 and 13. The treatment (method) sum of squares, *SSA*, was calculated in rows 14 and 15. The block (technician) sum of squares, *SSB*, was calculated in lines 16 and 17. The residual sum of squares, *SSR*, was calculated in rows 18 and 19. Degrees of freedom were calculated in rows 20:22. Mean squares were calculated in rows 23:25. Observed variance ratio was calculated in row 26.

On the basis of the simplest mathematical model for this case,

$$y_{ij} = \mu + \alpha_i + \beta_j + \varepsilon_{ij}$$

the null hypothesis and alternative hypothesis are as follows:

$H_{0,1}$:   The true effect of the method is zero, as opposed to
$H_{a,1}$:   the true effect of the method is not zero.

If the alternative hypothesis is true, the corresponding population variance ratio will be greater than 1.

We are now in a position to apply the *F*-test. Mean squares and observed *f*-ratios are shown in this case as part of the spreadsheet, rather than as a separate table. $f_{\text{observed A}}$ should be compared to f_{limit} for two degrees of freedom in the numerator and two degrees of freedom in the denominator at the 5% level of significance, which is 19.00. Therefore we do not have enough evidence to reject the null hypothesis that the true effect of the method is zero. However, the result for Method 3 is appreciably above the results for Methods 1 and 2. Thus, we may want to collect more evidence.

Unless this was only a preliminary experiment, we should probably use larger sample sizes from the beginning. Larger numbers of degrees of freedom would provide tests more sensitive to departures from the null hypotheses, as we have seen before. In this example the sample sizes have been kept small to make the calculations as simple as possible.

From these data there is no evidence at the 5% level of significance that one or two methods give higher results than the others or that the technicians really do affect the results.

12.4 Concluding Remarks

We have seen in Chapter 11 some of the chief strategies and considerations involved in designing industrial experiments. Chapter 12 has introduced the analysis of variance, one of the main methods of analyzing data from factorial designs. Both these chapters are introductory. Further information on both can be obtained from the books by Box, Hunter, and Hunter and by Montgomery (see the List of Selected References in section 15.2). For instance, both of these books have much more information on fractional factorial designs, including worked examples. Some persons find the book by Box, Hunter, and Hunter easier to follow, but the book by Montgomery is more up-to-date.

Introduction to Analysis of Variance

The worked examples on the analysis of variance that we've seen in this chapter were simple cases involving small amounts of data. They were chosen to make the calculations as easily understandable as possible. As the number of data increase, calculation using a calculator becomes more laborious and tedious, and the probability of mechanical error increases. There can be no doubt that the computer calculation is much quicker and more convenient, and the probability of error in calculation is much smaller. The advantage of the computer increases greatly as the size of the data set increases, and a typical set of data from industrial experimentation is much larger than the set of data used in Example 12.3. Thus, the great majority of practical analyses of variance are done nowadays using various types of software on digital computers.

There are two main approaches to computer calculations of ANOVA. One is the fundamental approach, in which the basic formulas of Excel or another spreadsheet are used to perform the calculations outlined in this chapter. That can be used from the beginning. The other is use of more complicated and specialized functions such as special software like SPSS and SAS. Those are very useful once the reader has a good grasp of the basic relations and their usefulness, but they should not be used in the learning phase.

We have noted also that the analysis of variance as introduced in this chapter does not give us all the information we want in many practical cases. If the independent variables are numerical quantities, rather than categories, we usually want to obtain a functional relationship between or among the variables. If a particular independent variable increases by ten percent, by how much does the dependent variable increase? The analysis of variance in the form discussed in this chapter may tell us that a certain independent variable has a significant effect on the dependent variable, but it cannot give a quantitative functional relationship between the variables. To obtain a functional relationship we must use a different mathematical model and a different analysis. That is the analysis called regression, which will be discussed in Chapter 14.

Problems

1. Three testing machines are used to determine the breaking load in tension of wire which is believed to be uniform. Nine pieces of wire are cut off, one after another. They are numbered consecutively, and three are assigned to each machine using random numbers. Random numbers are used also to determine the order in which specimens are tested on each machine. The breaking loads (in Newtons) found on each machine are shown in the table below.

	Testing Machine	
#1	#2	#3
1570	1890	1640
1750	1860	1760
1680	2390	2020

The diagnostic plots 1, 2, and 3 recommended in part (c) of section 12.1 were carried out and showed no significant discrepancies.
a) What further diagnostic plot should be made in this case? Why? If this plot is not satisfactory, how will that affect the subsequent analysis?
b) Assuming this further plot is also satisfactory, do the data indicate (at the 5% level of significance) that one or two of the machines give higher readings than others?

2. Two determinations were made of the viscosities of each of three polymer solutions. Viscosities were measured at the same flow rate in the same instrument. The order of the tests was determined using random numbers. The results were as follows.

	Solution	
#1	#2	#3
177	184	206
183	187	202
176	175	200

The diagnostic plots recommended in part (c) of section 12.1 were carried out and showed no significant discrepancies. Can we conclude (at the 5% level of significance) that the solutions have different viscosities?

3. A chemical engineer is studying the effects of temperature and catalyst on the percentage of undesired byproduct in the output of a chemical reactor. Orders of testing were determined using random numbers. Percentage of the byproduct is shown in the table below.

Catalyst	Temperature, °C	
	140	150
#1	2.3	1.6
	1.3	2.8
#2	3.6	3.0
	3.4	3.8

The diagnostic plots recommended in part (c) of section 12.1 were carried out and showed no significant discrepancies. Can we conclude (at the 5% level of significance) that catalyst or temperature or their interaction affects the percentage of byproduct?

4. A storage battery is being designed for use at low temperatures. Two materials have been tested at two temperatures. The orders of testing were determined using random numbers. The life of each battery in hours is shown in the following table.

	Temperature	
Material	−20 °C	−35 °C
#1	90	92
	119	86
#2	128	85
	150	103

The diagnostic plots recommended in part (c) of section 12.1 were carried out and showed no significant discrepancies. Can we conclude (at the 5% level of significance) that material or temperature or their interaction affects the life of the battery?

5. The copper sulfide solids from a unit of a metallurgical plant were sampled on March 3, March 10, and March 17. Half of each sample was dried and analyzed. The other half of each sample was washed with an experimental solvent and filtered, then dried and analyzed. The order of testing was determined using random numbers. The date of sampling was taken as a blocking factor. The percentage copper was reported as shown in the table below.

	March 3	March 10	March 17
Unwashed	64.48	68.67	68.34
Washed	68.22	72.74	74.54

The diagnostic plots showed no significant discrepancies. Is there evidence at the 5% level of significance that washing affects the percentage copper?

6. A test section in a fertilizer plant is used to test modifications in the process. A processing unit feeds continuously to three filters in parallel. A change is made in the processing unit. A sample is taken from the filter cake of each filter, both before and after the change. Percentage moisture is determined for each sample in an order determined by random numbers, and results are shown in the table below.

	Filter #1	Filter #2	Filter #3
Before change	2.14	2.31	2.32
After change	1.51	1.83	1.8

The diagnostic plots showed no significant discrepancies. Taking the filter number as a blocking factor, do these data give evidence at the 5% level of significance that the processing change affects the percentage moisture?

CHAPTER 13

Chi-squared Test for Frequency Distributions

For this chapter the reader should have a good understanding of statistical inference from Chapter 9, and of sections 2.2, 4.4, 5.3, and 5.4.

This is another case in which we set up a null hypothesis and then test the statistical significance of disagreement with it. But now we are concerned with frequency distributions. We compare *observed frequencies* with corresponding *expected frequencies* calculated on the basis of a null hypothesis with stated trial assumptions. Then we calculate a quantity which summarizes the disagreement between observed and expected frequencies, and we test whether it is so large that it would not likely occur by chance.

13.1 Calculation of the Chi-squared Function

Let the observed frequency for class i be o_i, and let the expected frequency for that same class be e_i. We must have $\sum o_i = \sum e_i =$ total frequency. Then we define

$$\chi^2_{\text{calculated}} = \sum_{\text{all classes}} \frac{(o_i - e_i)^2}{e_i} \tag{13.1}$$

This is a value of a random variable having approximately a chi-squared distribution, and the approximation generally gets better as the data set becomes larger. The theory behind that involves both the normal approximation to the binomial distribution and the mathematical relationship between the normal distribution and the chi-squared distribution. Notice that the summation in equation 13.1 must extend over all possible classes of a particular set rather than any selection of them.

To prevent the error of approximation in using the χ^2 distribution from becoming appreciable, each expected value of frequency should be at least 5. This is similar to the rough rule for the normal approximation to the binomial distribution, which requires that both np and $(n)(1-p)$ be greater than 5. Under some conditions a small proportion of the expected frequencies for the chi-squared test can be less than 5 without producing serious error (see the book by Barnes listed in section 15.2). However, a minimum expected frequency of 5 should be applied in solving problems in this book. If one or more expected frequencies are less than 5, it may be reasonable to combine adjacent cells or classes to get a combined expected frequency of at least 5. We will see that this is done frequently.

Chi-squared Tests for Frequency Distributions

However, like other tests of significance, the chi-squared test for frequency distributions becomes more sensitive as the number of degrees of freedom increases, and that increases as the number of classes increases. Thus, we should make the number of classes as large as we can, keeping the requirements of the last paragraph in mind.

The value of $\chi^2_{calculated}$ can be compared with theoretical values of χ^2 for appropriate number of degrees of freedom and level of significance. The χ^2 distribution to be used here is the same as the χ^2 distribution introduced in section 10.1 for comparing a sample variance with a population variance. Remember that χ is pronounced "kigh," like "high." The shape of the χ^2 distribution is always skewed; shapes of distributions for three different numbers of degrees of freedom are shown in Figure 10.1. Some tabulated values of χ^2 can be found in Table A3 in Appendix A.

If a computer with Excel or some alternative is available, it can be used instead of a table. Probabilities for particular values of χ^2 can be found from the Excel function CHIDIST. The arguments to be used with this function are the value of χ^2 and the number of degrees of freedom. The function then returns the upper-tail probability. For example, for $\chi^2 = 11.07$ at 5 degrees of freedom, we type in a cell for a work sheet the formula =CHIDIST(11.07,5), or else from the Formula menu, we choose Paste Function, Statistical Functions, CHIDIST(,), then type in the arguments and choose the OK button. The result is 0.05000962, the probability of obtaining a value of χ^2 greater than 11.07 completely by chance. If we have a value of the upper-tail probability and the number of degrees of freedom, we use the Excel function CHIINV to find the value of χ^2. Again, the function can be chosen using the Formula menu or it can be typed into a cell. For an upper-tail probability of 0.05 and 5 degrees of freedom, CHIINV(0.05,5) gives 11.0704826.

If the calculated value of χ^2 is greater than the corresponding tabulated or computer value of χ^2, the null hypothesis must be rejected at the level of significance equal to the stated upper-tail probability. The chi-squared test for frequency distributions is always a one-tailed or one-sided test.

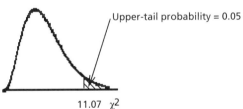

Figure 13.1: Upper-tail probability for Chi-squared Distribution

In general, the *number of degrees of freedom* for any statistical test is equal to the number of independent pieces of information in the data. For the chi-squared test for frequency distributions, the number of degrees of freedom is the number of classes or cells used in the comparison, less the number of linearly independent restrictions placed on those data. For example, if we make 100 tosses of a coin, we have two classes or cells, the number of heads and the number of tails, and one restriction, that the number of heads and the number of tails must add up to 100. Then the number of degrees of freedom in this case is 2 classes – 1 restriction = 1 degree of freedom. We always have at least one restriction, given by the total frequency for all classes or cells.

In some cases, which we will encounter in section 13.3, there are further restrictions. This is because one or more statistical parameters such as a mean or a standard deviation are estimated from the data. Each calculation of an estimated parameter from the data represents another independent restriction that reduces the number of degrees of freedom.

Another way of finding the number of degrees of freedom is to count the number of classes or cells to which frequencies could be assigned arbitrarily without changing total frequencies of any kind, and subtract the number of parameters (if any) which have been determined from the data. This is often the most practical approach.

If the number of degrees of freedom is 1, we should apply a correction for continuity (called the Yates correction). This correction for continuity is similar to the one used for a normal approximation to a discrete distribution. However, that will be omitted from this book. It is discussed in the book by Walpole and Myers (see reference in section 15.2) and other references.

The chi-squared test for frequency distributions appears in various forms depending on just what trial assumptions are used to give null hypotheses. In each case the expected frequency for any class or cell is the product of two quantities: the total frequency for all classes and the probability that a randomly chosen item will fall in that particular class.

13.2 Case of Equal Probabilities

If it is reasonable to make the trial assumption that all the classes or cells are equally probable, we can easily calculate the expected frequencies for the corresponding null hypothesis.

Example 13.1

A die was tossed 120 times with the observed frequencies shown below. Test whether the die shows evidence of bias at the 5% level of significance.

Result	1	2	3	4	5	6
Observed frequency	12	25	28	14	15	26

Answer:

If there is no bias, all the results are equally likely.

H_0: Pr [1] = Pr [2] = Pr [3] = Pr [4] = Pr [5] = Pr [6]

H_a: Not all the results are equally likely.

Chi-squared Tests for Frequency Distributions

On the basis of the null hypothesis, the probability of each of the six possible results is $\frac{1}{6}$, so the expected frequency of each result is $\frac{120}{6} = 20$. Then we have

$$\chi^2_{\text{calculated}} = \sum_{\text{all results}} \frac{(o_i - e_i)^2}{e_i}$$

$$= \frac{(12-20)^2}{20} + \frac{(25-20)^2}{20} + \frac{(28-20)^2}{20} + \frac{(14-20)^2}{20} + \frac{(15-20)^2}{20} + \frac{(26-20)^2}{20}$$

$$= 12.50$$

We have 6 classes or cells, and we have 1 restriction, that the sum of the frequencies must be equal to the total number of tosses. Then the number of degrees of freedom is given by

no. of classes or cells − no. of restrictions = 6 − 1 = 5.

From Table A3 for 5 d.f. and 0.05 upper-tail probability, $\chi^2_{\text{limit}} = 11.07$, or from Excel CHIINV(0.05,5) = 11.0704826 (quoting all the digits from Excel).

The calculated and limiting values of χ^2 are compared in Figure 13.2.

Since $\chi^2_{\text{calculated}} > \chi^2_{\text{limit}}$, we reject the null hypothesis.

Then there is evidence at the 5% level of significance that the die is biased.

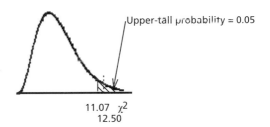

Figure 13.2: Comparison of Calculated and Limiting Values of χ^2

13.3 Goodness of Fit

We can use the chi-squared test for frequency distributions to compare experimental frequencies with the frequencies that would be expected if an assumed probability distribution applies. Are the differences between observed and expected frequencies small enough so that we can say that they could reasonably be due only to chance, or are they too large for that interpretation? We calculate the expected class frequencies on the basis of the assumed probability distribution, then use the chi-squared test to judge the significance of the differences. That is essentially what we did for a very simple probability distribution in section 13.2, but now we will use that approach for other, more complex distributions, such as the binomial, Poisson, and normal distributions, or generally for any case where probabilities for various categories are known. If the assumed probability distribution involves parameters that are estimated from the data, each estimated parameter will correspond to a further restriction, and that will have to be taken into account in determining the number of degrees of freedom.

Chapter 13

We should note that other tests are also used frequently for tests of goodness of fit and may have advantages in some cases. In particular, the Kolmogorov-Smirnov and Anderson-Darling tests are said to be better for small samples. See the book by Johnson (reference given in section 15.2).

Example 13.2

In Example 4.2 and Table 4.5 we had data on the thicknesses of 121 metal parts of an optical instrument. The histogram for these data was shown in Figure 4.4, and we saw later that its shape was similar to the shape we might expect for a normal frequency distribution. In Example 7.9 we plotted the data on normal probability paper and found good agreement. Now we will test the data for goodness of fit to a normal distribution at 5% level of significance.

Answer: The mean and standard deviation were estimated from the data of Example 4.2 to be $\bar{x} = 3.369$ mm and $s = 0.0629$ mm. These were used in Example 7.6 to calculate expected frequencies for the various class intervals according to the normal distribution, and these expected frequencies were compared to the observed frequencies. The comparison is shown in the table below:

Table 13.1: Expected and Observed Class Frequencies

Lower Class Boundary, mm	Upper Class Boundary, mm	Expected Class Frequency	Observed Class Frequency
—	3.195	0.3	0
3.195	3.245	2.6	2
3.245	3.295	11.4	14
3.295	3.345	28.2	24
3.345	3.395	37.2	46
3.395	3.445	27.6	22
3.445	3.495	10.9	10
3.495	3.545	2.4	2
3.545	3.595	0.3	1
3.595	—	0.0	0

Before we can apply the chi-squared test for frequency distributions to these data, some adjacent classes have to be combined so that the expected frequency for each revised cell is at least 5. Thus, the first three classes are combined to give a cell with expected cell frequency 14.3 and observed cell frequency 16, and the last four classes are combined to give a cell with expected cell frequency 13.6 and observed cell frequency 13. That leaves us with $10 - 2 - 3 = 5$ cells or classes.

Chi-squared Tests for Frequency Distributions

Now we can calculate

$$\chi^2_{calculated} = \sum_{all\ classes} \frac{(o_i - e_i)^2}{e_i} = \frac{(16-14.3)^2}{14.3} + \frac{(24-28.2)^2}{28.2} +$$

$$+ \frac{(46-37.2)^2}{37.2} + \frac{(22-27.6)^2}{27.6} + \frac{(13-13.6)^2}{13.6}$$

$$= 4.07$$

We have H_0: probabilities for the various cells are given by the normal distribution

and H_a: other factors affect probabilities.

The number of cells is 5, the total expected frequency has been made equal to the total observed frequency, and we have two statistical parameters, μ and σ, which have been determined from the data. Then the number of degrees of freedom is 5 – 1 – 2 = 2. For 2 degrees of freedom and 0.05 level of significance, Table A3 or the Excel function CHIINV gives $\chi^2_{critical}$ or χ^2_{limit} = 5.99. Since $\chi^2_{calculated} < \chi^2_{limit}$, we have no reason to reject the null hypothesis. The observed frequency distribution seems to be consistent with a normal distribution.

Example 13.3

In Example 5.15, a Poisson distribution was fitted to data for the numbers of cars crossing a bridge in forty successive 6-minute intervals of time. The sample mean was calculated from the data to be $\bar{x}$ = 2.875, and this value was used as an estimate of the population mean, μ, for calculation of Poisson probabilities. The comparison of frequencies for various values of the numbers counted, x, is as follows:

x	Observed Frequency	Expected Frequency
0	2	2.26
1	7	6.49
2	10	9.33
3	8	8.94
4	6	6.42
5	3	3.69
6	3	1.77
7	1	0.73
≥ 8	0	0.38

Is the goodness of fit satisfactory at the 5% level of significance?

Chapter 13

Answer: To make the minimum expected frequency in each cell at least 5, the first two cells should be combined, and also the last four. For counts of 0 or 1, the observed frequency becomes 9, and the expected frequency becomes 8.75. For counts of 5 or more, the observed frequency becomes 7, and the expected frequency becomes 6.57. After this modification, the new number of cells is 9 − 1 − 3 = 5.

Now we are ready to apply the chi-squared test.

H_0: The observed frequency distribution is consistent with a Poisson distribution.

H_a: The frequency distribution is not adequately fitted by a Poisson distribution.

$$\chi^2_{calculated} = \sum_{all\ classes} \frac{(o_i - e_i)^2}{e_i}$$

$$= \frac{(9-8.75)^2}{8.75} + \frac{(10-9.33)^2}{9.33} + \frac{(8-8.94)^2}{8.94} + \frac{(6-6.64)^2}{6.42} + \frac{(7-6.57)^2}{6.57}$$

$$= 0.007 + 0.048 + 0.099 + 0.027 + 0.028$$

$$= 0.21$$

We have 5 cells, and there are two restrictions, for the total frequency and estimation of μ from the data. Then there are 5 − 2 = 3 degrees of freedom. From Table A3 or the Excel function CHIINV we find for 0.05 upper tail probability and 3 degrees of freedom, $\chi^2_{critical}$ = 7.81. Since $\chi^2_{calculated} \ll \chi^2_{critical}$ there is no indication at all that the fit is not good enough.

In fact the fit is *too* good. You may remember from section 10.1.1 that for any number of degrees of freedom, the mean of the chi-squared distribution is equal to the number of degrees of freedom. In this case $\chi^2_{calculated}$ is smaller than the number of degrees of freedom and so smaller than the mean of the distribution. For 3 degrees of freedom and 0.95 upper tail probability, so at the other end of the distribution, Table A3 gives $\chi^2_{critical}$ = 0.35. Then the value of $\chi^2_{calculated}$ is even less than $\chi^2_{critical}$ for an upper-tail probability of 0.95. See Figure 13.3.

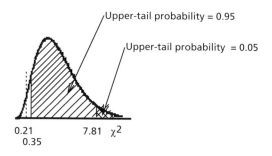

Figure 13.3: Comparison of Calculated and Limiting Values of χ^2

There is less than 5% probability of getting by chance a value of $\chi^2_{calculated}$ *smaller* than the reported value. This indicates that the reported data are too good to be true and may suggest that they were concocted rather than honestly observed.

Chi-squared Tests for Frequency Distributions

13.4 Contingency Tables

A contingency table involves two different factors in more than one row and more than one column, giving a two-dimensional array. Both factors are usually qualitative. We use the chi-squared distribution to test these two factors for independence: does one of the factors affect the other? or are they operating independently? If the factors are independent, then the simple form of the multiplication rule applies according to equation 2.2: the probability of a particular level of factor A and a particular level of factor B is simply the product of the probability of that level of factor A and the probability of that level of factor B. The best estimate we can make of the probability of a particular level of either factor is the total number of outcomes which occur at that level, divided by the total frequency for this set of data. On that basis the expected frequency for level i of factor A and level j of factor B is given by

Pr [level i of factor A $\cap$ level j of factor B] × total frequency =

$$= \Pr[\text{level } i \text{ of factor A}] \times \Pr[\text{level } j \text{ of factor B}] \times \text{total frequency}$$

$$\approx \left(\frac{\text{total number at level } i \text{ of factor A}}{\text{total frequency}}\right) \times \left(\frac{\text{total number at level } j \text{ of factor B}}{\text{total frequency}}\right) \times$$

$$\times \text{ total frequency}$$

$$\approx \frac{(\text{total number at level } i \text{ of factor A}) \times (\text{total number at level } j \text{ of factor B})}{\text{total frequency}}$$

The total numbers at particular levels are usually spoken of as column totals and row totals, and the total frequency for all conditions is called the grand total. Then the expected frequency for level i of factor A and level j of factor B is given by

$$\frac{(\text{row total})(\text{column total})}{\text{grand total}}.$$

This relationship for the expected frequency applies both for the case where all the total numbers at particular levels are random variables and for the case where some total numbers at particular levels (either for columns or for rows, not both) are fixed at chosen values. Thus, in Example 13.4 below, the total frequency for each shift is fixed at 300.

Example 13.4

The observed numbers of days on which accidents occurred in a factory on three successive shifts over a total of 300 days are as shown below. The numbers of days without accidents for each shift were obtained by subtraction.

Chapter 13

Shift	Days With Accidents	Days Without Accidents	Total
A	1	299	300
B	7	293	300
C	7	293	300
Total	15	885	900

Totals for all rows and all columns have been calculated. Is the difference in number of days with accidents between different shifts statistically significant? That is, is there evidence that the probability of accidents depends on the shift? Use the 5% level of significance.

Answer:

H_0: The numbers of days with accidents are independent of the shift.

H_a: Some shifts have greater probability of accidents than others.

The analysis will use the chi-squared test for frequency distribution with

$$\chi^2 = \sum_{\text{all classes}} \frac{(o_i - e_i)^2}{e_i}.$$

The expected frequencies are found using the null hypothesis and the column and row totals. Overall, the best estimate of the probability that there will be at least one accident on a randomly chosen shift and day is $\frac{15}{900}$, and the best estimate of the probability of no accident on a randomly chosen shift and day is $\frac{885}{900}$. (With these figures the probability of more than one accident on any particular shift and day is small enough to be neglected.) Similarly, the probability that any randomly chosen shift is A shift (or B shift, or C shift) is $\frac{300}{900}$. On the basis of the null hypothesis the expected number of days with accidents on A shift or B shift or C shift is then $\left(\frac{15}{900}\right)\left(\frac{300}{900}\right)(900) = 5$, or $\frac{(\text{row total})(\text{column total})}{\text{grand total}} = \frac{(15)(300)}{900} = 5$. Similarly, the expected number of days without accidents on A shift or B shift or C shift is $\frac{(\text{row total})(\text{column total})}{\text{grand total}} = \frac{(885)(300)}{900} = 295$. We use these expected frequencies and the corresponding observed frequencies to find $\chi^2_{\text{calculated}}$.

$$\chi^2_{\text{calculated}} = \sum_{\text{all classes}} \frac{(o_i - e_i)^2}{e_i}$$

$$= \frac{(1-5)^2}{5} + \frac{(7-5)^2}{5} + \frac{(7-5)^2}{5} + \frac{(299-295)^2}{295} + \frac{(293-295)^2}{295} + \frac{(293-295)^2}{295}$$

$$= 4.88$$

Chi-squared Tests for Frequency Distributions

We have 6 classes or cells. The restrictions are the totals for each shift, the total number of accidents, and the total number of days without accidents, but these are not all linearly independent. The number of degrees of freedom for a contingency table is best found as the number of class frequencies which could be varied arbitrarily without changing any of the row or column totals. In this problem that number of degrees of freedom is 2, since 2 cell frequencies could be varied arbitrarily without changing the totals. We can see that by removing the individual class frequencies from the contingency table, then marking x's in some cells until no more could be varied without affecting some of the totals:

Shift	Days With Accidents	Days Without Accidents	Total
A	x_1		300
B	x_2		300
C			300
Total	15	885	900

For instance, if the numbers of accidents for shifts A and B are fixed, then the values in all the other cells are determined by the totals. Therefore the number of degrees of freedom in this problem must be 2.

From Table A3 or the Excel function CHIINV, for upper-tail probability 0.05 and 2 degrees of freedom, $\chi^2_{critical} = 5.991$. This is shown on Figure 13.4.

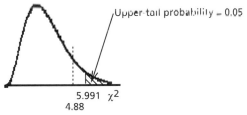

Figure 13.4: Calculated and Critical Values of Chi-squared

Since $4.88 < 5.991$, the calculated value of χ^2 is not significant at the 5% level of significance. Therefore we have insufficient evidence to reject the null hypothesis. We could gather more information. If further data continue to show more accidents on B and C shifts than on A shift, a later analysis might well show a significant value of χ^2.

Example 13.5

Results of a study of the repair records of three models of cars over the first three years of the cars' lives on the basis of a sample are shown below.

Car Model	Number Surveyed	Percentages Requiring		
		Major Repairs	Minor Repairs	No Repairs
A	60	20	50	30
B	30	40	40	20
C	40	30	60	10

Chapter 13

Test the hypothesis that all models perform equally well, so probabilities are independent of the model. The level of significance to be used will be 5%.

Answer: Before we can apply the chi-squared test we have to convert percentages to observed frequencies and find column and row totals.

Observed Frequencies

Car Model	Major Repairs	Minor Repairs	No Repairs	Total
A	12	30	18	60
B	12	12	6	30
C	12	24	4	40
Total	36	66	28	130

The corresponding expected frequencies are calculated using the null hypothesis.

H_0: Probabilities for repair are independent of the model.

H_a: At least one model has different probabilities of repair.

Expected frequencies are then calculated using totals: $\left(\dfrac{\text{row total}}{\text{grand total}}\right)$(column total).

For example, expected frequency of major repairs for model A is $\left(\dfrac{60}{130}\right)(36) = 16.6$. Expected frequencies are shown in the table below.

Expected Frequencies

Car Model	Major Repairs	Minor Repairs	No Repairs	Total
A	16.6	30.5	12.9	60
B	8.3	15.2	6.5	30
C	11.1	20.3	8.6	40
Total	36	66	28	130

$$\chi^2_{\text{calculated}} = \sum_{\text{all classes}} \frac{(o_i - e_i)^2}{e_i} = \frac{(12-16.6)^2}{16.6} + \frac{(30-30.5)^2}{30.5} + \frac{(18-12.9)^2}{12.9}$$

$$+ \frac{(12-8.3)^2}{8.3} + \frac{(12-15.2)^2}{15.2} + \frac{(6-6.5)^2}{6.5}$$

$$+ \frac{(12-11.1)^2}{11.1} + \frac{(24-20.3)^2}{20.3} + \frac{(4-8.6)^2}{8.6}$$

$$= 8.87$$

Chi-squared Tests for Frequency Distributions

The number of degrees of freedom is the number of class frequencies which could be changed arbitrarily without changing any of the totals for rows and columns. If the frequencies of, say, major repairs for Models A and B and minor repairs for Models A and B are chosen arbitrarily, all other frequencies are fixed if the totals are to stay the same. Then the number of degrees of freedom is 4.

(This last calculation can be reduced to a simple formula, but the reader will obtain better understanding during the learning process by reasoning from the underlying ideas, as we have done here. The formula for number of degrees of freedom for contingency tables can be found in a number of reference books, including the book by Walpole and Myers for which a citation is given in section 15.2.)

From Table A3 or the Excel function CHIINV at the 0.05 level of significance, $\chi^2_{critical} = 9.488$. Since $\chi^2_{calculated} < \chi^2_{critical}$, we cannot reject the null hypothesis. We do not have sufficient evidence to say that probabilities of the various categories of repairs depend on the model.

Problems

1. Numbers of people entering a commercial building by each of four entrances are observed. The resulting sample is as follows:

Entrance	1	2	3	4
No. of People	49	36	24	41

   a) Test the hypothesis that all four entrances are used equally. Use the 0.05 level of significance.
   b) Entrances 1 and 2 are on a subway level while 3 and 4 are on ground level. Test the hypothesis that subway and ground-level entrances are used equally often. Use again the 0.05 level of significance.

2. Two dice are rolled 100 times and the results are tabulated below according to the specified categories:

Value of roll	2 to 4	5 or 6	7	8 or 9	10 to 12
No. of rolls	21	21	18	28	12

   At the 5% level of significance, can we say that the dice are unbiased?

3. A robot-operated assembly line is developed to produce a range of new products, which are color-coded black, white, red and green. The assembly line is programmed to produce 11.76% black, 29.41% white, 7.06% red and 51.76% green items. A sample of 180 items was taken and the following distribution was observed:

Color	Black	White	Red	Green
Frequency	26	43	15	96

   a) Can you conclude at the 5% level of significance that the assembly line needs adjustment?

Chapter 13

b) What is the lowest level of significance at which you could conclude that the system needs adjustment?

4. When four pennies were tossed 160 times, the frequencies of occurrence of 0, 1, 2, 3 and 4 heads were 9, 48, 53, 44 and 6, respectively. Is there evidence at the 5% level of significance that the coins are not fair?

5. Consider the average daily yields of coke from coal in a coke oven plant summarized by the grouped frequency distribution shown below.

Lower Bound	Upper Bound	Class Midpoint	Frequency
67.95	68.95	68.45	1
68.95	69.95	69.45	8
69.95	70.95	70.45	22
70.95	71.95	71.45	22
71.95	72.95	72.45	9
72.95	73.95	73.45	8
73.95	74.95	74.45	2

The estimated mean and standard deviation from the data are 71.25 and 1.2775, respectively. Is the frequency distribution given above significantly different from a normal distribution at the 5% level of significance?

6. Consider the hourly labor costs (in dollars) for a random sample of small construction projects summarized in the frequency table below.

Lower Bound	Upper Bound	Class Midpoint	Frequency
18.505	19.505	19.005	6
19.505	20.505	20.005	24
20.505	21.505	21.005	17
21.505	22.505	22.005	16
22.505	23.505	23.005	7
23.505	24.505	24.005	3
24.505	25.505	25.005	2

The mean and standard deviation estimated from these data are $21.15 and $1.42, respectively. Are the above data significantly different from a normal distribution? Use .05 level of significance.

7. Scores made in the final exam by an elementary statistics section can be summarized in the following grouped frequency distribution:

Class No.	Class Midpoint	Frequency
1	14.5	3
2	24.5	2
3	34.5	3

Chi-squared Tests for Frequency Distributions

4	44.5	4
5	54.5	5
6	64.5	11
7	74.5	14
8	84.5	14
9	94.5	4

The mean and standard deviation calculated from these data are 65.48 and 20.957, respectively. At a 5% level of significance do the above data differ from a Normal Distribution?

8. A company has set up a production line for cans of carrots. The numbers of breakdowns on the production line over 49 shifts are summarized as follows:

No. of breakdowns in one shift	No. of shifts
0	18
1	12
2	8
3	6
4	3
5	2
> 5	0

Is this distribution significantly different from a Poisson distribution? Use the 5% level of significance.

9. A section of an oil field has been divided into 48 equal sub-areas. Counting the oil wells in the 48 sub-areas gives the following frequency distribution:

Number of oil wells	0	1	2	3	4	5	6	7
Frequency	5	10	11	10	6	4	0	2

Fitting the data to a Poisson Distribution gives the following estimated frequencies:

3.94   9.85   12.31   10.25   6.41   3.21   1.34   0.47

Test at the 5% level of significance the null hypothesis that the data fit a Poisson distribution.

10. A study of four block faces containing 52 one-hour parking spaces was carried out. Frequencies of vacant spaces were as follows:

No. of vacant parking spaces	0	1	2	3	4	5	>6
Observed frequency	30	45	20	15	7	3	0

From these data the mean number of vacant spaces was calculated to be 1.442. At the 5% level of significance, can you conclude that the distribution of vacant one-hour parking spaces follows a Poisson distribution?

Chapter 13

11. The number of weeds in each 10 m² square of lawn was recorded by a team of second-year students for a random sample of 220 lawns.

Number of weeds per 10 m²	Frequency
0	19
1	44
2	68
3	48
4	18
5	7
6	6
>6	10

  a) At the 5% level of significance, is this distribution significantly different from a Poisson Distribution?
  b) Is there any reason to suggest that the data may not have been reported honestly?

12. A factory buys raw material from three suppliers. All raw materials are made into products by the same workers using the same machines. An engineer thinks there is a difference in the likelihood of defects in products made from raw materials from different suppliers and collects the following information.

	Source of Raw Materials		
	Smith Co.	Jones Co.	Roberts Co.
No. of defective products	11	5	4
No. of satisfactory products	54	71	62

  Is there evidence at the 5% level of significance that the discrepancies are not due to chance?

13. A particular type of small farm machinery is produced by four different companies. The proportions of machines requiring repairs in the first year after sale to the farmers are as follows:

Company	Total Number of Machines	Proportion Requiring Repairs
A	145	0.1034
B	140	0.0429
C	120	0.0333
D	105	0.1143

  Is the distribution regarding requiring and not requiring repairs independent of the company? Use the 1% level of significance.

Chi-squared Tests for Frequency Distributions

14. An industrial engineer collected data on the frequency and severity of accidents in the mining industry and summarized her findings as follows:

Severity of Accident	Days of Week			Total
	Monday & Friday	Tuesday & Thursday	Wednesday	
Severe	22	9	4	35
Minor	283	254	128	665
Total	305	263	132	700

   a) Can you conclude at the 5% level of significance that the severity of accidents is independent of the day of the week?
   b) What is the lowest level of significance at which you could conclude that the frequency of *severe* accidents depends upon the day of the week?

15. In testing the null hypothesis that the level of heavy equipment usage and the owner's maintenance policy are independent variables, a mechanical engineer received replies to her questionnaire from a random sample of users. The following summary applies:

Equipment Usage	Maintenance Policy			Total
	By Calendar	By Hours of Operation	As Required	
Light	12	8	13	33
Moderate	7	15	22	44
Heavy	3	22	15	40
Total	22	45	50	117

At the 1% level of significance, should the engineer reject the null hypothesis?

16. The following data have been obtained by an automotive engineer interested in estimating owner preferences. From a sample of 163 automobiles the following data on engine size and transmission type were obtained.

Transmission	Engine Size		
	small	medium	large
4-speed	34	19	12
5-speed	24	28	5
Automatic	7	12	22

   a) He wishes to test the null hypothesis that transmission type and engine size chosen by the car-owning population are independent. Using a 5% level of significance, do the above data support this hypothesis?
   b) In all of Canada, statistics for cars equipped with automatic transmissions show that 21% have small engines, 23% have medium size engines and the remainder have large engines. Are the data in the above table consistent with the Canadian statistics?

Chapter 13

17. The tread life of a particular brand of tire was evaluated by recording kilometers traveled before wearout for a random sample of 500 cars. The cars were classified as subcompacts, compacts, intermediates, and full-size cars. The grouped frequency distribution is shown in the following table.

Treadwear, km		Class of Car			
Lower Bound	Upper Bound	Subcompact	Compact	Intermediate	Full Size
0	30,000	26	55	46	23
30,001	60,000	95	171	99	55
60,001	90,000	120	205	115	60

At the 1% level of significance, can you conclude that treadwear and class of car are independent?

18. Four alternative methods of loading a machine are tried to see whether the loading method has any effect on the likelihood that cycles will end in stoppages. The results are as follows:

Method of loading	A	B	C	D
Observed frequency of cycles with stoppages	8	4	9	3
Observed frequency of cycles without stoppages	10	16	12	18

Use the chi-squared test for frequencies to see whether these data show a significant effect of the method of loading on the probability of a stoppage. Use the 5% level of significance.

CHAPTER 14

Regression and Correlation

For this chapter the reader should have a good understanding of the material in sections 3.1 and 3.2 and in Chapter 9.

In previous chapters we have investigated frequency distributions, probability distributions, and central values such as means, all at fixed values of the independent variables. Now we want to see how the distributions and means change as one or more independent variables change. We will look at samples of data taken over a range of an independent variable or variables and use those data to obtain information regarding the relation between the dependent and independent variables.

In a simple case we have only one independent variable, x, and one dependent variable, y. *Regression analysis* assumes that there is no error in the independent variable, but there is random error in the dependent variable. Thus, all the errors due to measurement and to approximations in the modeling equations appear in the dependent variable, y. If other independent variables have an effect but are kept only approximately constant, effects of their variation may inflate the errors in the dependent variable. In some cases other independent variables may be varying appreciably and may affect the dependent variable, but the effect of a chosen independent variable may be examined by itself, as though it were the only independent variable, to obtain a preliminary indication of its effect. In any example of regression, the expectation or expected value of y varies as a function of x, and errors cause measured values of y to deviate from the expected value of y at any particular value of x. If there are several measured values of y at one value of x, the mean of the measured values of y will give an approximation of the expected value of y at that value of x.

Engineers often encounter situations where an independent variable affects the value of a dependent variable, and errors of measurement produce random fluctuations about the expected values. Thus, change in stress produces change in strain plus variation in measured strain due to error. The output of a stirred chemical reactor changes as the temperature within the reactor varies with time, and the measured concentration of any component in the output shows an additional variation caused by error. The power produced by an electric motor changes with variation of the input voltage, and measurements of output include measuring errors.

Chapter 14

Correlation involves a different approach and a different set of assumptions but some of the same quantities. Those will be discussed in section 14.6.

The methods of regression are used to summarize sets of data in a useful form. The values of *x* and *y* and any other quantities are already known from measurements and are therefore fixed, so it is not quite right to speak of them in this development as variables. The true variables will be the coefficients that are adjusted to give the best fit. Therefore, in sections 14.1 to 14.5 we will refer to *x* and the other "independent" pieces of data as *inputs* or regressors. A quantity such as *y*, which is a function of the inputs, will be called a *response*.

14.1 Simple Linear Regression

The simplest situation is a linear or straight-line relation between a single input and the response. Say the input and response are *x* and *y*, respectively. For this simple situation the mean of the probability distribution is

$$E(Y) = \alpha + \beta x \tag{14.1}$$

where α and β are constant parameters that we want to estimate. They are often called *regression coefficients*. From a sample consisting of *n* pairs of data (x_i, y_i), we calculate estimates, *a* for α and *b* for β. If at $x = x_i$, $\hat{y}_i$ is the estimated value of $E(Y)$, we have the fitted regression line

$$\hat{y}_i = a + b x_i \tag{14.2}$$

where the "hat" on $\hat{y}$ indicates that this is an *estimated* value.

(a) Method of Least Squares

The problem now is to determine *a* and *b* to give the best fit with the sample data. If the points given by (x_i, y_i) are close to a perfect straight line, it might be satisfactory to plot the points and draw the line by eye. However, for the present analysis we need a systematic recipe or algorithm. The reader may remember from section 3.2 (g) that the sum of squares of deviations from the *mean* of a sample is less than the sum of squares of deviations from any other constant value. We can adapt that requirement to the present case as follows. Let $e_i = y_i - \hat{y}_i$ be the deviation in the *y*-direction of any data point from the fitted regression line. Then the estimates *a* and *b* are chosen so that the sum of the *squares of deviations* of all the points, $\sum_{\text{all } i} e_i^2$, is smaller than for any other choice of *a* and *b*. Thus, *a* and *b* are chosen so that $\sum_{\text{all } i} e_i^2 = \sum_{\text{all } i} (y_i - \hat{y}_i)^2$ has a *minimum* value. This is called the *method of least squares* and the resulting equation is called the *regression line* of *y* on *x*, where *y* is the response and *x* is the input.

Say the points are as shown in Figure 14.1. This is called a *scatter plot* for the data. We can see that the points seem to roughly follow a straight line, but there are appreciable deviations from any straight line that might be drawn through the points.

Regression and Correlation

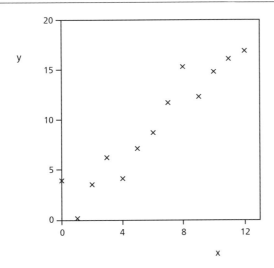

Figure 14.1: Points for Regression

Now let us consider the method of least squares in more detail. If the points or pairs of values are (x_i, y_i) and the estimated equation of the line is taken to be $\hat{y} = a + bx$, then the errors or deviations from the line in the y-direction are $e_i = [y_i - (a + bx_i)]$. These deviations are often called *residuals*, the variations in y that are not explained by regression. The squares of the deviations are $e_i^2 = [y_i - (a + bx_i)]^2$, and the sum of the squares of the deviations for all n points is $\sum_{i=1}^{n} e_i^2 = \sum_{i=1}^{n} [y_i - (a + bx_i)]^2$. This sum of the squares of the deviations or errors or residuals for all n points is abbreviated as SSE.

The quantity we want to minimize in this case is SSE = $\sum_{i=1}^{n} e_i^2 = \sum_{i=1}^{n} (y_i - \hat{y}_i)^2 = \sum_{i=1}^{n} [y_i - (a + bx_i)]^2$. Remember that the n values of x and the n values of y come from observations and so are now all fixed and not subject to variation. We will minimize SSE by varying a and b, so a and b become the independent variables at this point in the analysis. You should remember from calculus that to minimize a quantity we take the derivative with respect to the independent variable and set it equal to zero. In this case there are two independent variables, a and b, so we take partial derivatives with respect to each of them and set the derivatives equal to zero. Omitting some of the algebra we have

$$\frac{\partial}{\partial a}(\text{SSE}) = \frac{\partial}{\partial a} \sum_{i=1}^{n} [y_i - (a + bx_i)]^2 = -2 \left(\sum_{i=1}^{n} y_i - na - b \sum_{i=1}^{n} x_i \right) = 0$$

and

$$\frac{\partial}{\partial b}(\text{SSE}) = \frac{\partial}{\partial b} \sum_{i=1}^{n} [y_i - (a + bx_i)]^2 = -2 \left(\sum_{i=1}^{n} x_i y_i - a \sum_{i=1}^{n} x_i - b \sum_{i=1}^{n} x_i^2 \right) = 0.$$

Chapter 14

These are called the least squares equations (or normal equations) for estimating the coefficients, a and b. The right-hand equalities of these two equations give equations that are linear in the coefficients a and b, so they can be solved simultaneously. The results are

$$b = \frac{\sum_{i=1}^{n} x_i y_i - \frac{1}{n}\left(\sum_{i=1}^{n} x_i\right)\left(\sum_{i=1}^{n} y_i\right)}{\sum_{i=1}^{n} x_i^2 - \frac{1}{n}\left(\sum_{i=1}^{n} x_i\right)^2} \tag{14.3}$$

$$= \frac{\sum_{i=1}^{n}(x_i - \bar{x})(y_i - \bar{y})}{\sum_{i=1}^{n}(x_i - \bar{x})^2} \tag{14.3a}$$

and

$$a = \frac{\sum_{i=1}^{n} y_i - b\sum_{i=1}^{n} x_i}{n} = \bar{y} - b\bar{x} \tag{14.4}$$

The two forms of equation 14.3 for b are equivalent, as can be shown easily. The first form is usually used for calculations. The second form, equation 14.3a, is preferred when rounding errors in calculations may become appreciable. The second form indicates that the numerator is the sum of certain products and the denominator is the sum of similar squares.

These sums of products and squares are used repeatedly and so should be defined at this point. The quantity $\sum_{i=1}^{n}(x_i - \bar{x})^2$ is sometimes called the *sum of squares for x* and abbreviated S_{xx}. Similarly, the quantity $\sum_{i=1}^{n}(y_i - \bar{y})^2$ is sometimes called the *sum of squares for y* and abbreviated S_{yy}, and the quantity $\sum_{i=1}^{n}(x_i - \bar{x})(y_i - \bar{y})$ is sometimes called the sum of products for x and y and abbreviated S_{xy}. Then we have

$$S_{xx} = \sum_{i=1}^{n}(x_i - \bar{x})^2 = \sum_{i=1}^{n} x_i^2 - \frac{1}{n}\left(\sum_{i=1}^{n} x_i\right)^2 \tag{14.5}$$

$$S_{yy} = \sum_{i=1}^{n}(y_i - \bar{y})^2 = \sum_{i=1}^{n} y_i^2 - \frac{1}{n}\left(\sum_{i=1}^{n} y_i\right)^2 \tag{14.6}$$

$$S_{xy} = \sum_{i=1}^{n}(x_i - \bar{x})(y_i - \bar{y}) = \sum_{i=1}^{n} x_i y_i - \frac{1}{n}\left(\sum_{i=1}^{n} x_i\right)\left(\sum_{i=1}^{n} y_i\right) \tag{14.7}$$

Regression and Correlation

Equation 14.3 can be written compactly as

$$b = \frac{S_{xy}}{S_{xx}} \tag{14.8}$$

These abbreviations will be used also in later equations.

From equation 14.4 we have

$$a = \bar{y} - b\bar{x} \tag{14.4a}$$

Substituting for a in equation 14.2 with a little rearrangement gives

$$(\hat{y}_i - \bar{y}) = b(x_i - \bar{x}) \tag{14.9}$$

This indicates that the best-fit line passes through the point $(\bar{x}, \bar{y})$, which is called the centroidal point and is the center of mass of the data points. After the slope, b, is found from equation 14.8, the intercept, a, is usually calculated from equation 14.4a.

Equations 14.3 and 14.4 are called *regression equations*. The name "regression" arose because an early example of its use was in a study of heredity, which showed that under certain conditions some physical characteristics of offspring tended to revert or regress from the characteristics of the parents toward average values. The name "regression" has become well established for all uses for such equations and for the process of finding best-fit equations by the method of least squares.

Illustration

Now let's apply these equations to the points that were plotted in Figure 14.1. The data are given in Table 14.1.

Table 14.1: Data for Simple Linear Regression

x	0	1	2	3	4	5	6	7	8	9	10	11	12
y	3.85	0.03	3.50	6.13	4.07	7.07	8.66	11.65	15.23	12.29	14.74	16.02	16.86

We have 13 points, so $n = 13$. The data can be summarized by the following sums: $\sum_{i=1}^{13} x_i = 78$, $\sum_{i=1}^{13} y_i = 120.10$, $\sum_{i=1}^{13} x_i^2 = 650$, $\sum_{i=1}^{13} y_i^2 = 1483.0828$, $\sum_{i=1}^{13} x_i y_i = 968.95$

The centroidal point is given by $\bar{x} = \frac{78}{13} = 6$, $\bar{y} = \frac{120.10}{13} = 9.23846$.

The sums of squares and the sum of products are

$$S_{xx} = 650 - \frac{1}{13}(78)^2 = 650 - 468 = 182$$

$$S_{yy} = 1483.0828 - \frac{1}{13}(120.10)^2 = 1483.0828 - 1109.5392 = 373.5436$$

Chapter 14

$$S_{xy} = 968.95 - \frac{1}{13}(78)(120.10) = 968.95 - 720.60 = 248.35$$

Then $b = \dfrac{S_{xy}}{S_{xx}} = \dfrac{248.35}{182} = 1.36456$,

and using the values of $\bar{x}$ and $\bar{y}$ in equation 14.4a we find

$$a = 9.23846 - (1.36456)(6.000) = 9.23846 - 8.18736 = 1.0511$$

The best-fit regression equation of y as a function of x (often called the regression equation of y on x) by the method of least squares is

$$y = 1.0511 + 1.36456\, x$$

Notice that this calculation involves taking differences between numbers that are often of similar magnitude, so rounding the numbers too early could greatly reduce the accuracy of the results. As usual, rounding should be left to the end of the calculation.

The calculations for regression, especially for large sets of data, can be done much more quickly using a spreadsheet rather than a pocket calculator. Excel is suitable for such calculations.

The resulting regression equation of y on x is compared with the original points in Figure 14.2. The centroidal point is also shown. To emphasize that deviations in the y-direction are minimized, lines have been drawn in that direction between the points and the line.

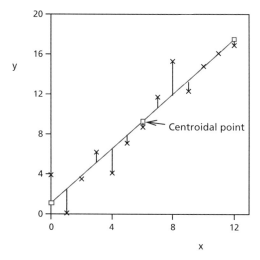

Figure 14.2: Comparison of Points and Regression Line

(b) Comparison of Regressions for Different Assumptions of Error

The derivation for the regression of y on x assumed that the values of x were known without error and that only values of y contained error. Is the result different if this assumption is not correct? The opposite assumption would be that values of y are known without error and only values of x contain error. In that case deviations from the line would be taken in the x-direction at constant y. The roles of y and x would be reversed. The equation of the new regression line would be $x = a' + b'y$, so $y = \dfrac{x - a'}{b'}$. Derivation for minimum sum of squares of deviations in this case would give

Regression and Correlation

$$b' = \frac{\sum_{i=1}^{n} x_i y_i - \frac{1}{n}\left(\sum_{i=1}^{n} x_i\right)\left(\sum_{i=1}^{n} y_i\right)}{\sum_{i=1}^{n} y_i^2 - \frac{1}{n}\left(\sum_{i=1}^{n} y_i\right)^2} = \frac{\sum_{i=1}^{n}(x_i - \bar{x})(y_i - \bar{y})}{\sum_{i=1}^{n}(y_i - \bar{y})^2} = \frac{S_{xy}}{S_{yy}} \quad (14.10)$$

and

$$a' = \frac{\sum_{i=1}^{n} x_i - b'\sum_{i=1}^{n} y_i}{n} = \bar{x} - b'\bar{y} \quad (14.11)$$

Again the regression line would pass through the centroidal point. If the equation of the new line is solved for y, it becomes

$$y = \frac{a'}{b'} + \frac{1}{b'}x \quad (14.12)$$

Thus its slope is S_{yy}/S_{xy}, instead of S_{xy}/S_{xx} for the slope of the regression of y on x. The new regression line is called the *regression of x on y*. Then the assumption concerning which variable contains the error does make a difference. The only case in which the lines for the regression of y on x and the regression of x on y would coincide is when the points form a perfect straight line. The more the data points depart from a straight line, the more the two regression lines will differ. Figure 14.3 shows the regression line of y on x and the regression line of x on y for the illustration of Figures 14.1 and 14.2.

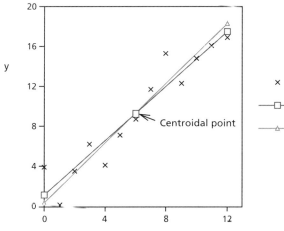

Figure 14.3: Comparison of Regression Lines

(c) Variance of Experimental Points Around the Line

Now we need to estimate the variance of points from the least-squares regression line for y on x. This must be found from the *residuals*, deviations of points from the least-squares line in the y-direction. As we discussed in part (b) of this section, these

Chapter 14

residuals can be calculated as $e_i = y_i - \hat{y}_i = y_i - (a + bx_i) = y_i - a - bx_i$. The *error sum of squares* abbreviated as *SSE*, is given by

$$SSE = \sum_{i=1}^{n}(y_i - a - bx_i)^2$$

or, since $\bar{y} = a + b\bar{x}$ and thus $a = \bar{y} - b\bar{x}$,

$$SSE = \sum_{i=1}^{n}\left[(y_i - \bar{y}) - b(x_i - \bar{x})\right]^2$$

$$= \sum_{i=1}^{n}(y_i - \bar{y})^2 - 2b\sum_{i=1}^{n}(x_i - \bar{x})(y_i - \bar{y}) + b^2\sum_{i=1}^{n}(x_i - \bar{x})^2$$

$$= S_{yy} - 2bS_{xy} + b^2 S_{xx}$$

But from equation 14.8, $b = \dfrac{S_{xy}}{S_{xx}}$, so $b^2 S_{xx} = \dfrac{(S_{xy})^2}{(S_{xx})^2} S_{xx} = \dfrac{(S_{xy})(S_{xy})}{S_{xx}} = b S_{xy}$, and $-2bS_{xy} + b^2 S_{xx} = -bS_{xy}$. Then we have

$$SSE = S_{yy} - b S_{xy} \tag{14.14}$$

This estimate of the error sum of squares, *SSE*, must be divided by the number of degrees of freedom. The number of degrees of freedom available to estimate the variance $\sigma^2_{y|x}$ is the number of points or pairs of values for *x* and *y*, less one degree of freedom for each of the independent coefficients estimated from the data. In this case we have n points and we have estimated from the data two independent coefficients, b and $\bar{y}$, or a and b. The available number of degrees of freedom is $(n - 2)$. The estimate of the variance of the points about the line is

$$s^2_{y|x} = \frac{SSE}{n-2} = \frac{S_{yy} - b S_{xy}}{n-2} \tag{14.15}$$

This quantity is a measure of the scatter of experimental points around the line. The square root of this quantity is, of course, the estimated standard deviation or standard error of the points from the line. The subscript, $y|x$, is meant to emphasize that the estimated variance around the line is found from deviations in the *y*-direction and at fixed values of *x*. The subscript is omitted in some books. The quantity $s^2_{y|x}$ can be used also to obtain estimates of the standard errors of other parameters, such as the slope of the best-fit line. These will be discussed in section 14.3 for statistical inferences.

14.2 Assumptions and Graphical Checks

Let us look first at the assumptions required for finding the best-fit lines. After that, we'll look at additional assumptions required for statistical inferences such as confidence limits and statistical significance. Then let's see how we can examine a plot of the data to see whether some assumptions are reasonable for any particular case.

Regression and Correlation

For simple linear regression of y on x in the simplest case we assume the data points are related by an equation of the form

$$y_i = a + b\,x_i + e_i \tag{14.16}$$

where the e_i represent errors or deviations or *residuals*. This involves certain assumptions. The first is that a linear relation between y and x represents the data adequately, so that the model represented by equation 14.16 is satisfactory. The second assumption is that the errors e_i are entirely in the y-direction and so independent of x; thus, there are assumed to be *no* errors in the values of x. Regression calculations also assume that the individual residuals, e_i, are essentially independent of one another, so that equation 14.16 is the only relation affecting y over the region in which measurements have been taken. Similar assumptions apply for the regression of x on y.

In order to apply statistical tests of significance and confidence limits we must also assume that the variance is constant and not varying as a function of the variables, and that the statistical distribution of errors or residuals is at least approximately normal. In particular, positive and negative deviations from the line should be equally likely at all values of x within the range of experimental data. Any outliers, points for which residuals are much larger than the others in absolute value, may make statistical inferences useless.

The reasonableness of the assumption that the values of x are known without error and all the error is in y must be tested by knowledge of the quantities and how they are measured. Because the line for regression of y on x and the line for x on y come closer together as data approach a perfect straight line, the effects of this assumption become less significant as the data come closer to a perfect correlation.

Now consider the assumption of a simple linear relation between y and x. Is a linear relation between y and x a satisfactory representation of the data? Or is there reason to think that some other form of relation would represent the data better? In many cases the underlying relation may be more complex, but a straight-line relation between y and x may represent the data satisfactorily for a particular range of values.

To examine these and other questions, we need to calculate the residuals from the best-fit straight line (or some more complex alternative), plot them against x or y, and examine the results. If we find some systematic relation between the residuals and either x or y, then apparently the straight-line relation of equation 14.13 is not adequate and we should try a different form for the relation between y and x.

We can obtain an indication of whether the variance is constant from the plot of residuals against x or y. If the residuals show markedly more or markedly less scatter (or first one, then the other) as x or y increases, so that the scatter shows a systematic pattern, then the variance is probably not constant. Of course, residuals vary randomly besides any systematic variation, so we have to be careful not to jump to conclusions. It is often desirable to obtain more data to confirm a tentative conclusion that the variance is not constant.

Chapter 14

We should consider the residuals to see whether the assumption of a normal distribution is reasonable. Are there about as many positive residuals as negatives? Are small deviations considerably more frequent than larger ones? Are there any outstanding outliers? We can answer these questions by examining the plot of residuals against x or y (usually x). These tests are adequate for relatively small sets of data. There are other, more sophisticated tests for a normal distribution that are useful for larger data sets, but they are beyond the scope of this book. For them the reader is referred to the book by Ryan (see reference in section 15.2).

Some examples of graphical checks

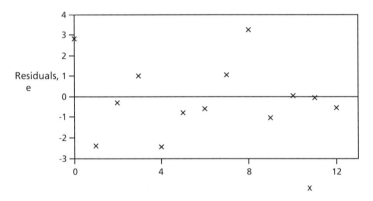

Figure 14.4: Residuals plotted against x

Figure 14.4 shows a plot of residuals versus x for the same data as in Figures 14.1, 14.2, and 14.3. There does not seem to be any strong pattern among the points in this plot, so we have no reason to discard the straight-line relation. Furthermore, there doesn't seem to be any systematic change beyond random variation in the scatter as x increases, so we have no reason to believe that the variance is anything other than constant.

The data points in Figure 14.4 show about the same number of positive and negative residuals and no pronounced outliers. Residuals of small absolute value are considerably more frequent than larger deviations. Therefore, the assumption that deviations from the line are normally distributed seems to be satisfactory.

For comparison, Figure 14.5 shows a residual plot for a case in which residuals plotted against x show systematic deviations. They are systematically less than or equal to zero in the middle of the diagram, and greater than or equal to zero on either side.

Regression and Correlation

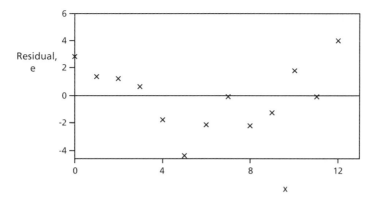

Figure 14.5: Another set of residuals plotted against *x*

This indicates that a linear relation between *y* and *x* does not represent the data adequately. In this case a quadratic relation between *y* and *x*, such as $y = a + bx^2$, might be more appropriate.

Figure 14.6 shows another plot of residuals against the input quantity. In this case we can see that the scatter of the residuals gradually increases as the independent variable increases. This indicates that the variance is not constant but appears to be increasing as *x* increases. Before confidence limits or tests of significance can be applied to these data, we must do something to make the variance about the line at least approximately constant. One possibility is to find a suitable transformation to give a quantity which will have a more constant variance about the line. That will be explored in section 14.4. Draper and Smith (see reference in section 15.2) suggest also a method of weighting the residuals in such a way that the variance becomes constant.

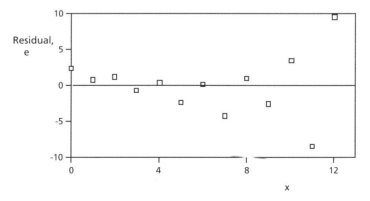

Figure 14.6: Residuals plotted against input or regressor

Chapter 14

Draper and Smith suggest other graphical checks of the data. In particular, they suggest a plot of residuals against the order in which measurements were made. This is done to check for any relation between errors and the time sequence, such as drift in calibration of instruments.

Lack of independence among the y-values in regression is often due to a correlation between these y-values and some other independent variable that is changing appreciably. If the various runs and measurements have not been properly randomized, as described in Chapter 11, bias can enter so that results are misleading.

14.3 Statistical Inferences

If the graphical checks are satisfactory, we can look at confidence intervals and tests of significance. Equation 14.15 has given us an expression for the estimated variance of points about the regression line, $s_{y|x}^2$.

(a) Inferences for Slope

Let us first look at the variance of the estimated slope, b. From equation 14.8,

$$b = \frac{S_{xy}}{S_{xx}} = \frac{\sum_{i=1}^{n}(x_i - \bar{x})(y_i - \bar{y})}{\sum_{i=1}^{n}(x_i - \bar{x})^2}.$$

Remembering that the x's are assumed to be known without any error, applying the rules for combining variables, which were discussed briefly in section 8.2, gives the following:

$$s_b^2 = \frac{s_{y|x}^2}{\sum_{i=1}^{n}(x_i - \bar{x})^2} = \frac{s_{y|x}^2}{S_{xx}} \qquad (14.17)$$

The estimated standard error (or standard deviation) of the slope is the square root of this. The standard error of the slope is multiplied by the value of t for the appropriate probability and the appropriate number of degrees of freedom to find confidence intervals. Hypothesis tests are done in the same way as in section 9.2.

Notice from equation 14.17 that the estimated variance of the slope becomes smaller as $S_{xx} = \sum_{i=1}^{n}(x_i - \bar{x})^2$ increases if the variance $s_{y|x}^2$ remains constant. Then if the x-values are more widely separated from the mean, we will have a better estimate of the slope of the line because the standard error of the slope will be smaller. That seems reasonable: if we base an estimate of the slope on more widely separated points while s^2, the variance of points about the line, remains constant, the estimate of the slope should be more reliable.

(b) Variance of the Mean Response

Next, we look at the variance of $\hat{y}$, which represents the least-squares estimate of y as a function of x. This is sometimes called the variance of the mean response. We will use as its symbol $s_{\hat{y}|x}^2$. (Remember that the circumflex or "hat" in $\hat{y}$ indicates an estimate from data.) $s_{\hat{y}|x}^2$ represents the estimated random variation of the best-fit line if the experiment is repeated with the same number of points at the same values of x. The position of the line is determined by three independent quantities: the position of the centroidal point $(\bar{x}, \bar{y})$, the slope of the line, b, and the difference between the x-coordinate of a particular point and the x-coordinate of the centroidal point. From this we get

$$s_{\hat{y}|x}^2 = \frac{s_{y|x}^2}{n} + \frac{s_{y|x}^2 (x-\bar{x})^2}{S_{xx}} = s_{y|x}^2 \left[\frac{1}{n} + \frac{(x-\bar{x})^2}{S_{xx}} \right] \quad (14.18)$$

Thus, we see that the variance of the mean response is smallest at the centroidal point and increases to the left and to the right by an amount proportional to the estimated variance of the points around the line and to the square of the x-distance from the centroidal point. Equation 14.18 can be applied at any value of x, including at $x = 0$, where it would give the variance of the intercept, a. Remember, however, that it is dangerous to extrapolate outside the region in which measurements have been taken.

(c) Variance of a Single New Observation

If we want the variance of a single new observation rather than the mean response, the variance shown in equation 14.18 will be larger by the variance around the line, $s_{y|x}^2$. Then the

$$\text{variance of a single new observation} = s_{y|x}^2 \left[1 + \frac{1}{n} + \frac{(x-\bar{x})^2}{S_{xx}} \right] \quad (14.19)$$

Corresponding standard errors are obtained from the variances of equations 14.18 and 14.19 by taking square roots. These standard errors are then multiplied by appropriate values of t to find confidence intervals (in the case of the mean response) and prediction intervals (in the case of single new observations). Tests of significance are performed similarly.

Illustration (continued)

Continuing with the previous numerical illustration, which was used in Figures 14.1 to 14.4 and for which data were shown in Table 14.1, we have:

$$s_b^2 = \frac{s_{y|x}^2}{S_{xx}} = \frac{3.15046}{182} = 0.01731, \ s_b = 0.1316$$

From equation 14.18 $s_{\hat{y}|x}^2 = s_{y|x}^2 \left[\frac{1}{n} + \frac{(x-\bar{x})^2}{S_{xx}} \right]$, and at $x = 11$ we have

$$s_{\hat{y}|x}^2 = (3.15046) \left[\frac{1}{13} + \frac{(11-6)^2}{182} \right] = 0.67510 \text{ and } s_{\hat{y}|x} = 0.8216.$$

Chapter 14

From equation 14.19 the estimated variance of a single new observation is

$$S_{y|x}^2 \left[1 + \frac{1}{n} + \frac{(x-\bar{x})^2}{S_{xx}}\right]; \text{ at } x = 11 \text{ this becomes } (3.15046)\left[1 + \frac{1}{13} + \frac{(11-6)^2}{182}\right]$$

= 3.8256, and the corresponding standard error is 1.956.

The 95% confidence limits for the mean response and 95% prediction intervals for a new individual observation are shown in Figure 14.7 for the same relationship as in Figures 14.1 to 14.4.

The arrows in Figure 14.7 emphasize that errors are assumed to be only in the y direction.

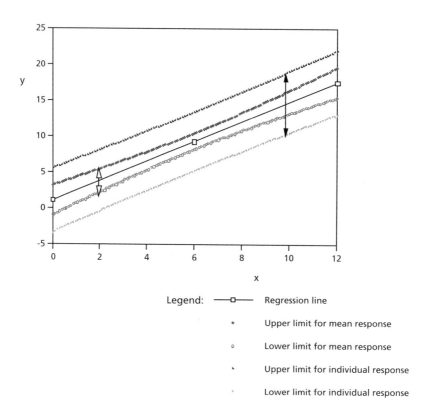

Figure 14.7: 95% Confidence intervals and prediction intervals around the regression line

Regression and Correlation

Example 14.1

The shear resistance of soil, y kN m^{-2}, is determined by measurements as a function of the normal stress, x kN m^{-2}. The data are as shown below:

x_i	10	11	12	13	14	15	16	17	18	19	20	21
y_i	14.08	15.57	16.94	17.68	18.49	19.55	20.68	21.72	22.80	23.84	24.79	25.67

Find the regression line of y on x. Plot the data, the regression line, and the centroidal point.

Answer: Calculations to find the regression line of y on x are shown below. Intermediate results are not rounded until final results are obtained to minimize rounding errors, but numbers are reported here to only three decimals.

	x	y	x^2	y^2	xy
	10	14.08	100	198.161	140.770
	11	15.57	121	242.506	171.299
	12	16.94	144	286.844	203.238
	13	17.68	169	312.646	229.863
	14	18.49	196	341.710	258.795
	15	19.55	225	382.179	293.241
	16	20.68	256	427.479	330.809
	17	21.72	289	471.698	369.216
	18	22.80	324	519.880	410.416
	19	23.84	361	568.520	453.029
	20	24.79	400	614.473	495.771
	21	25.67	441	659.118	539.139
Totals	186	241.80	3026	5025.214	3895.587

$$\bar{x} = \frac{186}{12} = 15.5 \quad \bar{y} = \frac{241.80}{12} = 20.151$$

$$S_{xx} = \sum_{i=1}^{n}\left(x_i^2\right) - \frac{\left(\sum_{i=1}^{n} x_i\right)^2}{n} = 3026 - \frac{(186)^2}{12}$$

$$= 3026 - 2883 = 143$$

Chapter 14

$$S_{yy} = \sum_{i=1}^{n} y_i^2 - \frac{\left(\sum_{i=1}^{n} y_i\right)^2}{n} = 5{,}025.214 - \frac{(241.80)^2}{12}$$

$$= 5{,}025.214 - 4{,}872.399 = \quad 152.815$$

$$S_{xy} = \sum_{i=1}^{n} x_i y_i - \frac{\left(\sum_{i=1}^{n} x_i\right)\left(\sum_{i=1}^{n} y_i\right)}{n} = 3895.587 - \frac{(186)(241.80)}{12}$$

$$= 3895.587 - 3{,}747.950 = \quad 147.637$$

Then $b = \dfrac{S_{xy}}{S_{xx}} = \dfrac{147.637}{143} = 1.032$ and $a = \bar{y} - b\bar{x} = 20.151 - (1.032)(15.5) = 4.148$.

Because the calculations of S_{yy}, S_{xy}, and a all involve differences of numbers of similar magnitudes, it is particularly important not to round the numbers too soon. The regression line of y on x is $y = 4.148 + 1.032\ x$.

The data, centroidal point and regression line of y on x are shown in Figure 14.8.

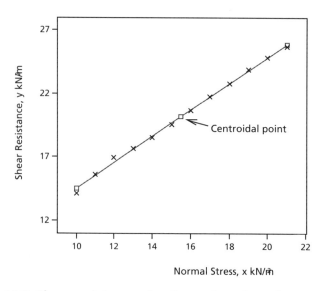

Figure 14.8: Shear resistance of soil as a function of normal stress

Regression and Correlation

Example 14.2

For the data of Example 14.1 calculate the standard deviation of points about the regression line, then plot residuals against the input, and comment on the results.

Answer: Calculations for these plots are shown in Table 14.1.

Table 14.1: Calculation of residuals

x	y	$\hat{y}$	Residuals
10	14.08	14.47	−0.39
11	15.57	15.50	+0.07
12	16.94	16.54	+0.40
13	17.68	17.57	+0.11
14	18.49	18.60	−0.12
15	19.55	19.63	−0.08
16	20.68	20.67	+0.01
17	21.72	21.70	+0.02
18	22.80	22.73	+0.07
19	23.84	23.77	+0.08
20	24.79	24.80	−0.01
21	25.67	25.83	−0.16

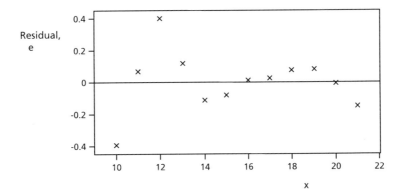

Figure 14.9: Residuals plotted against the regressor

Chapter 14

The residuals are just the differences between the measured values of y and the corresponding values on the regression line, $\hat{y}$. They are plotted against the input or regressor in Figure 14.9.

From equation 14.14 we have

$$SSE = S_{yy} - b\, S_{xy}$$
$$= 152.815 - (1.032)(147.637)$$
$$= 152.815 - 152.425$$
$$= 0.3896$$

From equation 14.15, $s_{y|x} = \text{SQRT}(SSE / (n-2))$
$$= \text{SQRT}(0.3896 / 10)$$
$$= 0.1974$$

Figure 14.9 shows no systematic effect of the input on the residuals, either in average or in variability. Thus there is no reason to think that the shear resistance of the soil is not well represented for this range of values by a linear function of the normal stress. Furthermore, there is no reason to think that the variance is a function of x. The distribution of the residuals is consistent with a normal distribution. Thus, we can legitimately use the calculated data to find confidence intervals and prediction intervals, and apply tests of significance.

Example 14.3

For the data given in Examples 14.1 and 14.2:

a) Find the 90% confidence interval for the slope of the regression line of shear resistance on normal stress.

b) Is the slope significantly larger than 1.000 at the 5% level of significance?

c) Find the 95% confidence interval for the mean response of shear resistance at a normal stress of 12 kN/m².

d) Is the mean response for the shear resistance at a normal stress of 12 kN/m² significantly more than 16.5 kN/m² at the 1% level of significance?

e) Find the 95% prediction interval for a single new observation at a normal stress of 20 kN/m².

Answer: a) By the method of least squares we found in Example 14.1 that the best estimate of the slope or regression coefficient is $b = 1.0324$. In Example 14.2 we calculated the estimated standard error of the points around the best-fit line to be $s_{y|x} = 0.1974$, and $s_{y|x}^2 = 0.03896$. From equation 14.17 the estimated variance of the slope is $s_b^2 = \dfrac{s_{y|x}^2}{S_{xx}}$, and from Example 14.1 $S_{xx} = 143$. Then the estimated standard error of the slope is $s_b = \sqrt{\dfrac{0.03896}{143}} = 0.01651$. For the 90% confidence interval we need the value of t corresponding to probability 0.05 in each tail with $12 - 2 = 10$ degrees of freedom. This is shown in Figure 14.10. From Table A2 we find for a one-tail probability of 0.05 and for 10 degrees of freedom, $t_1 = 1.812$. Then the 90% confidence interval for the slope is from $1.0324 - (1.812)(0.01651) = 1.0025$ to $1.0324 + (1.812)(0.01651) = 1.0623$.

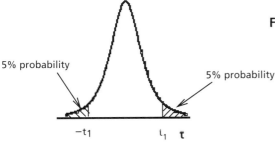

Figure 14.10: *t*-distribution for 90% confidence interval

b) $H_0: \beta = 1.000$

  $H_a: \beta > 1.000$ (one-tailed test)

Test statistic: $t = \dfrac{b - 1.000}{s_b}$.

Large values of $|t|$ will indicate that the null hypothesis is unlikely to be correct.

$b = 1.0324$

$s_b = 0.01651$ (from part a)

$t_{\text{calculated}} = \dfrac{1.0324 - 1.000}{0.01651} = 1.962$

For 10 d.f., one-tail probability 0.05, $t_{\text{critical}} = 1.796$ as in Figure 14.10.

Since $|t_{\text{calculated}}| > t_{\text{critical}}$, the slope is significantly larger than 1.000 at the 5% level of significance.

However, for 10 d.f. and one-tail probability of 0.025, Table A2 shows that $t_{\text{critical}} = 2.228$. Then at 2.5% level of significance, $|t_{\text{calculated}}| < t_{\text{critical}}$, so the slope is not significantly larger than 1.000. But we have to answer the problem as it was stated.

c) The variance of the mean response at $x = 12$ is given by equation 14.18 as

$$s^2_{\hat{y}|x} = s^2_{y|x}\left[\dfrac{1}{n} + \dfrac{(x - \bar{x})^2}{S_{xx}}\right] = 0.03896\left[\dfrac{1}{12} + \dfrac{(12 - 15.5)^2}{143}\right]$$

$$= 0.03896 [0.0833 + 0.0857]$$
$$= 0.00658$$

Then the standard error of the mean response is 0.0811.

For the 95% confidence interval we require the value of t for 10 d.f. and a one-tail probability of 0.025. This is shown in Figure 14.11. From Table A2 we find $t_1 = 2.228$.

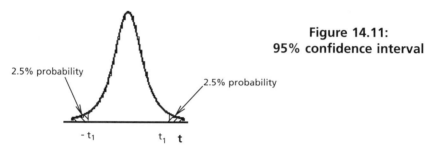

Figure 14.11: 95% confidence interval

At a normal stress of 12 kN/m² the prediction from the regression line is $y = 4.148 + 1.032\, x = 4.148 + (1.032)(12) = 16.532$. Then the 95% confidence interval for the mean response at 12 kN/m² is $16.532 \pm (2.228)(0.0811) = 16.351$ to 16.713.

Then we can have 95% confidence that the mean response for the shear resistance at a normal stress of 12 kN/m² is between 16.35 and 16.71 kN/m².

d) H_0: $\mu_{\hat{y}|x=12} = 16.5$

   H_a: $\mu_{\hat{y}|x=12} > 16.5$ (one-tailed test)

Test statistic: $t = \dfrac{\hat{y}_{x=12} - 16.5}{s_{\hat{y}|x=12}}$. Large values of $|t_{\text{calculated}}|$ will make H_0 unlikely to be correct.

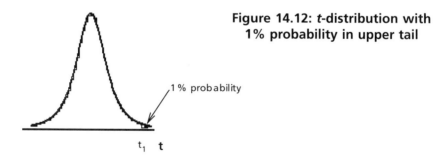

Figure 14.12: t-distribution with 1% probability in upper tail

From Table A2 we find for an upper-tail probability of 0.01, 10 d.f., $t_{critical}$ = 2.764. We found in part (c) that the standard error of the mean response is 0.0811, and that at a normal stress of 12 kN/m² the mean response is 16.532 kN/m². Then we have

$$t_{calculated} = \frac{16.532 - 16.5}{0.0811} = 0.395$$

Since $t_{calculated} < t_{critical}$, the mean response for the shear resistance at a normal stress of 12 kN/m² is not significantly more than 16.5 kN/m² at the 1% level of significance.

e) At a normal stress of 20 kN/m² the predicted shear resistance of the soil is 4.148 + (1.032)(20) = 24.788. From part (c) and Figure 14.13 for the 95% prediction interval we have t_1 = 2.228. The standard error for a single new observation is (from equation 14.19)

$$s_{y|x}\sqrt{1 + \frac{1}{n} + \frac{(x - \bar{x})^2}{S_{xx}}} = 0.1974\sqrt{1 + \frac{1}{12} + \frac{(20 - 15.5)^2}{143}} = 0.2185.$$ The 95% prediction interval for a single new observation is from 24.788 − (2.228)(0.2185) to 24.788 + (2.228)(0.2185), or from 24.30 to 25.27 kN/m².

14.4 Other Forms with Single Input or Regressor

(a) Other Forms Linear in the Coefficients

With an extra step of calculation an important group of equations can be fitted to data by the method of least squares. For instance, equations of the form, log $y = a + b x$, where a and b are coefficients to be determined by least squares, can be handled easily. Remember that x and y are known quantities, numbers. Then we can calculate without difficulty the value of log y for each data point. Then log y can be used in place of y, and so the regression coefficients can be calculated as before.

Example 14.4

We want to fit the following set of data to an equation of the form, ln $y = a + b x$, by the method of least squares.

x	0	1	2	3	4	5	6	7	8	9	10	11
y	1.178	1.142	1.273	1.354	1.478	1.737	1.842	1.778	2.160	2.418	2.339	2.931

Answer: The first step is to calculate ln y for each value of y. Then x^2, (ln y)², and the product $(x)(\ln y)$ are calculated.

Chapter 14

Table 14.2: Calculations for regression using ln y

x	ln y	x^2	$(\ln y)^2$	$(x)(\ln y)$
0	0.164	0	0.027	0.000
1	0.133	1	0.018	0.133
2	0.241	4	0.058	0.482
3	0.303	9	0.092	0.910
4	0.391	16	0.153	1.562
5	0.552	25	0.305	2.760
6	0.611	36	0.373	3.666
7	0.575	49	0.331	4.027
8	0.770	64	0.593	6.161
9	0.883	81	0.780	7.946
10	0.850	100	0.722	8.496
11	1.075	121	1.157	11.830
Totals 66	6.548	506	4.607	47.973

Then $\bar{x} = \dfrac{66}{12} = 5.5$ and $\overline{\ln y} = \dfrac{6.548}{12} = 0.5457$.

$$S_{xx} = \sum_{i=1}^{12} x^2 - \dfrac{\left(\sum_{i=1}^{12} x\right)^2}{12} = 506 - \dfrac{(66)^2}{12} = 506 - 363 = 143$$

$$S_{\ln y, \ln y} = \sum_{i=1}^{12} (\ln y)^2 - \dfrac{\left(\sum_{i=1}^{12} \ln y\right)^2}{12} = 4.607 - \dfrac{(6.548)^2}{12} = 4.607 - 3.573 = 1.034$$

$$S_{x, \ln y} = \sum_{i=1}^{12} (x)(\ln y) - \dfrac{\left(\sum_{i=1}^{12} x\right)\left(\sum_{i=1}^{12} \ln y\right)}{12} = 47.973 - \dfrac{(66)(6.548)}{12} = 47.973 - 36.014 = 11.958$$

Then $b = \dfrac{S_{x, \ln y}}{S_{xx}} = \dfrac{11.958}{143} = 0.08362$

and $a = \overline{\ln y} - (b)(\bar{x}) = 0.5457 - (0.08362)(5.5) = 0.5457 - 0.4599 = 0.0858$

Then the fitted equation is

$$\ln y = 0.0858 + 0.08362\, x$$

The residuals, fitted values of ln y less observed values of ln y, are plotted versus x in Figure 14.13. In fact these data correspond to the data plotted in Figure 14.6 as residuals of y vs. x. If these two plots are compared, it will be seen that the logarithmic transformation has made the variance much more constant.

Regression and Correlation

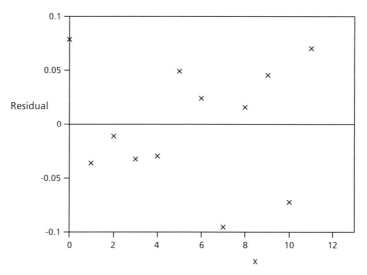

Figure 14.13: Residuals of ln y vs. x

This modified method works for a considerable number of cases. The requirement is that the equation to which we fit data must be of the form $f_1(y) = a + b f_2(x)$, where x is the only input quantity. The two functions, $f_1(y)$ and $f_2(x)$, can be of any form and do not have to be linear, but both a and b must be coefficients to be determined by the method of least squares. Thus the fitting equation must be *linear in the coefficients* so that it is easy to solve for a and b. The modified method is sometimes still considered to be simple linear regression, but then the word "simple" means that there is only one input, and the word "linear" means that the equation is linear in the *coefficients*. Fitting equations amenable to the modified method include the following types:

$$y = a + b x^3$$
$$y = a + b \sqrt{x}$$
$$y = a + b\, e^{x^2}$$
$$y = a + \frac{b}{x}$$
$$y = a + b x^{-3}$$
$$\frac{1}{y} = a + \frac{b}{x}$$
$$\frac{1}{y} = a + b \ln x$$
$$e^{-y} = a + bx$$
$$\log y = a + b \log x$$

and many others. Sometimes the nonlinear form for y or x is suggested by theory or previous experience, and sometimes it is suggested by consideration of the pattern of the residuals and by trial-and-error.

The graphical checks for constant variance, for fit to the chosen equation, and for normality can be done as before. If the checks are satisfactory, statistical inferences regarding confidence limits and tests of hypothesis can be applied. However, remember that if y is normally distributed, $\ln y$, y^{-1}, e^{-y} and so on are unlikely to be satisfactorily normal. The assumption that the input, x, is known without error still applies.

(b) Other forms transformable to give equations linear in the coefficients

Various common forms of equations involving one input can be transformed easily to give forms of equations which are linear in the coefficients.

(1) The exponential function, $y = a\,b^x$, can be modified suitably by taking logarithms of both sides. This gives $\log y = \log a + x \log b$. Notice that this is the form that gives straight lines on semi-log graph paper.

(2) The power function, $y = a\,x^b$, can also be treated by taking logarithms of both sides. The result is $\log y = \log a + b \log x$. Notice that this form would give straight lines on log-log graph paper.

(3) The function, $y = \dfrac{x}{a+b\,x}$, can be inverted to give $\dfrac{1}{y} = \dfrac{a}{x} + b$. An alternative is to multiply the inverted form by x to give $\dfrac{x}{y} = a + b\,x$.

It is important to note that the squares of the deviations are minimized in the *transformed* response variable ($\log y$ or $1/y$ or x/y in the cases above) rather than y, and the graphical tests need to be applied to the transformed response variable. It is possible in some cases to apply a simple weighting function to make the variance approximately constant (see the book by Draper and Smith, reference given in section 15.2).

(c) Extension: Nonlinear Forms

Equations that cannot be transformed into forms linear in the coefficients can still be treated by least squares. However, now instead of applying the relations discussed to this point, iterative numerical methods must be used to minimize the sum of squares of the deviations from the fitted line. The Excel feature called Solver can be used for that calculation.

14.5 Correlation

Correlation is a measure of the association between two random variables, say X and Y. We do not have for this calculation the assumption that one of these variables is known without error: *both variables are assumed to be varying randomly*. We do assume for this analysis that X and Y are related *linearly*, so the usual *correlation*

Regression and Correlation

coefficient gives a measure of the linear association between X and Y. Although the underlying correlation is defined in terms of variances and covariance, in practice we work with the *sample correlation coefficient*. This is calculated as

$$r_{xy} = \frac{S_{xy}}{\sqrt{S_{xx}S_{yy}}} \tag{14.20}$$

where S_{xx}, S_{yy}, and S_{xy} are defined in equations 14.5 to 14.7. This correlation coefficient is often denoted simply by r.

If the points (x_i, y_i) are in a perfect straight line and the slope of that line is positive, $r_{xy} = 1$. If the points are in a perfect straight line and the slope is negative, $r_{xy} = -1$. If there is no systematic relation between X and Y at all, $r_{xy} \approx 0$, and r_{xy} differs from zero only because of random variation in the sample points. If X and Y follow a linear relation affected by random errors, r_{xy} will be close to $+1$ or -1. These cases are illustrated in Figure 14.14. In all cases, because of the definitions $-1 \leq r_{xy} \leq +1$.

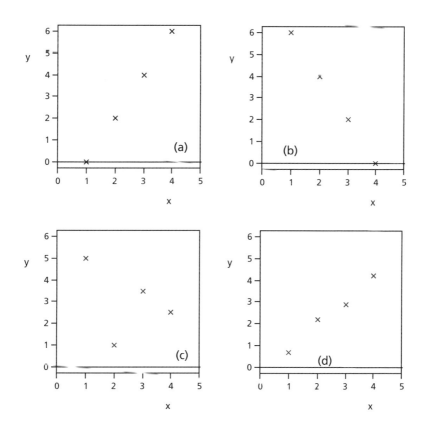

Figure 14.14: Illustrations of various correlation coefficients
(a) $r_{xy} = +1$, (b) $r_{xy} = -1$, (c) $r_{xy} \approx 0$, (d) $r_{xy} \approx 1$

Chapter 14

Examination of equations 14.8, 14.7 and 14.20 will indicate that r_{xy} has the same sign as the slope or regression coefficient, b. Furthermore, $r_{xy} = \pm\sqrt{b\,b'}$. Then as r_{xy} becomes closer to −1 or +1, the equations of the linear regression line of y on x and the linear regression line of x on y come closer to coinciding with one another.

An approximate expression for the standard error of the sample correlation coefficient has been derived and is available. It is therefore possible to test the hypothesis that the underlying correlation is zero and r_{xy} differs from it only because of random fluctuations. However, this gives the same information as testing the hypothesis that the slope or regression coefficient b differs from zero only because of random fluctuations. We have already seen (Example 14.3) how to test a hypothesis concerning the underlying value of the linear slope, b. Therefore this book will not be concerned with a test of hypothesis for r_{xy}.

A common misunderstanding is to assume that a strong correlation between two variables is evidence that one causes the other. This *may* be correct, but often there is another explanation. Huff (*How to Lie with Statistics,* see section 15.2 for reference) cites a number of examples in which some other explanation is more likely. One example he quotes is that for certain years there was a close relationship between the salaries of Presbyterian ministers in Massachusetts and the price of rum in Havana, Cuba. Closer examination indicates that both were likely due to a common factor, a marked and widespread inflation of salaries and prices over those years.

If the data come from a well planned, carefully randomized experiment, rather than from accumulated routine data, there is much less probability that other factors are responsible. In that case, it is considerably more likely that a correlation indicates some sort of causal relation between the variables.

The square of the correlation coefficient, r_{xy}^2, is called the *coefficient of determination*. The sum of squares of deviations from $\bar{y}$ in the y-direction is $\sum_{i=1}^{n}(y_i - \bar{y})^2$. The coefficient of determination is the fraction of this sum of squares which is explained by the linear relation between $\hat{y}$ and x given by regression of y on x. Then the coefficient is given by the ratio of $\sum_{i=1}^{n}(\hat{y}_i - \bar{y})^2$ to $\sum_{i=1}^{n}(y_i - \bar{y})^2$. A closely related quantity is the coefficient of multiple determination, which is useful in multiple linear regression.

If the correlation coefficient or the coefficient of determination becomes larger for the same algebraic forms, that indicates that the relationship between the variables has become stronger. However, if an algebraic form changes, say from x to $\ln x$, comparing the values of the coefficients is not useful. Instead, in that case we should compare the variances of the points around the regression lines.

Regression and Correlation

Example 14.5

a) Calculate a correlation coefficient for the data of Example 14.1.

b) What fraction of the sum of squares of deviations in the y-direction from $\bar{y}$ is explained by the linear relation between y and x given by regression?

Answer: a) $r_{xy} = \dfrac{S_{xy}}{\sqrt{S_{xx} S_{yy}}} = \dfrac{147.637}{\sqrt{(143)(152.815)}} = 0.9987$

b) The fraction of the sum of squares of deviations in the y-direction which is explained by the linear regression of y on x is equal to the coefficient of determination, $r^2 = (0.9987)^2 = 0.997$. Thus, in this case only 0.3% of the sum of squares of deviations in the y-direction from $\bar{y}$ is not explained by the regression.

For comparison, the data shown in Table 14.1 and plotted with the regression line of y on x in Figure 14.2 give a correlation coefficient of 0.952, and the regression explains 90.7% of the sum of squares of deviations in the y-direction from $\bar{y}$.

14.6 Extension: Introduction to Multiple Linear Regression

If there is more than one input or regressor, the basic ideas of linear regression still apply, but the algebra becomes considerably more complicated.

Let us look briefly at three simple cases in which there are only two or three inputs. If the two independent inputs are x and z, which enter only as first powers, the relation for point i with residual e_i becomes

$$y_i = a + b x_i + c z_i + e_i \tag{14.21}$$

If the two inputs are x and x^2, the relation still comes under the heading of multiple linear regression:

$$y_i = a + b x_i + c x_i^2 + e_i \tag{14.22}$$

This could be extended to a longer power series. If both x and z affect y, but now second-order terms are included, the relation for point i becomes

$$y_i = a + b x_i + c x_i^2 + d z_i + e z_i^2 + f x_i z_i + e_i \tag{14.23}$$

Notice that the term involving $x_i z_i$ represents an interaction of the sort discussed in section 11.3.

Multiple linear regression can include terms of the type discussed in part (a) of section 14.4. These are forms nonlinear in x or y or both, but linear in the *coefficients*. Thus, the term "linear" in "multiple linear regression" refers to fitting equations which are linear in the coefficients. For example, data for vapor pressure of pure components are sometimes related to temperature by expressions of the following form:

Chapter 14

$$\ln y = a - \frac{b}{x} + c \ln x + d\, x^6$$

This equation can be fitted to data by multiple linear regression.

As the number of terms increases, the complexity of the algebra increases. For each additional term there is another coefficient to be determined by the method of least squares. The algebra of the theoretical development becomes simpler if we use matrix notation, but the resulting expressions still have to be expanded into scalar algebra for calculations.

Furthermore, the analysis of multiple linear regression frequently involves re-calculation with more or fewer terms. We add terms to try to describe the relationship more fully, or we remove terms that do not contribute significantly to a useful description.

Thus, present-day calculations of multiple linear regression are almost always done on a digital computer using specialized software. Various computer packages, such as SAS and SPSS, are extremely useful once the reader has a good grasp of the fundamentals.

If the data for multilinear regression come from routine operating data rather than from a designed experiment, we have to worry about possible correlation among the inputs. That is eliminated if data are from a *designed experiment* with appropriate randomization.

Problems

1. Scraps of iron were selected on the basis of their densities, x_i, and their iron contents, y_i, were measured. The results were as follows:

x_i	2.8	2.9	3.0	3.1	3.2	3.2	3.2	3.3	3.4
y_i	27	23	30	28	30	32	34	33	30

   Find the regression equation of y on x by the method of least squares.

2. For the data given in problem 1 above, use a graph to check whether the form of the equation represents the data adequately, whether the variance appears to be independent of x, and whether the residuals appear to be normally distributed.

3. For the data given in problem 1 above, assume that graphical checks for fit, constancy of variance, and normality are satisfied. Find:
   a) the standard error of the slope of the regression equation, and
   b) the 95% confidence limits of the slope.

4. For the data given in problem 1 above,
   a) find the sample correlation coefficient
   b) what percentage of the variation of y_i about $\bar{y}$ is explained by regression?

Regression and Correlation

5. The density of molten salt mixtures, y g/cm^3, was measured at various temperatures x°C. The results were:

x_i	250	270	290	310	330	350
y_i	1.955	1.935	1.890	1.920	1.895	1.865

   a) Plot a graph of y vs. x showing these points (the graph is called a scatter diagram).
   b) Calculate Σx, Σy, Σx^2, Σy^2, Σxy, $\bar{x}$, $\bar{y}$.
   c) Calculate the regression equation of y on x in the form $y = a + bx$. Plot this line and the centroidal point on the graph.
   d) Calculate the regression of x on y in the form $x = a' + b'y$. Plot on the same graph as for parts (a) and (c).

6. For the data given in problem 5 above, use a graph to check whether the form of the equation in part (c) represents the data adequately, whether the variance appears to be independent of x, and whether the residuals appear to be normally distributed.

7. Assume that the graphical checks of problem 6 above are satisfactory. Then for the data given in problem 5 above:
   a) Calculate the estimated variance around the regression line, $s_{y|x}^2$.
   b) Is the estimated regression coefficient, b, significantly different from zero at the 1% level of significance? Can we conclude that temperature has a significant effect on density in this case?
   c) What is the 95% confidence interval for ß, the true slope or regression coefficient?
   d) Calculate the 95% confidence interval for the mean value of y at each of $x = 250, 300, 350$.
   e) Suppose we repeat the experiment. What is the 95% prediction interval for individual values of y at each of $x = 250, 300, 350$?

8. For the data given in problem 5 above,
   a) Calculate the correlation coefficient for this data set.
   b) Calculate the coefficient of determination.

9. A physical measurement, such as intensity of light of a particular wavelength transmitted through a solution, can often be calibrated to give the concentration of a particular substance in the solution. 9 pairs of values of intensity (x) and concentration (y) were obtained and can be summarized as follows:

   $\Sigma x = 30.3$, $\Sigma y = 91.1$, $\Sigma xy = 345.09$, $\Sigma x^2 = 115.11$, $\Sigma y^2 = 1036.65$.
   (a) Find the regression equation for y on x.
   (b) Find the correlation coefficient between x and y.
   (c) Assuming that graphical checks are satisfactory, test the null hypothesis that the slope of the regression line of y on x is not significantly different from zero, using the 1% level of significance, and
   (d) Find the 99% confidence limits of the slope of the straight line relation giving y as a function of x.

Chapter 14

10. An engineering student has a summer job with the forestry service. He measured the tree trunk diameters (x) and related them to the age of the tree (y). The following information was obtained:
    $n = 6$, $\Sigma x = 21$, $\Sigma x^2 = 91$, $\Sigma y = 26.4$, $\Sigma y^2 = 142.52$, $\Sigma xy = 113.8$
    a) Find:
       i) the regression equation of y on x
       ii) the correlation coefficient between x and y.
    b) Assuming that graphical checks are satisfactory, test whether the regression coefficient is significantly different from 0 at the 1% level of significance.

11. Gasoline consumption of a test automobile was recorded at speeds (x) ranging from 56 to 112 km/hr. The observed gasoline consumptions were converted to distance traveled per liter of gasoline (y). The following information was compiled:
    $\Sigma x = 984$, $\Sigma x^2 = 84416$, $\Sigma xy = 13418.4$, $\Sigma y = 165.3$, $\Sigma y^2 = 2282.45$, $n = 12$.
    a) Find the regression equation for y on x.
    b) Find the correlation coefficient.

12. Assuming that the graphical checks are satisfactory, for the data given in problem 11 above:
    a) Test the hypothesis that slope (regression coefficient) is equal to -0.045, using 5% level of significance.
    b) Find the 90% confidence limits for the mean distance traveled per liter of gasoline at 90 km/hr speed.

13. It is much easier to measure diameters of spot welds than to measure their shear strengths, and under some conditions they are related. Corresponding values were obtained for 10 welds, with shear strengths expressed by y p.s.i. and weld diameters, x, expressed as thousandths of an inch. For these 10 pairs of data, $\Sigma y = 22\,860$, $\Sigma x = 2325$, $\Sigma y^2 = 67{,}719{,}400$, $\Sigma x^2 = 697{,}425$, $\Sigma xy = 6{,}872{,}250$.
    a) Find the regression equation for y on x.
    b) Find the correlation coefficient.
    c) Assuming that graphical checks are satisfactory:
       i) Is the slope of the regression equation significantly different from 0 at the 5% level of significance?
       ii) Test the hypothesis that the slope (regression coefficient) is equal to 10, using 1% level of significance.
       iii) Find the 95% confidence limits for the true shear strength of a weld 0.210 inches in diameter.

14. Ms. Patsy Knowlet, a water quality engineer, noted that there seemed to be a close connection between an important streamflow water quality parameter, Y, and the flow, X m³/s. She found that 9 pairs of observations yielded the following data: $\Sigma x = 15.2$; $\Sigma x^2 = 57.6$; $\Sigma y = 45.6$; $\Sigma y^2 = 518.3$; $\Sigma xy = 172.6$. She would like to develop an equation that would allow her to predict Y knowing X.

a) Find the best estimate of the linear regression line of y on x.
b) Find the correlation coefficient between x and y.

15. Assume that the appropriate graphical check is satisfactory. Then for the data given in problem 14 above:
a) Find the 95% confidence limits of the slope or regression coefficient, β.
b) What are the 95% confidence limits for the predicted water quality parameter, Y, at a flow of 4.6 m³/s?

16. Shear stress (y) and rate of shear (x) can be measured for a liquid in a viscometer. For 12 pairs of values the data can be summarized as follows: $\Sigma x = 132.0$, $\Sigma y = 151.7$, $\Sigma x^2 = 1944.0$, $\Sigma y^2 = 2570.48$, $\Sigma xy = 2233.2$
a) Find the linear regression of y on x.
b) What is the correlation coefficient?
c) What fraction of the variance of y is explained by the correlation?

17. Assume that the appropriate graphical check is satisfactory. Then for the data given in problem 16 above:
a) Find the standard error of the slope.
b) Find the 95% confidence interval of the slope.
c) Is the slope significantly smaller than a slope of 1.210 at the 5% level of significance?

18. The number of errors per hour of radio telegraphists (y) as a function of the temperature (x) is determined. The relevant data about the relationship are as follows:

$\Sigma x = 118$ $\Sigma x^2 = 3510$
$\Sigma y = 56.1$ $\Sigma y^2 = 809.63$
$\Sigma xy = 1679.2$ $n = 4$

a) Find the regression line of errors/hour on temperature.
b) Find the correlation coefficient between temperatures and errors/hour.
c) Assuming that graphical checks are satisfactory:
   i) Is the slope of the regression line significantly different from zero? What does this imply about the relation between y and x?
   ii) Find the 99% confidence limits of the slope of the regression line.

19. The relationship between the temperature of a rocket engine (t) and the thrust force (f) was investigated in a series of tests. Pairs of data for t and f (in suitable units) were collected and can be summarized as follows: $n = 15$; $\Sigma t = 540$; $\Sigma f = 33.00$; $\Sigma t^2 = 21426$; $\Sigma f^2 = 77.08$; $\Sigma tf = 1267.10$
a) Find the regression equation of f on t.
b) Assuming that graphical checks are satisfactory:
   i) Find the 95% confidence limits of the intercept and slope.
   ii) According to Snooker's theory, the slope of the straight line relating f and t in this range of values should be 0.0500. Do the data disagree significantly with Snooker at the 5% level of significance?

Chapter 14

20. The speed (rpm) of a Danor ventilation fan was varied and the airflow capacity (cubic meters per second) of the fan was measured. The data pairs of capacity (y) versus speed (x) were collected. The following information was obtained:
   $n = 12$, $\Sigma x = 118.98$, $\Sigma y = 30.09$, $\Sigma x^2 = 1251.06$, $\Sigma y^2 = 80.57$, $\Sigma xy = 317.34$.
   (a) Find the regression equation of Y on X, and
   b) Find the correlation coefficient between X and Y.
   c) Assuming that graphical checks are satisfactory, test whether the slope of the regression line is significantly different from zero at the 1% level of significance.

21. The value of Y, the percentage decrease of volume of leather from the value at one atmosphere pressure, was measured for certain fixed values of high pressure, x atmospheres. The relevant data about the relationship are as follows:
   $\Sigma x = 28{,}000$, $\Sigma y = 19.0$, $\Sigma xy = 148{,}400$, $\Sigma x^2 = 216{,}000{,}000$, $\Sigma y^2 = 102.2$, $n = 4$.
   a) Find the regression line of y on x.
   b) Find the correlation coefficient between decrease of volume and pressure.
   c) Assuming that graphical checks are satisfactory:
      i) Test whether the slope of the regression line is significantly different from zero at the 1% level of significance.
      ii) Find the 99% confidence limits of the slope of the regression line.

CHAPTER **15**

Sources of Further Information

15.1 Useful Reference Books

Many reference books are available in the area of probability and statistics for engineers. Some that I have found useful will be mentioned here. Detailed references are given in section 15.2.

Some readers may want books that are a little more theoretical and advanced than this one. The book by Walpole and Myers is such a book. It is clearly written and contains such topics as Baysian and maximum likelihood approaches to estimation. It also contains a chapter on nonparametric statistics and one on statistical quality control.

The book by Burr focuses on statistics while providing a good background in probability. That author uses to good effect his experience as a statistical consultant.

The book by Vardeman concentrates on statistics and is very good in that general area, including considerable discussion on design of experiments and analysis of experimental results. It contains a large number of reports on study projects done by undergraduate students.

On the other hand, the book by Ziemer concentrates on probability and its applications in electrical engineering, rather than on statistics. Some readers of the present book will want now or later a more mathematically rigorous development of probability; they should consider the book by Ziemer.

The book by Ross also takes a more rigorous approach to probability and statistics.

The book by Barnes is notable because it takes an approach strongly based on using a computer. It uses diskettes of specially formulated software for calculations involving probability and statistics.

The book by Kennedy and Neville has been popular with engineering students because it includes many problems from practical engineering situations, both solved problems and problems for the student to solve. The present book tries to follow the example of Kennedy and Neville in that regard.

Finally, there are four areas for which specialized books should be recommended. One is design and analysis of experiments, for which mention has already been made in Chapters 11 and 12 of books by Box, Hunter, and Hunter and by Montgomery. Another area is regression, both simple and multiple, for which I have found the book by Draper and Smith very useful. A more up-to-date reference, and an excellent source of information, is the book by Ryan. A third area is the application of probability to the theory of electrical communication systems, for which the book by Haykin is suitable. The fourth is engineering reliability, for which the book by Billinton and Allan is suggested.

15.2 List of Selected References

Barnes, J. Wesley. *Statistical Analysis for Engineers and Scientists, A Computer-based Approach.* New York: McGraw-Hill, 1994

Billinton, Roy, and Ronald N. Allan. *Reliability Evaluation of Engineering Systems: Concepts and Techniques, Second Edition.* New York: Plenum Press, 1996

Box, G.E.P., W.E.G. Hunter, and J.S. Hunter. *Statistics for Experimenters.* New York: Wiley, 1978

Burr, Irving W. *Applied Statistical Methods.* New York: Academic Press, 1974

Haykin, Simon. *Communications Systems, Fourth Edition.* New York: Wiley, 2000

Huff, Darrell. *How to Lie with Statistics.* New York: Norton, 1954, 1993 (reissue)

Johnson, Richard A. *Miller & Freund's Probability and Statistics for Engineers, Sixth Edition.* New Jersey: Prentice Hall, 2000

Kennedy, John B., and Adam N. Neville. *Basic Statistical Methods for Engineers and Scientists, Third Edition.* New York: Harper & Row, 1986

Mendenhall, William, Dennis D. Wackerly and Richard L. Scheaffer. *Mathematical Statistics with Applications, Fourth Edition.* Boston: PWS-Kent, 1995

Montgomery, Douglas C. *Design and Analysis of Experiments, 5th Edition.* New York: Wiley, 2000

Ross, Sheldon M. *Introduction to Probability and Statistics for Engineers and Scientists, Second Edition.* Academic Press, 2000

Ryan, Thomas P., *Modern Regression Methods,* New York: Wiley, 1997

Vardeman, Stephen B. *Statistics for Engineering Problem Solving.* Boston: PWS Publishing, 1994

Walpole, Ronald E., & Raymond H. Myers. *Probability and Statistics for Engineers and Scientists, 7th Edition.* New Jersey: Prentice Hall, 2002

Ziemer, Rodger E. *Elements of Engineering Probability & Statistics.* New Jersey: Prentice Hall, 1997

Appendices

Appendix A contains probability tables for use in statistical calculations. These are for the normal distribution, the t-distribution, the chi-squared distribution, and the F-distribution. The numbers in the tables were calculated using MS Excel.

Appendix B describes some of the abilities and properties of MS Excel which are useful in statistical calculations. This appendix gives brief instruction in using Excel for such a purpose, but the reader is assumed to have a basic knowledge of Excel already.

Appendix C describes some functions of Excel not recommended for use while the reader is learning the fundamentals of probability and statistics. These can save time in calculations once the reader has fully understood the fundamentals.

Appendix D contains answers to some of the problem sets.

Appendix A: Tables

Table A1
Cumulative Normal Probability
$\Phi(z) = \Pr[Z < z]$

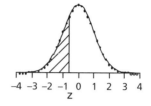

$\Delta z = $	−0.09	−0.08	−0.07	−0.06	−0.05	−0.04	−0.03	−0.02	−0.01	−0.00	
z_0											z_0
−3.7	0.0001	0.0001	0.0001	0.0001	0.0001	0.0001	0.0001	0.0001	0.0001	0.0001	−3.7
−3.6	0.0001	0.0001	0.0001	0.0001	0.0001	0.0001	0.0001	0.0001	0.0002	0.0002	−3.6
−3.5	0.0002	0.0002	0.0002	0.0002	0.0002	0.0002	0.0002	0.0002	0.0002	0.0002	−3.5
−3.4	0.0002	0.0003	0.0003	0.0003	0.0003	0.0003	0.0003	0.0003	0.0003	0.0003	−3.4
−3.3	0.0003	0.0004	0.0004	0.0004	0.0004	0.0004	0.0004	0.0005	0.0005	0.0005	−3.3
−3.2	0.0005	0.0005	0.0005	0.0006	0.0006	0.0006	0.0006	0.0006	0.0007	0.0007	−3.2
−3.1	0.0007	0.0007	0.0008	0.0008	0.0008	0.0008	0.0009	0.0009	0.0009	0.0010	−3.1
−3.0	0.0010	0.0010	0.0011	0.0011	0.0011	0.0012	0.0012	0.0013	0.0013	0.0013	−3.0
−2.9	0.0014	0.0014	0.0015	0.0015	0.0016	0.0016	0.0017	0.0018	0.0018	0.0019	−2.9
−2.8	0.0019	0.002	0.0021	0.0021	0.0022	0.0023	0.0023	0.0024	0.0025	0.0026	−2.8
−2.7	0.0026	0.0027	0.0028	0.0029	0.0030	0.0031	0.0032	0.0033	0.0034	0.0035	−2.7
−2.6	0.0036	0.0037	0.0038	0.0039	0.0040	0.0041	0.0043	0.0044	0.0045	0.0047	−2.6
−2.5	0.0048	0.0049	0.0051	0.0052	0.0054	0.0055	0.0057	0.0059	0.0060	0.0062	−2.5
−2.4	0.0064	0.0066	0.0068	0.0069	0.0071	0.0073	0.0075	0.0078	0.0080	0.0082	−2.4
−2.3	0.0084	0.0087	0.0089	0.0091	0.0094	0.0096	0.0099	0.0102	0.0104	0.0107	−2.3
−2.2	0.0110	0.0113	0.0116	0.0119	0.0122	0.0125	0.0129	0.0132	0.0136	0.0139	−2.2
−2.1	0.0143	0.0146	0.0150	0.0154	0.0158	0.0162	0.0166	0.017	0.0174	0.0179	−2.1
−2.0	0.0183	0.0188	0.0192	0.0197	0.0202	0.0207	0.0212	0.0217	0.0222	0.0228	−2.0
−1.9	0.0233	0.0239	0.0244	0.025	0.0256	0.0262	0.0268	0.0274	0.0281	0.0287	−1.9
−1.8	0.0294	0.0301	0.0307	0.0314	0.0322	0.0329	0.0336	0.0344	0.0351	0.0359	−1.8
−1.7	0.0367	0.0375	0.0384	0.0392	0.0401	0.0409	0.0418	0.0427	0.0436	0.0446	−1.7
−1.6	0.0455	0.0465	0.0475	0.0485	0.0495	0.0505	0.0516	0.0526	0.0537	0.0548	−1.6
−1.5	0.0559	0.0571	0.0582	0.0594	0.0606	0.0618	0.063	0.0643	0.0655	0.0668	−1.5
−1.4	0.0681	0.0694	0.0708	0.0721	0.0735	0.0749	0.0764	0.0778	0.0793	0.0808	−1.4
−1.3	0.0823	0.0838	0.0853	0.0869	0.0885	0.0901	0.0918	0.0934	0.0951	0.0968	−1.3
−1.2	0.0985	0.1003	0.1020	0.1038	0.1056	0.1075	0.1093	0.1112	0.1131	0.1151	−1.2
−1.1	0.1170	0.1190	0.1210	0.1230	0.1251	0.1271	0.1292	0.1314	0.1335	0.1357	−1.1
−1.0	0.1379	0.1401	0.1423	0.1446	0.1469	0.1492	0.1515	0.1539	0.1562	0.1587	−1.0
−0.9	0.1611	0.1635	0.1660	0.1685	0.1711	0.1736	0.1762	0.1788	0.1814	0.1841	−0.9
−0.8	0.1867	0.1894	0.1922	0.1949	0.1977	0.2005	0.2033	0.2061	0.2090	0.2119	−0.8
−0.7	0.2148	0.2177	0.2206	0.2236	0.2266	0.2296	0.2327	0.2358	0.2389	0.2420	−0.7
−0.6	0.2451	0.2483	0.2514	0.2546	0.2578	0.2611	0.2643	0.2676	0.2709	0.2743	−0.6
−0.5	0.2776	0.2810	0.2843	0.2877	0.2912	0.2946	0.2981	0.3015	0.3050	0.3085	−0.5
−0.4	0.3121	0.3156	0.3192	0.3228	0.3264	0.3300	0.3336	0.3372	0.3409	0.3446	−0.4
−0.3	0.3483	0.352	0.3557	0.3594	0.3632	0.3669	0.3707	0.3745	0.3783	0.3821	−0.3
−0.2	0.3859	0.3897	0.3936	0.3974	0.4013	0.4052	0.4090	0.4129	0.4168	0.4207	−0.2
−0.1	0.4247	0.4286	0.4325	0.4364	0.4404	0.4443	0.4483	0.4522	0.4562	0.4602	−0.1
−0.0	0.4641	0.4681	0.4721	0.4761	0.4801	0.4840	0.4880	0.4920	0.4960	0.5000	−0.0

Table A1 (continued)

Table A1 (continued)
Cumulative Normal Probability
$\Phi(z) = \Pr[Z < z]$

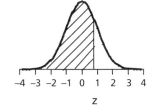

$\Delta z =$	0.00	0.01	0.02	0.03	0.04	0.05	0.06	0.07	0.08	0.09	
z_0											z_0
0.0	0.5000	0.5040	0.5080	0.5120	0.5160	0.5199	0.5239	0.5279	0.5319	0.5359	0.0
0.1	0.5398	0.5438	0.5478	0.5517	0.5557	0.5596	0.5636	0.5675	0.5714	0.5753	0.1
0.2	0.5793	0.5832	0.5871	0.5910	0.5948	0.5987	0.6026	0.6064	0.6103	0.6141	0.2
0.3	0.6179	0.6217	0.6255	0.6293	0.6331	0.6368	0.6406	0.6443	0.6480	0.6517	0.3
0.4	0.6554	0.6591	0.6628	0.6664	0.6700	0.6736	0.6772	0.6808	0.6844	0.6879	0.4
0.5	0.6915	0.6950	0.6985	0.7019	0.7054	0.7088	0.7123	0.7157	0.7190	0.7224	0.5
0.6	0.7257	0.7291	0.7324	0.7357	0.7389	0.7422	0.7454	0.7486	0.7517	0.7549	0.6
0.7	0.7580	0.7611	0.7642	0.7673	0.7704	0.7734	0.7764	0.7794	0.7823	0.7852	0.7
0.8	0.7881	0.7910	0.7939	0.7967	0.7995	0.8023	0.8051	0.8078	0.8106	0.8133	0.8
0.9	0.8159	0.8186	0.8212	0.8238	0.8264	0.8289	0.8315	0.8340	0.8365	0.8389	0.9
1.0	0.8413	0.8438	0.8461	0.8485	0.8508	0.8531	0.8554	0.8577	0.8599	0.8621	1.0
1.1	0.8643	0.8665	0.8686	0.8708	0.8729	0.8749	0.8770	0.8790	0.8810	0.8830	1.1
1.2	0.8849	0.8869	0.8888	0.8907	0.8925	0.8944	0.8962	0.8980	0.8997	0.9015	1.2
1.3	0.9032	0.9049	0.9066	0.9082	0.9099	0.9115	0.9131	0.9147	0.9162	0.9177	1.3
1.4	0.9192	0.9207	0.9222	0.9236	0.9251	0.9265	0.9279	0.9292	0.9306	0.9319	1.4
1.5	0.9332	0.9345	0.9357	0.937	0.9382	0.9394	0.9406	0.9418	0.9429	0.9441	1.5
1.6	0.9452	0.9463	0.9474	0.9484	0.9495	0.9505	0.9515	0.9525	0.9535	0.9545	1.6
1.7	0.9554	0.9564	0.9573	0.9582	0.9591	0.9599	0.9608	0.9616	0.9625	0.9633	1.7
1.8	0.9641	0.9649	0.9656	0.9664	0.9671	0.9678	0.9686	0.9693	0.9699	0.9706	1.8
1.9	0.9713	0.9719	0.9726	0.9732	0.9738	0.9744	0.9750	0.9756	0.9761	0.9767	1.9
2.0	0.9772	0.9778	0.9783	0.9788	0.9793	0.9798	0.9803	0.9808	0.9812	0.9817	2.0
2.1	0.9821	0.9826	0.983	0.9834	0.9838	0.9842	0.9846	0.9850	0.9854	0.9857	2.1
2.2	0.9861	0.9864	0.9868	0.9871	0.9875	0.9878	0.9881	0.9884	0.9887	0.989	2.2
2.3	0.9893	0.9896	0.9898	0.9901	0.9904	0.9906	0.9909	0.9911	0.9913	0.9916	2.3
2.4	0.9918	0.9920	0.9922	0.9925	0.9927	0.9929	0.9931	0.9932	0.9934	0.9936	2.4
2.5	0.9938	0.9940	0.9941	0.9943	0.9945	0.9946	0.9948	0.9949	0.9951	0.9952	2.5
2.6	0.9953	0.9955	0.9956	0.9957	0.9959	0.9960	0.9961	0.9962	0.9963	0.9964	2.6
2.7	0.9965	0.9966	0.9967	0.9968	0.9969	0.9970	0.9971	0.9972	0.9973	0.9974	2.7
2.8	0.9974	0.9975	0.9976	0.9977	0.9977	0.9978	0.9979	0.9979	0.9980	0.9981	2.8
2.9	0.9981	0.9982	0.9982	0.9983	0.9984	0.9984	0.9985	0.9985	0.9986	0.9986	2.9
3.0	0.9987	0.9987	0.9987	0.9988	0.9988	0.9989	0.9989	0.9989	0.9990	0.9990	3.0
3.1	0.9990	0.9991	0.9991	0.9991	0.9992	0.9992	0.9992	0.9992	0.9993	0.9993	3.1
3.2	0.9993	0.9993	0.9994	0.9994	0.9994	0.9994	0.9994	0.9995	0.9995	0.9995	3.2
3.3	0.9995	0.9995	0.9995	0.9996	0.9996	0.9996	0.9996	0.9996	0.9996	0.9997	3.3
3.4	0.9997	0.9997	0.9997	0.9997	0.9997	0.9997	0.9997	0.9997	0.9997	0.9998	3.4
3.5	0.9998	0.9998	0.9998	0.9998	0.9998	0.9998	0.9998	0.9998	0.9998	0.9998	3.5
3.6	0.9998	0.9998	0.9999	0.9999	0.9999	0.9999	0.9999	0.9999	0.9999	0.9999	3.6
3.7	0.9999	0.9999	0.9999	0.9999	0.9999	0.9999	0.9999	0.9999	0.9999	0.9999	3.7
3.8	0.9999	0.9999	0.9999	0.9999	0.9999	0.9999	0.9999	0.9999	0.9999	0.9999	3.8

Appendix A

Table A2: *t*-distribution

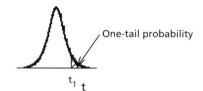

	One-tail Probability						
	0.1	0.05	0.025	0.01	0.005	0.001	
df							df
1	3.078	6.314	12.706	31.821	63.656	318.289	1
2	1.886	2.920	4.303	6.965	9.925	22.328	2
3	1.638	2.353	3.182	4.541	5.841	10.214	3
4	1.533	2.132	2.776	3.747	4.604	7.173	4
5	1.476	2.015	2.571	3.365	4.032	5.894	5
6	1.440	1.943	2.447	3.143	3.707	5.208	6
7	1.415	1.895	2.365	2.998	3.499	4.785	7
8	1.397	1.860	2.306	2.896	3.355	4.501	8
9	1.383	1.833	2.262	2.821	3.250	4.297	9
10	1.372	1.812	2.228	2.764	3.169	4.144	10
11	1.363	1.796	2.201	2.718	3.106	4.025	11
12	1.356	1.782	2.179	2.681	3.055	3.930	12
13	1.350	1.771	2.160	2.650	3.012	3.852	13
14	1.345	1.761	2.145	2.624	2.977	3.787	14
15	1.341	1.753	2.131	2.602	2.947	3.733	15
16	1.337	1.746	2.120	2.583	2.921	3.686	16
17	1.333	1.740	2.110	2.567	2.898	3.646	17
18	1.330	1.734	2.101	2.552	2.878	3.610	18
19	1.328	1.729	2.093	2.539	2.861	3.579	19
20	1.325	1.725	2.086	2.528	2.845	3.552	20
21	1.323	1.721	2.080	2.518	2.831	3.527	21
22	1.321	1.717	2.074	2.508	2.819	3.505	22
23	1.319	1.714	2.069	2.500	2.807	3.485	23
24	1.318	1.711	2.064	2.492	2.797	3.467	24
25	1.316	1.708	2.060	2.485	2.787	3.450	25
26	1.315	1.706	2.056	2.479	2.779	3.435	26
27	1.314	1.703	2.052	2.473	2.771	3.421	27
28	1.313	1.701	2.048	2.467	2.763	3.408	28
29	1.311	1.699	2.045	2.462	2.756	3.396	29
30	1.310	1.697	2.042	2.457	2.750	3.385	30
40	1.303	1.684	2.021	2.423	2.704	3.307	40
50	1.299	1.676	2.009	2.403	2.678	3.261	50
60	1.296	1.671	2.000	2.390	2.660	3.232	60
70	1.294	1.667	1.994	2.381	2.648	3.211	70
80	1.292	1.664	1.990	2.374	2.639	3.195	80
90	1.291	1.662	1.987	2.368	2.632	3.183	90
100	1.290	1.660	1.984	2.364	2.626	3.174	100
110	1.289	1.659	1.982	2.361	2.621	3.166	110
120	1.289	1.658	1.980	2.358	2.617	3.160	120
∞	1.282	1.645	1.960	2.326	2.576	3.090	∞

Table A3: Chi-squared Distribution

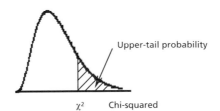

df	Upper Tail Probabilities					
	0.99	0.975	0.95	0.05	0.025	0.01
	Values of Chi-squared					
1	0.0002	0.0010	0.004	3.84	5.02	6.63
2	0.020	0.051	0.103	5.99	7.38	9.21
3	0.11	0.22	0.35	7.81	9.35	11.34
4	0.30	0.48	0.71	9.49	11.14	13.28
5	0.55	0.83	1.15	11.07	12.83	15.09
6	0.87	1.24	1.64	12.59	14.45	16.81
7	1.24	1.69	2.17	14.07	16.01	18.48
8	1.65	2.18	2.73	15.51	17.53	20.09
9	2.09	2.70	3.33	16.92	19.02	21.67
10	2.56	3.25	3.94	18.31	20.48	23.21
11	3.05	3.82	4.57	19.68	21.92	24.73
12	3.57	4.40	5.23	21.03	23.34	26.22
13	4.11	5.01	5.89	22.36	24.74	27.69
14	4.66	5.63	6.57	23.68	26.12	29.14
15	5.23	6.26	7.26	25.00	27.49	30.58
16	5.81	6.91	7.96	26.30	28.85	32.00
17	6.41	7.56	8.67	27.59	30.19	33.41
18	7.01	8.23	9.39	28.87	31.53	34.81
19	7.63	8.91	10.12	30.14	32.85	36.19
20	8.26	9.59	10.85	31.41	34.17	37.57
21	8.90	10.28	11.59	32.67	35.48	38.93
22	9.54	10.98	12.34	33.92	36.78	40.29
23	10.20	11.69	13.09	35.17	38.08	41.64
24	10.86	12.40	13.85	36.42	39.36	42.98
25	11.52	13.12	14.61	37.65	40.65	44.31
26	12.20	13.84	15.38	38.89	41.92	45.64
27	12.88	14.57	16.15	40.11	43.19	46.96
28	13.56	15.31	16.93	41.34	44.46	48.28
29	14.26	16.05	17.71	42.56	45.72	49.59
30	14.95	16.79	18.49	43.77	46.98	50.89
35	18.51	20.57	22.47	49.80	53.20	57.34
40	22.16	24.43	26.51	55.76	59.34	63.69
45	25.90	28.37	30.61	61.66	65.41	69.96
50	29.71	32.36	34.76	67.50	71.42	76.15

Appendix A

Table A4: F-Distribution

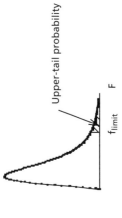

Values of F with df1 degrees of freedom in the numerator and df2 degrees of freedom in the denominator to give Upper-tail Probability of 0.05

df1	1	2	3	4	5	6	7	8	9	10	11	12	15	20	27	40	60	∞
df2																		
1	161	199	216	225	230	234	237	239	241	242	243	244	246	248	250	251	252	254
2	18.51	19.00	19.16	19.25	19.30	19.33	19.35	19.37	19.38	19.40	19.40	19.41	19.43	19.45	19.46	19.47	19.48	19.50
3	10.13	9.55	9.28	9.12	9.01	8.94	8.89	8.85	8.81	8.79	8.76	8.74	8.70	8.66	8.63	8.59	8.57	8.53
4	7.71	6.94	6.59	6.39	6.26	6.16	6.09	6.04	6.00	5.96	5.94	5.91	5.86	5.80	5.76	5.72	5.69	5.63
5	6.61	5.79	5.41	5.19	5.05	4.95	4.88	4.82	4.77	4.74	4.70	4.68	4.62	4.56	4.51	4.46	4.43	4.37
6	5.99	5.14	4.76	4.53	4.39	4.28	4.21	4.15	4.10	4.06	4.03	4.00	3.94	3.87	3.82	3.77	3.74	3.67
7	5.59	4.74	4.35	4.12	3.97	3.87	3.79	3.73	3.68	3.64	3.60	3.57	3.51	3.44	3.39	3.34	3.30	3.23
8	5.32	4.46	4.07	3.84	3.69	3.58	3.50	3.44	3.39	3.35	3.31	3.28	3.22	3.15	3.10	3.04	3.01	2.93
9	5.12	4.26	3.86	3.63	3.48	3.37	3.29	3.23	3.18	3.14	3.10	3.07	3.01	2.94	2.88	2.83	2.79	2.71
10	4.96	4.10	3.71	3.48	3.33	3.22	3.14	3.07	3.02	2.98	2.94	2.91	2.85	2.77	2.72	2.66	2.62	2.54
11	4.84	3.98	3.59	3.36	3.20	3.09	3.01	2.95	2.90	2.85	2.82	2.79	2.72	2.65	2.59	2.53	2.49	2.41
12	4.75	3.89	3.49	3.26	3.11	3.00	2.91	2.85	2.80	2.75	2.72	2.69	2.62	2.54	2.48	2.43	2.38	2.30
15	4.54	3.68	3.29	3.06	2.90	2.79	2.71	2.64	2.59	2.54	2.51	2.48	2.40	2.33	2.27	2.20	2.16	2.07
20	4.35	3.49	3.10	2.87	2.71	2.60	2.51	2.45	2.39	2.35	2.31	2.28	2.20	2.12	2.06	1.99	1.95	1.84
27	4.21	3.35	2.96	2.73	2.57	2.46	2.37	2.31	2.25	2.20	2.17	2.13	2.06	1.97	1.90	1.84	1.79	1.67
40	4.08	3.23	2.84	2.61	2.45	2.34	2.25	2.18	2.12	2.08	2.04	2.00	1.92	1.84	1.77	1.69	1.64	1.51
60	4.00	3.15	2.76	2.53	2.37	2.25	2.17	2.10	2.04	1.99	1.95	1.92	1.84	1.75	1.67	1.59	1.53	1.39
∞	3.84	3.00	2.60	2.37	2.21	2.10	2.01	1.94	1.88	1.83	1.79	1.75	1.67	1.57	1.49	1.39	1.32	1.01

Appendices

Table A4: F-Distribution (continued)

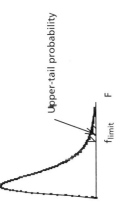

Values of F with df1 degrees of freedom in the numerator and df2 degrees of freedom in the denominator to give Upper-tail Probability of 0.01

df1 df2	1	2	3	4	5	6	7	8	9	10	12	15	20	27	40	60	∞
1	4052	4999	5404	5624	5764	5859	5928	5981	6022	6056	6107	6157	6209	6249	6286	6313	6366
2	98.5	99.0	99.2	99.3	99.3	99.3	99.4	99.4	99.4	99.4	99.4	99.4	99.4	99.5	99.5	99.5	99.5
3	34.1	30.8	29.5	28.7	28.2	27.9	27.7	27.5	27.3	27.2	27.1	26.9	26.7	26.5	26.4	26.3	26.1
4	21.2	18.0	16.7	16.0	15.5	15.2	15.0	14.8	14.7	14.5	14.4	14.2	14.0	13.9	13.7	13.7	13.5
5	16.3	13.3	12.1	11.4	11.0	10.7	10.5	10.3	10.2	10.1	9.39	9.72	9.55	9.42	9.29	9.20	9.02
6	13.7	10.9	9.78	9.15	8.75	8.47	8.26	8.10	7.98	7.87	7.72	7.56	7.40	7.27	7.14	7.06	6.88
7	12.2	9.55	8.45	7.85	7.46	7.19	6.99	6.84	6.72	6.62	6.47	6.31	6.16	6.03	5.91	5.82	5.65
8	11.3	8.65	7.59	7.01	6.63	6.37	6.18	6.03	5.91	5.81	5.67	5.52	5.36	5.23	5.12	5.03	4.86
9	10.6	8.02	6.99	6.42	6.06	5.8	5.61	5.47	5.35	5.26	5.11	4.96	4.81	4.68	4.57	4.48	4.31
10	10.0	7.56	6.55	5.99	5.64	5.39	5.20	5.06	4.94	4.85	4.71	4.56	4.41	4.28	4.17	4.08	3.91
12	9.33	6.93	5.95	5.41	5.06	4.82	4.64	4.50	4.39	4.30	4.16	4.01	3.86	3.74	3.62	3.54	3.36
15	8.68	6.36	5.42	4.89	4.56	4.32	4.14	4.00	3.89	3.80	3.67	3.52	3.37	3.25	3.13	3.05	2.87
20	8.10	5.85	4.94	4.43	4.10	3.87	3.70	3.56	3.46	3.37	3.23	3.09	2.94	2.81	2.69	2.61	2.42
27	7.68	5.49	4.60	4.11	3.78	3.56	3.39	3.26	3.15	3.06	2.93	2.78	2.63	2.51	2.38	2.29	2.10
40	7.31	5.18	4.31	3.83	3.51	3.29	3.12	2.99	2.89	2.80	2.66	2.52	2.37	2.24	2.11	2.02	1.80
60	7.08	4.98	4.13	3.65	3.34	3.12	2.95	2.82	2.72	2.63	2.50	2.35	2.20	2.07	1.94	1.84	1.60
∞	6.64	4.61	3.78	3.32	3.02	2.80	2.64	2.51	2.41	2.32	2.18	2.04	1.88	1.74	1.59	1.47	1

Appendix B

Appendix B: Some Properties of Excel Useful During the Learning Process

(a) Formulas

A formula combines values with operators such as a plus sign. We will be concerned at present with only the arithmetic operators, which are +, –, /, *, %, and ^. Values may be expressed as number constants, such as 34.7, or as references to the content of a cell, such as F28 (meaning the cell that is both in column F and in row 28). Instead of references we may use names, such as Cost, if those names have been defined. An Excel formula always begins with an equal sign, =. That sign indicates to the computer that the content of a cell is a formula that needs to be evaluated. In many cases the most convenient method of inserting an Excel function is to paste it into the appropriate cell. This is discussed briefly at the end of section 5.5.

As always in engineering calculations, we must make clear how we are performing a calculation. When the formula in a cell has been entered correctly, the corresponding cell on the computer screen will show the arithmetic result for that formula. For example, entering the formula =20+34 will give the result, 54, and that will show in the space for the cell on the screen. If we select that cell, the Formula Bar will show the formula. But if we print the work sheet, we will see in that cell only the result, 54, and the formula will not appear. Then to make the printed work sheet more understandable, a neighboring cell (usually to the left or right of the cell in question or above it) should give a clear statement of the formula being used. That statement must not begin with an equal sign, or Excel will interpret it as the formula itself. Instead, it should (for purposes of this book) *end* with an equal sign—e.g., 20 + 34 =. We see instances of this in the body of the text, such as Examples 3.4, 4.4, and 4.5.

(b) Array Formulas

An ordinary formula, as in part (a) above, produces a result in just one cell. Often we want to produce results simultaneously in two or more cells. For that we use an *array formula*. For example, we may want to calculate the deviation of each measurement from the mean of those measurements. Say the measurements are in rows 18 to 88 of column B, which we show as B18:B88, and the mean of the measurements is in cell B90. We could calculate each deviation separately, as =B18-B90 in cell C18, =B19-B90 in cell C19, and so on. A faster alternative is to calculate them all together by the array formula =B18:B88-B90 in cells C18:C88. It is clear that the array formula can be applied much more quickly than the 71 individual formulas.

To apply the array formula we first select the cells in which we want the answer to appear, cells C18:C88 in this case. Then we type in the formula, which is =B18:B88-B90 for this example. Then to indicate to the computer that we are applying an array formula, we press not just Enter but simultaneously CONTROL+SHIFT+ENTER (note: three keys) in Excel for Windows, or

COMMAND+ENTER in Excel for Macintosh. The array formula is shown in the Formula Bar inside braces, { }, but do not type the braces yourself.

(c) Sorting

The Sort command can save a good deal of effort in developing a frequency distribution and in finding quantiles from the distribution. The Sort command is on the Data menu of Excel. For example, say the data we want to sort are in columns A and B, and we want to sort according to the numbers or letters in column B, which is headed Thickness in row 1. We select columns A and B, then from the Data menu we choose Sort. The Sort dialog appears, and we click the button indicating that the list has a header row. We select sorting by the heading Thickness in ascending order, increasing in magnitude from the smallest to the largest, then click the OK button. (Note that if the first results are not in the form desired, we can immediately afterward choose Undo Sort from the Edit menu, then try again.) After the data have been sorted from the smallest to the largest we can number them in order, say in column C, by entering 1 in the first row and 2 in the second row, then completing the series by selecting these first two cells in the column, then dragging the fill handle down and releasing the mouse button when all the data have been numbered.

(d) Summing

Of course we can add up a column of figures and put the result in cell B6 by selecting that cell and typing (say) =B1+B2+B3+B4+B5, then pressing enter or return. It is usually faster and more convenient to sum a column or row of data by selecting the cell at the end of that column or row of data and clicking the AutoSum tool on the standard toolbar. The AutoSum tool is marked with the Greek letter sigma, Σ. After it is clicked, the cells which will be added are surrounded by dotted lines, and the formula bar shows =SUM(B1:B5) if cells B1 to B5 are the ones in question. If there are possible cells to the left of the selected cell as well as above it, either set of cells might have been chosen. Then we need to make sure that we have the right ones. If we select the cells we want just before we select the cell for the sum, it seems to come out right, but that may not always be so.

(e) Functions

An Excel function is a special prewritten formula that uses a value or values as input, performs an operation, and returns a value or values as a result. Excel functions vary greatly in complexity, from simple functions that add up input quantities to complex functions that perform a multitude of tasks in a particular sequence. Excel functions can be inserted by either of two methods. One is to use the Insert menu and choose Function. Then the Paste Function dialog appears, and we choose a category such as Math & Trig or Statistical, then the required function such as Sum or Frequency. A dialog box appears to prompt us to choose values of the argument of the function. The alternative method is to type the equals sign (since the function is a type of formula), the name of the function, and then the arguments within parentheses and separated by commas [for example, =SUM(A1, A2)].

Appendix B

Many of the Excel functions are not recommended for use while a person is learning the fundamentals of probability and statistics. That is because they act as "black boxes" that perform calculations without requiring any thinking on the part of the person using them; the person has only to supply the input values and the computer supplies the logic. Thus, these functions do not help the process of learning the fundamentals. Some of these functions which are useful at a later stage, when a person has already gained a good fundamental knowledge of probability and statistics, will be listed in Appendix C.

However, a few functions can be recommended for use even when the fundamentals are being learned. They are as follows.

(i) Sum Function

The SUM function simply adds up the arguments or, if the arguments are references, the contents of the cells. The Sum function in Excel is found in the Mathematics and Trigonometry category. The arguments for this function may be arrays, such as B1:B5; references to individual cells, such as A26; or numbers, such as 5. Thus, =SUM(16,13) gives 29. If A1 contains 11 and B1 contains 7, =SUM(A1,B1) gives 18. If A2 contains 14, B2 contains 9, and C2 contains 6, =SUM(A2:C2) gives 29. In fact, the AutoSum tool which we saw above uses the SUM function.

(ii) Frequency Function

The FREQUENCY function counts the numbers of values within given class boundaries and returns a frequency array. It is found in the category of Statistical functions. In most cases the arguments of the FREQUENCY function are, first, the reference to the array of cells containing values of the data which are to be counted; then, second, the reference to the array of cells giving the upper class boundaries in *ascending* order (that is, beginning with the smallest and working up). As usual, these arguments are separated by commas. The number of values less than the *lowest* upper class boundary appears in the first cell, and the numbers of values more than previous upper class boundaries but less than successive upper class boundaries appear in subsequent cells, ending with the number of values *larger* than the largest class boundary. Thus, the number of cells for class frequencies is one more than the number of upper class boundaries.

The procedure is to select the vertical or horizontal array of cells where we want the class frequencies to appear, then to enter =FREQUENCY(input reference, class boundaries reference). An illustration of the use of the frequency function is given in Example 4.4, where the array formula =FREQUENCY(B2:B122,B135:B143) was entered in cells D135:D144. Then the values of the data were taken from cells B2:B122, the upper class boundaries were in cells B135:B143, and the resulting frequencies were placed in cells D135:D144.

Appendices

(f) Making Histograms or Other Charts or Graphs

As we see in Chapter 4, histograms are used frequently to show graphically the class frequencies for various classes or intervals of the variate. The information for a histogram is contained in a grouped frequency table. The ChartWizard provides a convenient way to produce a histogram or other chart or graph.

A chart can be produced from a table of data (for a histogram, that would be a grouped frequency table) by selecting Chart from the Insert menu. Modifications, major or minor, of the chart are produced using the Chart menu. The procedure for histograms is discussed in more detail in section 4.5, particularly in Example 4.4.

Appendix C: Functions Useful Once the Fundamentals Are Understood

There are a number of Excel functions which should *not* be used during the learning process but can be very useful later on. The following statistical functions fall in this category:

AVEDEV( ) calculates the mean of the absolute deviations from the mean (see section 3.3.4).

AVERAGE( ) returns the arithmetic mean of the arguments.

COUNT ( ) counts the numbers in the list of arguments.

COUNTA( ) counts the number of nonblank values.

DEVSQ( ) calculates the sum of squares of deviations of data points from their sample mean, e.g. $\sum(x_i - \bar{x})^2$.

GEOMEAN( ) returns the geometric mean of the arguments.

HARMEAN( ) gives the harmonic mean of the arguments.

LARGE(array,k) returns the kth largest value in the array.

MAX( ) gives the maximum value in a list of arguments.

MEDIAN( ) returns the median of the stated numbers.

MIN( ) gives the minimum value in a list of arguments.

MODE( ) returns the mode of the data set.

PERCENTILE(array,k) returns the *k*th percentile of numbers in the array.

PERCENTRANK(array,x,) returns the percentage rank of *x* among the values in the array.

QUARTILE(array,) returns the minimum, maximum, median, lower quartile, or upper quartile from the array.

RANK( ) gives the rank (order in a sorted list) of a number.

STDEV( ) gives the sample standard deviation, *s*, of a set of numbers.

STDEVP( ) calculates the standard deviation, σ, of a set of numbers taken as a complete population.

TRIMMEAN(array,) calculates the mean after a certain percentage of values are removed at the top and the bottom of the set of numbers.

VAR( ) returns the sample variance, s^2, of a set of numbers.

VARP( ) finds the variance, σ^2, of a set of numbers taken as a complete population.

Appendix D: Answers to Some of the Problems

The following answers are believed to be correct, but if you find different answers which seem right, you should check with your instructor.

In chapter 2, section 2.1, problem set beginning on page 10:
1. (a) 3/14, (b) 9/14, (c) 11/14
2. (i) 0.644, (ii) 0.689, (iii) 0.089, (iv) 0.267
4. (a) 0.0909, (b) 0.143
6. (a) 64, (b) 84, (c) 52, (d) 0.619
8. (a) (i) 1/6, (ii) 5 to 1, (iii) 1 to 5
   (b) (i) 1/26, (ii) 25 to 1, (iii) 1 to 25

In chapter 2, section 2.2, problem set beginning on page 25:
1. (a) 0.045, (b) 0.955, (c) 21 to 1
3. (i) 80, (ii) 0.750, (iii) 0.340
5. (a) 26, (b) 0.308
8. 6
9. 3
11. (a) 0.216, (b) 0.324, (c) 0.216
13. For C-F-C 0.512. For F-C-F 0.384. Then choose C-F-C.
16. (a) 0.904, (b) 0.0475, (c) 0.0250
19. (a) 0.192, (b) 0.344, (c) 0.757
21. (a) (i) 0.526, (ii) 0.0526, (b) 0.0093, (c) 1.53×10^{-9}

In chapter 2, section 2.3, problem set beginning on page 32:
1. 5040
3. (i) 36, (ii) 15, (iii) 26
7. 10 combinations
9. (a) 6.1×10^{-4}, (b) 4.95×10^{-4}, (c) 1.54×10^{-6}
11. 56
15. (a) 0.067, (b) 0.333

In chapter 2, section 2.4, problem set beginning on page 38:
1. (a) 0.261, (b) 0.652
3. (a) 0.907, (b) 0.118, (c) 0.282
7. (a) 0.28, (b) 0.755, (c) 0.371
9. (a) 1.82%, (b) 29.6%, (c) 26.7%

In chapter 3, sections 3.1 to 3.4, problem set beginning on page 60:
1. 21.575 mm, 21.57 mm
2. (a) 0.0746 mm^2, 0.273 mm, 1.27%
   (b) 0.0895 mm^2, 0.299 mm, 1.39%

Appendix D

In chapter 4, sections 4.1 to 4.5, problem set beginning on page 80:
- 3   (e) 79, 75, 84, (f) 79.4, (i) 80%

In chapter 5, sections 5.1 and 5.2, problem set beginning on page 91:
- 1   (b) 1.23
- 3   (a) 1.50, 0.583, (b) 0.0917
- 5   (b) 2.333, (c) 0.556, 0.745
- 9   (a) 0.162, (b) 57%

In chapter 5, section 5.2, problem set beginning on page 97:
- 1 $1425
- 7   (a) 9.875, 10.12, (b) 0.830, (c) 0.059
- 9   (a) 0.717, (b) $350, (c) 8.47
- 11   (a) $2.25 million, (b) –$0.30 million

In chapter 5, section 5.3, problem set beginning on page 111:
- 3   0.264
- 5   (b) 2/3, 1/3, (d) 0.812, (e) 28.9%
- 7   (a) 0.0137, (b) 0.0152

In chapter 5, section 5.4, problem set beginning on page 126:
- 5   (a) 1.20, (b) 1.22
- 7   (b) 0.36, (c) 0.20, (d) 1.13
- 9   (a) 0.717, (b) 0.14, (c) 0.036
- 11   (a) 0.45, (b) 0.19, (c) 0.14, (d) 0.05

In chapter 5, section 5.6, problem set beginning on page 138:
- 1   0.0575, 0.6227 vs 0.600 etc.
- 5   (b) 1.225, 1.2297, (c) 1.107, (d) 0.0014

In chapter 6, section 6.1, problem set beginning on page 147:
- 3   (c) (i) 0.393, (ii) 0.368, (iii) 0.238

In chapter 6, section 6.2, problem set beginning on page 153:
- 1   (a) 1.5, (b) 0.25, (c) 0, (d) 1.65, (e) 0.533
- 3   (b) 1.5, (c) 0.2887, (d) 0.5774
- 5   (a) 1/3 month, (b) 1/3 month, (c) 0.865, (d) 0.950

In chapter 7, sections 7.1 to 7.4, problem set beginning on page 170:
- 1   (a) 95.2%, (b) 0.5%
- 3   (a) 0.4%, (b) 98.6%
- 7   (a) 50.89 kg, (b) 39.8%, (c) 51.2 kg
- 13   (a) 0.215 cm, (b) 0.826
- 17 $13,660
- 21   (a) (i) 0.115, (ii) 0.576, (iii) 0.309
  (b) 332 L/min, (c) $29.0/hr

Appendices

In chapter 7, sections 7.5 to 7.7, problem set beginning on page 193:
- 1 (b) 0.3125, (d) 0.308
- 5 (a) (i) 0.002, (ii) 0.005, (iii) 0.01
  (b) (i) 0.370, (ii) 0.371, (iii) 0.390
- 7 0.0015, 0.0015, 0.003

In chapter 8, sections 8.1 to 8.4, problem set beginning on page 208:
- 1 (i) 0.147 kg, (ii) 2.94 kg
- 3 (a) 12.6%, (b) 100.63 kg, (c) 5.008 kg
- 5 (a) $55.46, (b) 0.064
- 13 0.019

In chapter 9, section 9.1.1, problem set beginning on page 218:
- 1 $z = 2.14 > 1.96$. Adjustment required.
- 3 Observed level of significance is $< 0.1\%$. Significant at 1% level.
- 5 (a) 11.1%, (b) 0.3%
- 9 (a) 37.21 kg, (b) 0.019, (c) 37.5 kg, 0.031

In chapter 9, section 9.1.2, problem set beginning on page 223:
- 5 (a) 106, (b) 74%, (c) 0.725%
- 7 (a) 0.28, (b) 31
- 11 (b) 17.4%, (c) 14
- 13 (a) 17, (b) 98.9%
- 15 (a) $z = -2.08$, adjust, (b) 0.046, (c) 42

In chapter 9, section 9.2, problem set beginning on page 240:
- 1 (a) $t = -2.67$, yes, (b) 0.57 to 0.95 ppm
- 7 $t_{calculated} = 1.686$, $t_{critical} = 2.201$, not significant
- 9 $t_{calculated} = 1.745$, $t_{critical} = 2.571$, no significant difference
- 13 (a) $t_{calculated} = 1.673$, $t_{critical} = 2.365$, difference not significant
  (b) $t_{calculated} = 1.038$, $t_{critical} = 1.761$, difference not significant

In chapter 9, sections 9.1 and 9.2, problem set beginning on page 245:
- 3 $2.06 < 2.33$ so no
- 7 (a) $4.41 > 1.701$ so yes
  (b) $1.63 < 1.701$ so no

In chapter 10, section 10.1, problem set beginning on page 257:
- 7 (a) 0.10 level of signif. gives limit 3.18. $2.04 < 3.18$, so not significant
  (b) 0.05 level of signif. gives limit 1.88, $2.66 > 1.88$, so significantly more
- 13 $2.59 < 2.71$, so not significantly higher

In chapter 10, section 10.2, problem set beginning on page 276:
- 1 (c) 3
- 5 (b) 5.6%, (c) 8.1%

Appendix D

In chapter 14, sections 14.1 to 14.5, problem set beginning on page 368:

5   (c) $y = 2.141 - 7.71 \times 10^{-4} x$
     (d) $x = 2210 - 1000\, y$

7   (a) 3.09×10^{-4}
     (b) $3.67 < 4.60$ so not significant. No
     (c) -1.88×10^{-4} to -13.54×10^{-4}
     (d) At $x = 250$, 1.913 to 1.984. At $x = 300$, 1.890 to 1.930
     (e) At $x = 300$, 1.857 to 1.963. At $x = 350$, 1.811 to 1.932

8   (a) 0.878, (b) 0.771

9   (a) $y = 0.257 + 2.930\, x$
     (b) 0.991
     (c) $19.686 > 3.499$ so significant
     (d) 2.41 to 3.45

Engineering Problem-Solver Index

This handy index shows all of the solved example problems arranged by engineering application.

A

Analysis of data using ANOVA (Analysis of Variance)
 Example 12.1, p. 299
 Example 12.2, p. 308
 Example 12.3, p. 312
 Example 12.4, p. 318
Analysis of data using chi-squared test
 Example 13.2, p. 328
 Example 13.3, p. 329
 Example 13.4, p. 331
 Example 13.5, p. 333

C

Chemical process control
 Example 9.1, p. 215
Choosing a distribution type for a particular application
 Section 6.3, p. 155
Correlation
 Example 14.5, p. 367

E

Estimating demand using Poisson distribution
 Example 5.14, p. 121
 Example 5.17, p. 136
Experiment design, testing effectiveness
 Example 9.9, p. 237
 Example 9.10, p. 238

Example 10.3, p. 254
Example 10.4, p. 254
Example 11.1, p. 276, 281
Example 11.2, p. 281
Example 11.3, p. 282
Example 11.4, p. 183
Example 11.5, p. 284
Example 11.6, p. 286
Example 11.7, p. 287
Example 11.8, p. 288

M

Metal analysis
 Example 9.8, p. 235

O

Ore sample analysis
 Example 9.3, p. 221
 Example 9.4, p. 222

P

Particle size distribution
 Example 7.11, p. 191
Plotting and analyzing data sets
 Example 4.1, p. 63
 Example 4.2, p. 68
 Example 4.3, p. 72
 Example 4.4, p. 75

Engineering Problem-Solver Index

Process control
    Example 10.1, p. 250, 256
Production line quality
    Example 3.2, p. 51
    Example 3.5, p. 58

R

Random sampling
    Example 8.2, p. 201
    Example 8.4, p. 203
    Example 9.2, 216
    Example 10.7, p. 262
    Example 10.8, p. 263
    Example 10.9, p. 264
Regression analysis of data set
    Example 14.1, p. 355
    Example 14.2, p. 357
    Example 14.3, p. 358
    Example 14.4, p. 361
Reliability, time to failure
    Example 7.2, p. 156, 163
    Example 7.3, p. 166
    Example 8.5, p. 204

S

Sampling components on production line
    Example 2.4, p. 12
    Example 2.14, p. 30
    Example 2.16, p. 34
    Example 5.8, p. 106
    Example 5.10, p. 110
    Example 5.15, p. 125
    Example 5.16, p. 135
    Example 8.6, p. 204
    Example 9.7, p. 233
    Example 10.2, p. 251

T

t-distribution
    Example 10.10, p. 266
    when to use over normal distribution, p. 228-229
Testing for defective components
    Example 2.16, p. 34
    Example 5.5, p. 103
    Example 5.9, p. 108

Index

A

addition rule, 11
alternative hypothesis, 213
analysis of variance, 255, 294 321
  one-way, 295-304
  two-way, 304-316
ANOVA, 294
applications, 4
arithmetic mean, 41-42
axioms of probability, 9

B

Bayes' Rule, 34-38
Bernoulli distribution, 132
beta distribution, 156
bias, 285
binomial distribution, 101-111
  nested, 110
blocking, 285
block, randomized analysis, 316
box plots, 65

C

Central Limit Theorem, 205-208
central location, 41
chance, 2
chi-squared distribution, 249
chi-squared function, 324-340
circular permutations, 31
class boundaries, 67
coefficient of determination, 366
coefficient of variation, 50

combinations, 29-32
computer, 3, 55, 249, 325
  binomial distribution, 264
  equivalent to Normal Probability paper, 185
  F-distribution, 253
  normal distribution, 173
  plotting individual points to compare with normal distribution, 188
  random numbers, 284
conditional probability, 17
confidence interval, 221, 251, 256, 266
  for variance, 251
confidence limits
  proportion, 266
contingency tables, 331
continuous random variable, 141
correction for continuity, 179
correlation, 364-367
correlation coefficient, 365
cumulative distribution function, 86, 142
cumulative frequency, 67
cumulative frequency diagram, 72
cumulative probabilities, 184

D

deciles, 51
degrees of freedom, 228, 325
descriptive statistics, 41
design
  sequential or evolutionary, 278
design of experiments, 272-290
deterministic, 2

Index

diagnostic plots, 298
discrete random variable, 84

E

empirical approach to probability, 7
error sum of squares, 348
estimate, 221
   interval, 221
estimate of variance
   combined or pooled, 234
event, 9
evolutionary operation, 274
Excel, 4, 55, 75-80
expectation, 88
expected mean, 105
expected value, 149
experimentation, 273-290
   factorial design, 274-276
   randomization in, 280
exponential distribution, 155
extensions, 4

F

F-distribution, 252
F-test, 252, 253
factorial design, 274, 288-290
fair odds, 9
fitting
   normal distribution to frequency data, 175
   fitting binomial, 135, 136
fractional factorial design, 288-290
frequency distribution, 133
   characteristics, 157
frequency graphs, 66

G

gamma distribution, 156
geometric distribution, 132
geometric mean, 43
goodness of fit, 327
graphical checks, 349
grouped frequency, 66

H

harmonic mean, 43
histogram, 70
hypergeometric probability distribution, 132
hypothesis testing, 213

I

inference
   mean
      known variance, 213
      with estimated variance, 228
inference, for variance, 248
inferences for the mean, 228
inference, statistical, 212
inputs, 342
interaction, 274, 275
interfering factors, 236, 272
interquartile range, 45

L

least squares, 342
level of confidence, 221
level of significance
   critical, 215
   observed, 214
linear combination of independent variables, 198
linear regression, 342
list
   references, 374
logarithmic mean, 43
lognormal distributions, 192
loss of significance, 49, 69
lurking factors, 272

M

mean
   arithmetic, 41
   geometric, 43
mean deviation from the mean, 45
median, 43-44
mode, 44

Index

multinomial distribution, 111
multiple linear regression, 367
multiplication rule, 16
mutually exclusive, 12

N

negative binomial distribution, 131
normal approximation to a binomial distribution, 178
normal distribution, 155, 157-192
    approximation to binomial distribution, 178-183
    fitting to frequency data, 175
    tables, 161
normal probability paper, 184
null hypothesis, 213-215

O

one-tailed test, 217
operating data, routine, 273

P

p-value, 214
percentiles, 51
permutations, 29-32
permutations into classes, 30
planned experiments, 273
Poisson approximation to binomial distribution, 124
Poisson distribution, 117-125
population, 2, 197
probabilistic, 2
probability, 6
    classical or a priori approach, 7
    distributions, 84-140
    empirical or frequency approach, 7
    subjective estimate, 7
probability density function, 141, 158
probability distributions, 84-140
probability functions, 85
probability plotting, 190
proportion, 261
    binomial distribution, 108

Q

quantile, 53
quantile-quantile plotting, 190
Quartiles, 51

R

random numbers, 133, 284
random sample, 197
random variable, 84
randomization, 280
randomizing, 236
range
    interquartile, 45
reference books, 373
regression, 341-368
    evidence of cause, 366
    multiple linear, 367
    non-linear, 364
    simple linear, 342
    transformable forms, 364
    x on y, 347
regression coefficients, 342, 361
regression equations, 345
regression line
    y on x, 342
relative cumulative frequency, 68
relative frequency, 67
reliability, 156
replication, 279
residuals, 348
response, 342
Rough Rule, 181
rounding, 10
rules of probability
    addition, 11
    multiplication, 16

S

sample, 1
    random, 2
sample correlation coefficient, 365
sample range, 45

Index

sample size, 202
   proportion, 269
sample standard deviation, 47, 105-106
sample variance, 47
sampling, 197-211
sampling with replacement, 200
sampling without replacement, 201
scale of experimentation, 273
scatter plot, 342
significance test
   paired measurements, 238
   sample mean vs population mean, 233
   unpaired sample means, 234
simple linear regression, 342
spread, data, 44-51
statistical inferences, 212
   slope, 352
standard deviation, 46, 105-106
   estimation from a sample, 46
standard error of the mean, 200
statistical inference, 1
   proportion, 261
   two sample proportions, 269
statistical significance, 215
   sample variance vs population variance, 250, 256
statistics, 1
stem-and-leaf displays, 63-64
stochastic relations, 2
Student's t-test, 229
Sturges' Rule, 67
sum of products, 344

T

t-distribution, 229
t-test, 233
   paired, 238
   unpaired, 234
test of hypothesis, 213
test statistic, 214
transformation of variables, 190
tree diagram, 8, 19
two-tailed test, 213
Type I Error, 217
Type II Error, 217

U

uniform distribution, 155
unpaired t-test, 234

V

variability, 44-51
variance, 45
   discrete random variable, 89
   estimation from a sample, 46
   points about line, 348
   of a difference, 199
   of a single new observation, 353
   of a sum, 199
   of sample means, 199
variance of the mean response, 352
variance ratio, 295
variance-ratio test, 252
Venn Diagram, 12

W

Weibull distribution, 156

Y

Yates correction, 326

LIMITED WARRANTY AND DISCLAIMER OF LIABILITY

[[NEWNES.]] AND ANYONE ELSE WHO HAS BEEN INVOLVED IN THE CREATION OR PRODUCTION OF THE ACCOMPANYING CODE ("THE PRODUCT") CANNOT AND DO NOT WARRANT THE PERFORMANCE OR RESULTS THAT MAY BE OBTAINED BY USING THE PRODUCT. THE PRODUCT IS SOLD "AS IS" WITHOUT WARRANTY OF ANY KIND (EXCEPT AS HEREAFTER DESCRIBED), EITHER EXPRESSED OR IMPLIED, INCLUDING, BUT NOT LIMITED TO, ANY WARRANTY OF PERFORMANCE OR ANY IMPLIED WARRANTY OF MERCHANTABILITY OR FITNESS FOR ANY PARTICULAR PURPOSE. [[NEWNES.]] WARRANTS ONLY THAT THE MAGNETIC CD-ROM(S) ON WHICH THE CODE IS RECORDED IS FREE FROM DEFECTS IN MATERIAL AND FAULTY WORKMANSHIP UNDER THE NORMAL USE AND SERVICE FOR A PERIOD OF NINETY (90) DAYS FROM THE DATE THE PRODUCT IS DELIVERED. THE PURCHASER'S SOLE AND EXCLUSIVE REMEDY IN THE EVENT OF A DEFECT IS EXPRESSLY LIMITED TO EITHER REPLACEMENT OF THE CD-ROM(S) OR REFUND OF THE PURCHASE PRICE, AT [[NEWNES.]]'S SOLE DISCRETION.

IN NO EVENT, WHETHER AS A RESULT OF BREACH OF CONTRACT, WARRANTY OR TORT (INCLUDING NEGLIGENCE), WILL [[NEWNES.]] OR ANYONE WHO HAS BEEN INVOLVED IN THE CREATION OR PRODUCTION OF THE PRODUCT BE LIABLE TO PURCHASER FOR ANY DAMAGES, INCLUDING ANY LOST PROFITS, LOST SAVINGS OR OTHER INCIDENTAL OR CONSEQUENTIAL DAMAGES ARISING OUT OF THE USE OR INABILITY TO USE THE PRODUCT OR ANY MODIFICATIONS THEREOF, OR DUE TO THE CONTENTS OF THE CODE, EVEN IF [[NEWNES.]] HAS BEEN ADVISED OF THE POSSIBILITY OF SUCH DAMAGES, OR FOR ANY CLAIM BY ANY OTHER PARTY.

ANY REQUEST FOR REPLACEMENT OF A DEFECTIVE CD-ROM MUST BE POSTAGE PREPAID AND MUST BE ACCOMPANIED BY THE ORIGINAL DEFECTIVE CD-ROM, YOUR MAILING ADDRESS AND TELEPHONE NUMBER, AND PROOF OF DATE OF PURCHASE AND PURCHASE PRICE. SEND SUCH REQUESTS, STATING THE NATURE OF THE PROBLEM, TO ELSEVIER SCIENCE CUSTOMER SERVICE, 6277 SEA HARBOR DRIVE, ORLANDO, FL 32887, 1-800-321-5068. [[NEWNES.]] SHALL HAVE NO OBLIGATION TO REFUND THE PURCHASE PRICE OR TO REPLACE A CD-ROM BASED ON CLAIMS OF DEFECTS IN THE NATURE OR OPERATION OF THE PRODUCT.

SOME STATES DO NOT ALLOW LIMITATION ON HOW LONG AN IMPLIED WARRANTY LASTS, NOR EXCLUSIONS OR LIMITATIONS OF INCIDENTAL OR CONSEQUENTIAL DAMAGE, SO THE ABOVE LIMITATIONS AND EXCLUSIONS MAY NOT [[NEWNES.]] APPLY TO YOU. THIS WARRANTY GIVES YOU SPECIFIC LEGAL RIGHTS, AND YOU MAY ALSO HAVE OTHER RIGHTS THAT VARY FROM JURISDICTION TO JURISDICTION.

THE RE-EXPORT OF UNITED STATES ORIGIN SOFTWARE IS SUBJECT TO THE UNITED STATES LAWS UNDER THE EXPORT ADMINISTRATION ACT OF 1969 AS AMENDED. ANY FURTHER SALE OF THE PRODUCT SHALL BE IN COMPLIANCE WITH THE UNITED STATES DEPARTMENT OF COMMERCE ADMINISTRATION REGULATIONS. COMPLIANCE WITH SUCH REGULATIONS IS YOUR RESPONSIBILITY AND NOT THE RESPONSIBILITY OF [[NEWNES.]].